T0301783

Circular Economy

The reclamation of wastewater (and other essential materials) is among the major research areas for understanding the effects of implementing a circular economic model. The reuse and recycling of wastewater can greatly reduce the overall demand for freshwater for various industrial applications. Such concepts could potentially greatly reduce the overall water demands of our planet if implemented successfully. *Circular Economy: Applications for Water Remediation* will examine the current understanding of the circular economy in water remediation processes, its drawbacks, and relatively unexplored areas that require further research. This book:

- Provides an overview of the processes available to extract value-added materials from wastewater, such as clean water, nutrients, and energy.
- Explores the possibilities of re-using wastewater for agricultural uses.
- Provides an overview of the current policies and regulations concerning the implementation of circular economy concepts in wastewater remediation.

Maulin P. Shah has been an active researcher and scientific writer in his field for over 20 years. He received his BSc (1999) in Microbiology from Gujarat University, Godhra, Gujarat, India. He also earned his PhD (2005) in Environmental Microbiology from Sardar Patel University, Vallabh Vidyanagar, Gujarat, India. His research interests include biological wastewater treatment, environmental microbiology, biodegradation, bioremediation, and phytoremediation of environmental pollutants from industrial wastewaters. He has published more than 350 research papers in national and international journals of repute on various aspects of microbial biodegradation and the bioremediation of environmental pollutants. He is the editor of 150 books of international repute. He has edited 25 special issues specifically on industrial wastewater research, microbial remediation, and biorefinery of wastewater treatment area. He is associated as an Editorial Board Member for 25 highly reputed journals.

Suvendu Manna received his PhD in Materials Science in 2015. Currently, he is an assistant professor at the University of Petroleum and Energy Studies, Dehradun, India. He has authored or co-authored close to 45 international research and review articles in SCI- and Scopus-indexed journals. In addition, he has written some 21 book chapters and edited four books. His current research interests include nanomaterials for waste management, waste recycling, environmental microbiology, and water treatment.

Papita Das received her BTech in Chemical Engineering from the University of Calcutta, West Bengal, India, and ME PhD in Chemical Engineering from Jadavpur University, West Bengal, India. She is a professor in the Department of Chemical Engineering and Director, School of Advanced Studies in Industrial Pollution Control Engineering, Jadavpur University, India. She is known for her work in water treatment using different novel adsorbent materials. She also works on biomass-based energy production and the synthesis and degradation of polymeric-nanocomposites. She has published more than 175 international journal research articles and reviews and more than 40 book chapters in various SCI- and Scopus-indexed journals. She was also ranked among the top 2% of scientists (2020, 2021, and 2022) in the World Rankings published by Stanford University, representing the top 2% most-cited scientists in various disciplines. In the field of chemical engineering, in 2021, she was ranked 432 among 53,348 researchers based on career-long impact and 171 among single-year citations, and, in 2020, she was ranked 614 among 55,697 researchers based on career-long impact and 217 for single-year citations. She has also supervised ten (completed), two (submitted), and seven PhD students (ongoing). She has been the editor of two books, published by Elsevier and Springer, and she is an editorial board member, editor-in-chief, and associate editor of various international journals. She has completed 17 projects funded by government agencies and industry.

Circular Economy

Applications for Water Remediation

Edited by
Maulin P. Shah, Suvendu Manna, and
Papita Das

CRC Press
Taylor & Francis Group
Boca Raton London New York

CRC Press is an imprint of the
Taylor & Francis Group, an **informa** business

Designed cover image: Shutterstock

First edition published 2025
by CRC Press
6000 Broken Sound Parkway NW, Suite 300, Boca Raton, FL 33487-2742

CRC Press is an imprint of Taylor & Francis Group, LLC

ISBN: 978-1-032-55908-7 (hbk)
ISBN: 978-1-032-55909-4 (pbk)
ISBN: 978-1-003-43748-2 (ebk)

DOI: 10.1201/9781003432869

Typeset in Times
by Deanta Global Publishing Services, Chennai, India

Contents

Preface

In the past few decades, waste has been considered in terms of the resources that could be recycled and reused, if processed correctly. Reuse and recycling reduce the pressure of raw material supply for various industrial applications. Reclamation of clean water (and other essential materials) is one of the major research areas for understanding the effect of implementing circular economy concepts. Such a concept would change the overall water demands of our planet completely, if implemented successfully. This book revisits the current understanding of the circular economy in water remediation processes, its drawbacks, and relatively unexplored areas that need to be investigated by researchers. This book provides an overview of the available processes by which to extract value-added materials from wastewater, such as clean water, nutrients, and energy. This book also highlights the possibility of reusing wastewater for agricultural use and whether such an application would have an impact on the biomagnification of some of the pollutants present in the wastewater. Also, it is known that the overall composition of wastewater depends on the economic growth of a region, the types of industry present, climatic conditions, and the overall lifestyle of the citizens. Thus, implementation of the circular economy might need to be revised with some changes, as needed. This book presents our readers with a comprehensive overview of the development of such research. Researchers, academics, and industrial scientists share their research, reviews, and case studies in this book in the form of book chapters. Extraction of heavy metals, organic components, and energy from wastewater, using direct and indirect nano-technological interventions, are critically discussed in this book. The drawbacks, technical difficulties, and future research possibilities are also mentioned in many of the chapters.

We hope this book will provide guidance to all those budding researchers who want to start research on this innovative topic. We would like to thank all of the contributing authors for associating with this project. We express a deep sense of gratitude to UPES, Dehradun, and Jadavpur University, Kolkata, for allowing us to publish this book. We also would like to thank the Vice-Chancellor, Pro-Vice-Chancellor, and Registrar of both universities for their continuous support and motivation. We greatly appreciate the efforts of all of our reviewers for spending their valuable time to voluntarily review the manuscripts and provide their comments and suggestions in a timely manner. We would also like to thank CRC Press for accepting our proposal and publishing the book in a timely manner.

Dr. Maulin P. Shah
Dr. Papita Das
Dr. Suvendu Manna

Biography Including Experience

Dr. Maulin P Shah has been an active researcher and scientific writer in his field for over 20 years. He received a BSc in 1999 in Microbiology from Gujarat University, Godhra, Gujarat, India. He also earned his PhD in 2005 in Environmental Microbiology from Sardar Patel University, Vallabh Vidyanagar, Gujarat, India. His research interests include biological wastewater treatment, environmental microbiology, biodegradation, bioremediation, and phytoremediation of environmental pollutants from industrial wastewaters. He has published more than 350 research papers in national and international journals of repute on various aspects of microbial biodegradation and bioremediation of environmental pollutants. He is the editor of 150 books from publishers of international repute (RSC, Wiley, DeGruyter, Elsevier, Springer, and CRC Press). He has edited 25 special issues specifically in the industrial wastewater research, microbial remediation, and the biorefinery of wastewater treatment areas. He is associated as an Editorial Board Member in 25 highly respected journals, published by Elsevier, Springer, Taylor & Francis, and Wiley.

Dr. Papita Das was awarded her BTech in Chemical Engineering from the University of Calcutta, West Bengal, India, and ME and PhD in Chemical Engineering from Jadavpur University, West Bengal, India. She is a professor at the Department of Chemical Engineering and Director of the School of Advanced Studies in Industrial Pollution Control Engineering, Jadavpur University. She is known for her work on water treatment using different novel adsorbent materials. She also works on biomass-based energy production and polymeric-nanocomposite synthesis and their degradation. She has published more than 175 research articles and reviews in various SCI- and Scopus-indexed journals and more than 40 book chapters. She was also included in the World Ranking of the top 2% scientists (2020 and 2021), published by Stanford University, which represents the top 2% most-cited scientists in various disciplines. She was ranked 614 among 55,697 researchers in chemical engineering (2020) and 534 among 66,189 researchers in the field of chemical engineering (2021) based on career-long impact and ranked 217 in the discipline for the single-year (2020) impact. She has also supervised ten PhD students (completed), three submitted and seven ongoing. She is the editor of two books published by Elsevier and Springer and she is and Editorial Board Member, Editor-in-Chief, and Associate Editor of various international journals. She has completed 17 projects funded by government agencies and industrial partners.

Dr. Suvendu Manna was awarded his PhD in Materials Science in 2015 from the Indian Institute of Technology Kharagpur, West Bengal, India. Currently, he is an assistant professor in the UPES (University of Petroleum and Energy Studies), Dehradun, India. He has authored or co-authored almost 45 international research and review articles in SCI- and Scopus-indexed journals. Also, he has written some 21 book chapters and edited four books. His current research interests are nanomaterials for waste management, waste recycling, environmental microbiology, and water treatment.

1 Integrating Circular Economy and Sustainability Strategy in Water Remediation Applications

Kassian T.T. Amesho
Institute of Environmental Engineering,
National Sun Yat-Sen University, Kaohsiung 804, Taiwan
Center for Emerging Contaminants Research,
National Sun Yat-Sen University, Kaohsiung 804, Taiwan
Tshwane School for Business and Society, Faculty
of Management of Sciences, Tshwane University
of Technology, Pretoria, South Africa
The International University of Management,
Centre for Environmental Studies, Main Campus,
Dorado Park Ext 1, Windhoek, Namibia
Destinies Biomass Energy and Farming Pty Ltd,
P. O. Box 7387, Swakopmund, Namibia
Regent Business School, Durban 4001 South Africa

Abner Kukeyinge Shopati
Namibia Business School (NBS), Faculty of Commerce,
Management and Law, University of Namibia, Private
Bag 13301, Main Campus, Windhoek, Namibia

Sumarlin Shangdiar
Center for Emerging Contaminants Research,
National Sun Yat-Sen University, Kaohsiung 804, Taiwan
Tshwane School for Business and Society, Faculty
of Management of Sciences, Tshwane University
of Technology, Pretoria, South Africa

DOI: 10.1201/9781003432869-1

Timoteus Kadhila
School of Education, Department of Higher Education and Lifelong Learning, University of Namibia, Private Bag 13301, Windhoek, Namibia

Sioni Iikela
The International University of Management, Centre for Environmental Studies, Main Campus, Dorado Park Ext 1, Windhoek, Namibia

Nastassia Thandiwe Sithole
Department of Chemical Engineering, Faculty of Engineering and the Built Environment at the University of Johannesburg, South Africa

E.I. Edoun
Tshwane School for Business and Society, Faculty of Management of Sciences, Tshwane University of Technology, Pretoria, South Africa

Subrata Chowdhury
Department of Computer Science and Engineering, Sreenivasa Institute of Technology and Management Studies, Chittoor, Andra Pradesh, India

1.1 INTRODUCTION

The confluence of circular economy and sustainability has emerged as a compelling imperative in the realm of water remediation, where the preservation of environmental integrity and the well-being of societies converge. This chapter embarks on a profound exploration of their intersection, aiming to reveal the synergistic potential they hold when harmoniously integrated into water remediation strategies.

1.1.1 Significance of the Circular Economy and Sustainability in Water Remediation

Water, an indispensable and finite resource, assumes a pivotal role in maintaining ecological equilibrium and supporting human livelihoods. However, contemporary challenges, including water pollution, scarcity, and contamination, underscore the urgency of innovative approaches. Circular economy principles, with their emphasis on optimizing resource utilization, minimizing waste, and adopting closed-loop systems, offer a promising framework for addressing these complex issues. Through the lens of circular water management, the potential emerges to counteract resource depletion, alleviate waste burden, and recalibrate the ecological balance that conventional linear models often strain (Liu et al., 2018; Salminen et al., 2022).

The infusion of sustainability principles amplifies the transformative potential of circular strategies. Sustainability, encompassing economic, environmental, and social dimensions, accentuates the need for holistic, enduring solutions. In the context of water remediation, this implies not only safeguarding aquatic ecosystems but also ensuring equitable access to clean water resources for diverse communities (Mannina et al., 2021; ECA, 2019). The amalgamation of the circular economy and sustainability presents an integrated pathway toward environmental restoration, economic prosperity, and societal equity through resource recovery in wastewater treatment as demonstrated in Figure 1.1.

1.1.2 Importance of Integration: Toward a Holistic Paradigm

While circular economy and sustainability exhibit inherent merits, their synergistic convergence fosters a holistic paradigm that transcends individual tenets. This synergy begets an approach that synergizes ecological, economic, and social considerations, establishing a resilient foundation for water remediation efforts (Pradel et al., 2016; Gherghel et al., 2019).

The subsequent sections of this chapter will cover the intricate tapestry of the integration of the circular economy and sustainability in water remediation. With meticulous examination of theoretical underpinnings, empirical insights, and illustrative case studies, we navigate the multifaceted dimensions of this integration. By delving into policy frameworks, innovative strategies, and practical applications, we illuminate the dynamic interplay that propels water remediation toward a regenerative trajectory.

Our endeavor is to illuminate the transformative potential borne out by the harmonious alignment of circular economy and sustainability principles. As we embark on this exploratory journey, we invite readers to embark on a comprehensive exploration of the intricacies, complexities, and synergies that underpin this vital convergence.

1.2 CIRCULAR ECONOMY AND WATER REMEDIATION

1.2.1 Definition and Key Principles of the Circular Economy

The concept of a circular economy represents a profound departure from the conventional linear model of resource consumption and disposal (Ferrans et al., 2020). Grounded in principles of regeneration and sustainability, a circular economy aims to establish closed-loop systems where resources are continually cycled, waste is minimized, and the value of materials is retained for as long as possible (Maryam et al., 2019). This paradigm shift entails a departure from the linear "take-make-dispose" mindset to a holistic and cyclical approach, promoting the mantra of "reduce-reuse-recycle" and emphasizing the need for systemic change (World Bank, 2020; Kehrein et al., 2020).

1.2.2 Relevance of Circular Economy to Water Remediation

The interface between circular economy principles and water remediation is far from coincidental; it holds transformative potential to address critical environmental and social challenges. The scarcity and vulnerability of water resources underscore

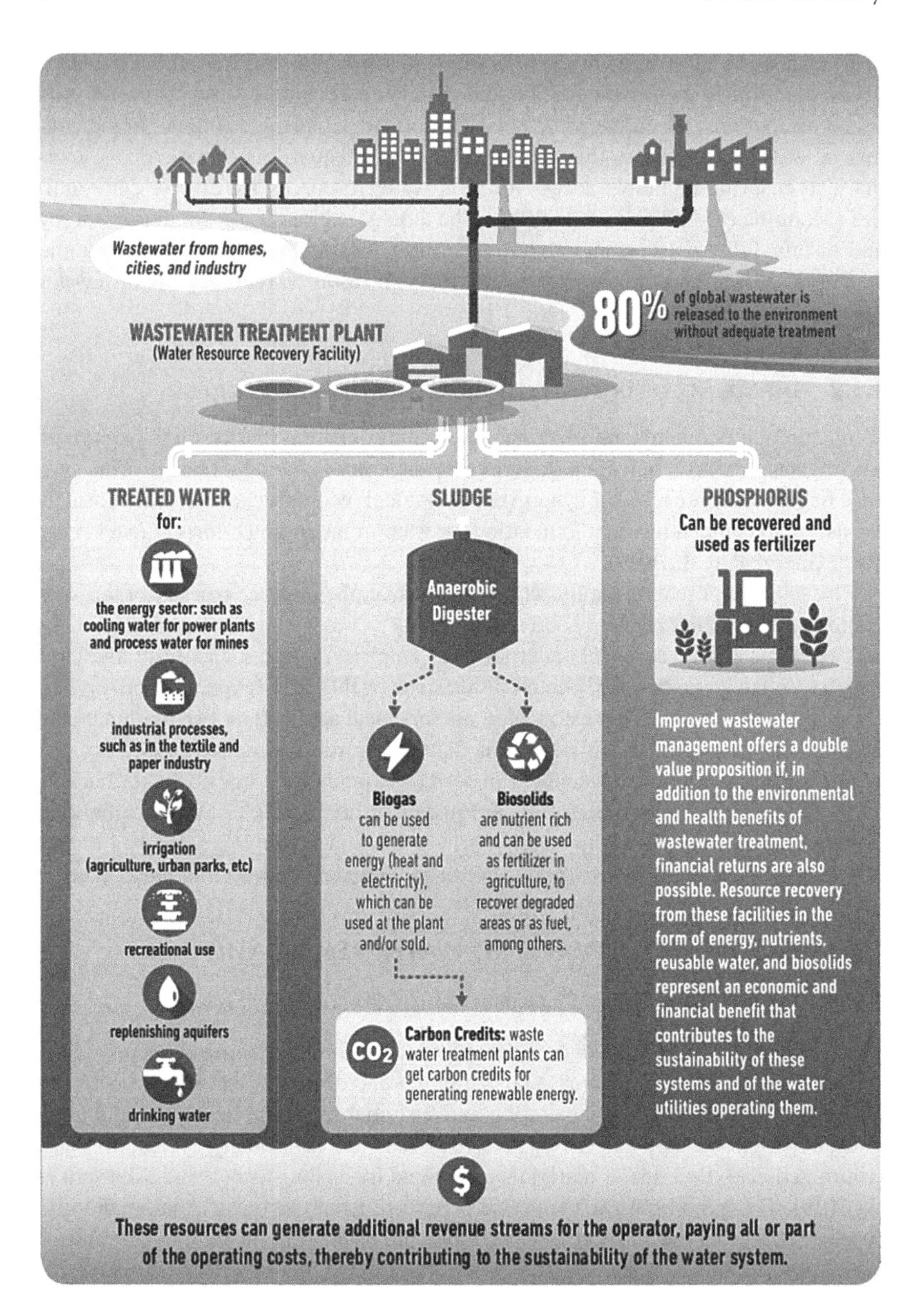

FIGURE 1.1 Resource recovery in wastewater treatment plants. (Source: Rodriguez et al., 2020.)

the urgency of adopting sustainable practices that conserve and optimize water use. Circular water management aligns seamlessly with the ethos of circular economy by advocating for resource efficiency, recovery, and reuse (IWA, 2018). This convergence between circular economy and water remediation is not merely theoretical; it holds the promise of offering pragmatic solutions to the interconnected problems of water pollution, scarcity, and degradation, as demonstrated in Figure 1.2.

The alignment of circular economy principles with the realm of water remediation goes beyond mere happenstance; it represents a convergence of ideologies that bears the potential to effect profound transformations, tackling pressing environmental and societal dilemmas head-on. In a world where water resources are beleaguered by

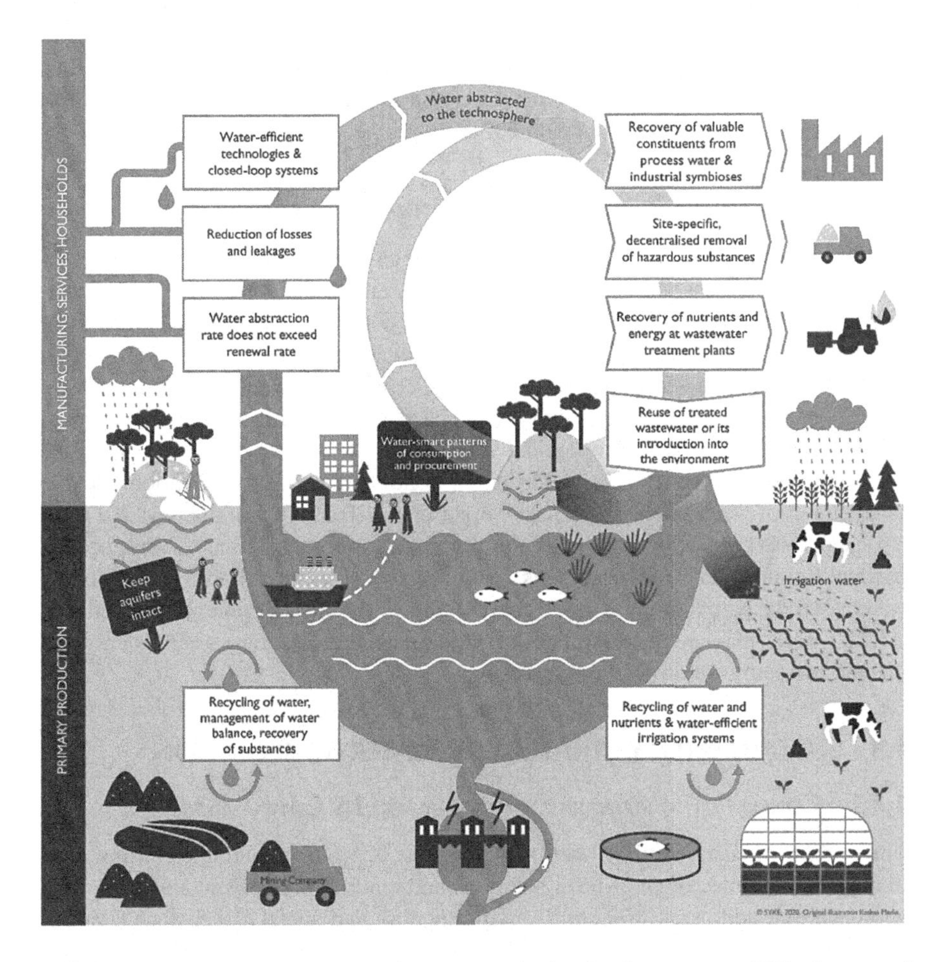

FIGURE 1.2 Water and water-related ecosystems in the circular economy (CE); elements of a water-smart CE. (Obtained with permission from Salminen et al., 2022.) [This is an open access article distributed under the terms of the Creative Commons CC-BY license, which permits unrestricted use, distribution, and reproduction in any medium, provided the original work is properly cited – https://doi.org/10.1016/j.jclepro.2021.130065.]

scarcity and vulnerability, the imperative for sustainable practices that safeguard and optimize water utilization has never been more urgent. This is where the symbiotic marriage of circular economy principles with water remediation enters the stage, poised to chart a trajectory toward resilience and rejuvenation.

Amid the backdrop of burgeoning urbanization and escalating industrialization, the discourse on water remediation becomes indelibly intertwined with considerations of ecological balance and social equity (PUB, 2016). The convergence of the circular economy and water remediation channels its energies toward the resolution of these multifaceted challenges. The integration of circularity into water management heralds the promise of rejuvenated ecosystems, as well as enhanced water quality that nurtures both human populations and biodiversity. The transformative potential of this amalgamation lies in its capacity to weave together intricate strands of sustainable practices, resource optimization, and environmental stewardship, thus unveiling an avenue to address the intricate tapestry of modern-day water dilemmas (Ganora et al., 2019; Cipolletta et al., 2021).

1.2.3 Potential Benefits of Circular Water Management Practices

The integration of circular water management practices harbors a spectrum of benefits that cascade across environmental, economic, and societal dimensions. The efficiency-driven principles of circularity inherently contribute to easing the ecological burden by curbing pollution and ecosystem degradation (Wang et al., 2017; Mannina et al., 2020). Concurrently, resource recovery from wastewater and other sources conserves precious freshwater resources and harnesses valuable materials that would otherwise be squandered. Economically, circular water management can stimulate economic growth through the establishment of new industries, job creation, and the fortification of water supply systems (Werker et al., 2020; Valentino et al., 2017). In a social context, the adoption of circular practices can foster equitable water access, particularly in water-stressed regions, and mitigate social inequities exacerbated by water scarcity (Cinà et al., 2019). Table 1.1 shows the diverse potential benefits of circular water management practices and the associated challenges across environmental, economic, and societal dimensions.

1.3 SUSTAINABILITY STRATEGY IN WATER REMEDIATION

1.3.1 Concept of Sustainability Strategy and Its Components

The pursuit of sustainability, within the context of water remediation, transcends the narrow confines of short-term solutions and embraces a holistic and integrated approach that addresses economic, environmental, and social dimensions (Papa et al., 2017). A sustainability strategy represents a structured framework that integrates these dimensions to ensure that actions taken today do not compromise the ability of future generations to meet their own needs (Hart et al., 2019). At its core, a sustainability strategy seeks equilibrium between economic prosperity, environmental stewardship, and social equity (Mukherjee and Jensen, 2020).

TABLE 1.1
Potential Benefits of Circular Water Management Practices and the Associated Challenges across Environmental, Economic, and Societal Dimensions

Dimension	Potential Benefits of Circular Water Management Practices	Challenges
Environmental	• **Resource Conservation:** Minimizes resource extraction and consumption, reducing pressure on ecosystems. • **Waste Reduction:** Reduces water pollution and minimizes waste generation, promoting cleaner water bodies. • **Biodiversity Conservation:** Preserves aquatic ecosystems and supports biodiversity through sustainable water use. • **Energy Efficiency:** Optimizes energy use through efficient water treatment processes, lowering carbon emissions. • **Climate Resilience:** Enhances water availability during drought and supports climate adaptation through sustainable practices.	• **Integration Complexity:** Adapting existing infrastructure to circular systems can be challenging. • **Behavior Change:** Encouraging public and industry adoption of circular practices may require significant behavior change. • **Technological Barriers:** Implementing advanced treatment technologies for resource recovery may face technical barriers. • **Policy and Regulation**: Inconsistent or inadequate regulations can hinder circular water management adoption. • **Monitoring and Measurement:** Accurate tracking and reporting of circular water practices' impact may pose challenges.
Economic	• **Cost Savings:** Reduces costs associated with resource procurement, treatment, and waste management. • **Revenue Generation:** Creates revenue streams through resource recovery, such as reclaimed water and valuable by-products. • **Market Opportunities:** Unlocks new markets for recycled water, reclaimed materials, and circular products. • **Long-Term Viability:** Ensures sustainable water supply, reducing business vulnerability to water scarcity and price fluctuations. • **Job Creation:** Stimulates job growth in water treatment, technology development, and circular supply chains.	• **Financial Investment:** Initial investments in circular infrastructure and technology can be substantial. • **Market Acceptance:** Developing markets for circular products may require overcoming consumer and industry skepticism. • **Regulatory Uncertainty:** Evolving policies and regulations can impact the economic feasibility of circular water practices. • **Return on Investment:** Demonstrating tangible economic benefits and payback periods may be challenging for circular projects. • **Value Chain Coordination:** Coordinating various stakeholders in circular value chains can be complex.

(Continued)

TABLE 1.1 (CONTINUED)
Potential Benefits of Circular Water Management Practices and the Associated Challenges across Environmental, Economic, and Societal Dimensions

Dimension	Potential Benefits of Circular Water Management Practices	Challenges
Societal	• **Public Health:** Improves water quality, reducing health risks associated with contaminated water. • **Community Engagement:** Fosters public participation in water conservation and sustainability efforts. • **Equity and Access:** Ensures equitable water distribution and access, benefiting marginalized communities. • **Resilience to Disasters:** Enhances disaster resilience by maintaining water availability during emergencies. • **Education and Awareness:** Raises awareness about water conservation and fosters a sense of environmental responsibility.	• **Cultural Barriers:** Cultural norms and practices may influence the acceptance of circular water management. • **Socioeconomic Disparities:** Ensuring equitable access to circular water benefits may be challenging in marginalized communities. • **Behavioral Change:** Encouraging individuals and communities to adopt circular practices can require behavior change efforts. • **Public Awareness:** Raising awareness about the benefits and importance of circular water management may be needed. • **Social Acceptance:** Garnering societal support for circular initiatives could face resistance or lack of understanding.

1.3.2 Role of Sustainability in Shaping Effective Water Remediation Approaches

In the realm of water remediation, the integration of sustainability principles is paramount for several compelling reasons. First, water is an invaluable global resource, indispensable for human well-being, ecosystem health, and industrial processes (Alamerew and Brissaud, 2019). Therefore, any endeavor to rectify water pollution and scarcity must rest upon a foundation that fosters long-term resilience. Second, the multifaceted nature of water-related challenges necessitates a comprehensive strategy that considers not only technical and scientific aspects but also societal and economic implications (Alamerew et al., 2020). Third, sustainability injects a sense of responsibility into water management practices, obliging stakeholders to consider the broader ramifications of their actions on ecosystems and societies.

1.3.3 Synergies Between the Circular Economy and Sustainability Goals

The symbiotic relationship between circular economy principles and sustainability objectives is a driving force in advancing effective water remediation strategies. Circular economy principles, rooted in waste minimization, resource optimization, and closed-loop systems, align seamlessly with sustainability's mandate to harmonize economic prosperity, environmental health, and social well-being (de Oliveira et al., 2021). Circular water management strategies, such as wastewater recycling and resource recovery, embody the essence of sustainable development by simultaneously addressing water pollution, water scarcity, and economic inefficiencies (Aydin and Badurdeen et al., 2019). These strategies offer an illustrative example of how circular economy approaches can amplify the impact of sustainability initiatives.

1.4 CLOSED-LOOP SYSTEMS AND RESOURCE RECOVERY

1.4.1 Closed-loop Systems in Water Treatment Processes

Closed-loop systems, emblematic of circular economy principles, stand as a paradigm shift in water treatment processes, characterized by their capacity to minimize waste generation and maximize resource utilization (Fatimah et al., 2020). Unlike the traditional linear approach, wherein water is used once and then discarded as an effluent, closed-loop systems emphasize the recirculation and reuse of water within a controlled cycle (Liu et al., 2018) as demonstrated in Figure 1.3. This entails the integration of innovative technologies that enable the purification and regeneration of water to suitable quality levels for specific applications, obviating the need for continuous freshwater intake (Ghisellini et al., 2016).

1.4.2 Showcase of Successful Resource Recovery Initiatives in Water Remediation

One of the salient achievements of closed-loop systems lies in their ability to achieve resource recovery from wastewater streams. Through advanced treatment processes,

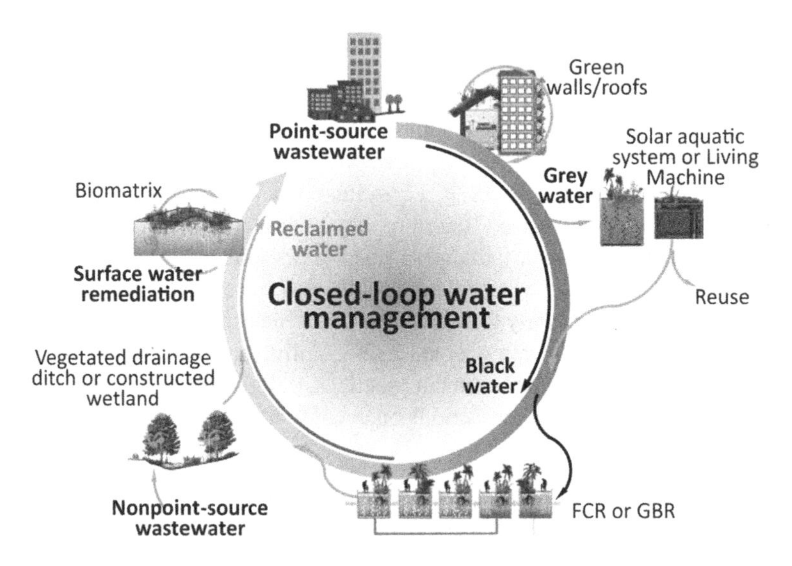

FIGURE 1.3 Closed-loop systems for water remediation processes. (Obtained with permission from Liu et al., 2018.) Copyright (2018), Elsevier [License Number: 5603330145179].

valuable resources, such as nutrients, metals, and energy, are extracted from wastewater, transforming it from a mere waste product into a reservoir of latent economic potential (Naidoo et al., 2021). Nutrient recovery, for instance, has gained considerable traction in recent years, with technologies such as struvite precipitation and algal cultivation offering avenues to convert wastewater-derived nutrients into agricultural fertilizers and biofuels (Hapuwatte et al., 2022). Moreover, the extraction of valuable metals from industrial effluents exemplifies the amalgamation of environmental stewardship and economic gain, aligning them closely with the circular economy ethos (Panchal et al., 2021).

1.4.3 Case Studies Illustrating the Implementation and Outcomes of Closed-loop Systems

Several compelling case studies illustrate the transformative potential of closed-loop systems in water remediation. The Copenhagen Water Recycling Plant in Denmark exemplifies the successful implementation of a closed-loop system, wherein treated wastewater is repurposed for cooling in industrial processes, reducing freshwater consumption and alleviating strain on local water resources (Sartal et al., 2020; Morseletto, 2020). In another instance, the NEWater initiative in Singapore underscores the viability of closed-loop systems in augmenting water supply. This initiative involves treating municipal wastewater to ultra-pure standards, rendering it fit for potable use and thereby enhancing water resilience in a water-scarce region (PUB, 2016; Schroeder et al., 2019).

As the push for sustainability gains momentum, the integration of closed-loop systems in water remediation offers a potent pathway to achieve water security, resource efficiency, and environmental integrity. The next section will delve further into the concept of product design for circularity and its pivotal role in advancing circular economy objectives within the context of water treatment.

1.5 INNOVATIVE PRODUCT DESIGN FOR CIRCULARITY

1.5.1 Product Design for Circularity in Water Treatment

In the pursuit of a sustainable and circular approach to water remediation, the concept of innovative product design emerges as a potent enabler. Product design for circularity embodies the ideology of creating goods with a lifespan that transcends the conventional linear model of production and disposal, instead emphasizing durability, repairability, and recyclability (Priyadarshini et al., 2020; Chia et al., 2021). Applied to water treatment technologies, circular design principles advocate for the development of systems that not only purify water effectively but also embody longevity and resource efficiency across their lifecycle.

1.5.2 Water Treatment Products Designed for Extended Use and Recycling

In the field of water remediation, remarkable strides have been made in aligning product design with circularity goals. Water filtration systems, for instance, have undergone transformational innovation, giving rise to modular and adaptable configurations that permit the replacement of worn components rather than the entire unit (Kristensen and Mosgaard, 2020). This not only diminishes electronic waste but also reduces the demand for virgin raw materials. Another exemplar is the advent of water treatment membranes engineered for enhanced durability, thereby elongating their service life and necessitating fewer replacements (Rocchi et al., 2021).

1.5.3 Benefits of Incorporating Circular Design Principles in Water Remediation Technologies

The incorporation of circular design principles in water remediation technologies accrues an array of benefits that reverberate across environmental, economic, and societal dimensions. Environmental gains stem from the reduction in raw material consumption and waste generation, thereby attenuating the ecological footprint of water treatment processes (Aditya et al., 2022; Bolognesi et al., 2021). Economically, circularly designed systems can engender substantial cost savings through reduced maintenance, lower replacement frequency, and the prospect of reusing recovered materials (Chai et al., 2021). Furthermore, the societal advantages include heightened resource resilience, where the preservation and regeneration of resources contribute to enhanced water security and sustained ecosystem health (Cheng et al., 2019).

As an illustration, the Watergy system, developed by American Water, underscores the potency of circular product design. This system integrates resource recovery into

TABLE 1.2
Benefits of Incorporating Circular Design Principles in Water Remediation Technologies

Benefits	Description
1. Resource efficiency	Circular design minimizes resource consumption by optimizing processes, reducing the need for raw materials and energy.
2. Reduced waste generation	Circular design minimizes resource consumption by optimizing processes, reducing the need for raw materials and energy.
3. Extended product lifespan	Circular design promotes the creation of products and systems with longer lifespans, reducing the need for frequent replacements and minimizing environmental impact.
4. Material selection	Circular design encourages the use of sustainable and eco-friendly materials, reducing the environmental footprint of the remediation technologies.
5. Modular and scalable	Modular designs allow for easier maintenance and upgrades leading to more adaptable and scalable solutions that can meet changing remediation needs.
6. Reduced energy consumption	Circular design optimizes energy use by promoting energy efficiency and integration of renewable energy sources.
7. Water conservation and reuse	Circular design focuses on minimizing water consumption and promoting water reuse, contributing to sustainable water management.
8. Ecological benefits	Circular design can have positive impacts on ecosystems, reducing pollution and supporting biodiversity.
9. Cost savings	Circular design principles can lead to cost savings in the long term through reduced resource and operational costs.
10. Innovation and collaboration	Circular design encourages innovative thinking and creative solutions, fostering collaboration among stakeholders.

water treatment, facilitating the extraction of energy and nutrients from wastewater while concurrently purifying water (Mohsenpour et al., 2021; Ferreira et al., 2019). This amalgamation of functionality epitomizes the synergy between circular design and water remediation, fostering not only environmental benefits but also economic viability. Table 1.2 shows the benefits of incorporating circular design principles in water remediation technologies.

In the subsequent section, the pivotal role of stakeholder engagement and collaboration in propelling circular economy initiatives within the realm of water remediation will be expounded upon.

1.6 STAKEHOLDER ENGAGEMENT AND COLLABORATION

1.6.1 Importance of Stakeholder Engagement in Circular Water Management

The harmonization of circular economy principles and sustainable water remediation necessitates an orchestration of efforts, involving stakeholders across industries, governments, communities, and academia. The multifaceted challenges associated with water remediation, coupled with the intricate interplay of ecological, societal, and economic dynamics, highlight the indispensability of stakeholder engagement. Stakeholders, as both recipients and custodians of water resources, possess insights, perspectives, and aspirations that are pivotal for crafting effective and contextually relevant circular strategies (Goswami et al., 2021; Liyanaarachchi et al., 2021). Their active participation instills a sense of shared ownership and commitment, underpinning the development of holistic and enduring solutions.

1.6.2 Collaborative Partnerships among Industries, Governments, and Communities

Collaboration emerges as the linchpin for advancing circular water management endeavors. The synergy between industries, governments, and communities presents a powerful amalgamation of expertise, resources, and authority, each essential for the materialization of circular visions. Industries contribute technological innovation, operational acumen, and financial resources, acting as catalysts for the implementation of circular practices within water treatment operations (Veleva et al., 2017). Governments wield regulatory influence and policy instruments that can incentivize circular behaviors while ensuring equitable access to water resources. Communities, as the primary beneficiaries, harbor contextual knowledge, local insights, and the potential to drive behavioral changes that support circular water management (Mesa et al., 2018).

1.6.3 Collaborative Initiatives in Sustainable Water Treatment

Examples abound wherein collaborative initiatives have underpinned the realization of sustainable water treatment goals. The Singaporean water reclamation initiative, NEWater, epitomizes the convergence of collaborative efforts. Orchestrated by the Public Utilities Board (PUB), NEWater demonstrates how interagency collaboration, research partnerships, and community awareness campaigns engendered the acceptance and adoption of treated wastewater for potable and non-potable uses (PUB, 2016). This collaborative triumph underscores the essential requirement of stakeholder unity and proactive engagement in reshaping water treatment paradigms.

Additionally, the partnership between the City of Philadelphia and the non-profit organization Philadelphia Water Green showcases the synergy between local governance and community action. Through collaborative planning, design, and implementation, the Green City, Clean Waters program integrates nature-based solutions to urban water management, mitigating stormwater runoff, and improving water quality (Schroeder et al., 2019). This harmonious interaction accentuates the role of local communities in fostering resilient and circular water systems. Table 1.3 lists collaborative partnerships between industries, governments, and communities in advancing circular water management, along with their pros and challenges.

As the subsequent section investigates the role of policy and regulations, it becomes evident that stakeholder engagement and collaboration serve as potent

TABLE 1.3
Collaborative Partnerships between Industries, Governments, and Communities in Advancing Circular Water Management

Collaborative Partnership	Description	Pros	Challenges
Public-Private Partnerships (PPP)	Collaborative efforts between governmental agencies and private companies to jointly develop, finance, and implement circular water management projects.	• Access to private sector expertise and funding. • Shared risks and responsibilities. • Faster project implementation. • Improved resource allocation.	• Complex negotiation and agreement processes. • Potential conflicts of interest. • Balancing profit motives with public interests.
Multi-Stakeholder Platforms	Inclusive forums involving industries, governments, communities, NGOs, and academia to collectively design and implement circular water initiatives.	• Diverse perspectives and expertise. • Enhanced innovation through collaboration. • Broader acceptance and support for initiatives.	• Potential for disagreements and conflicts among stakeholders. • Decision-making may be slower due to need for consensus-building.
Regulatory Collaboration	Governmental bodies and industries collaborating to develop and enforce regulations that promote circular water management practices.	• Alignment of policies and regulations with industry goals. • Clear guidelines for implementation. • Ensures compliance and accountability.	• Balancing regulatory stringency with industry viability. • Potential resistance to change from industries.
Community Engagement	Engaging local communities in circular water management decisions, ensuring their needs and concerns are considered.	• Local knowledge and insights. • Increased public acceptance and support. • Tailored solutions to community needs.	• Time-consuming community outreach and consultation. • Potential resistance to change from communities. • Balancing community interests with broader environmental goals.

(Continued)

TABLE 1.3 (CONTINUED)
Collaborative Partnerships between Industries, Governments, and Communities in Advancing Circular Water Management

Collaborative Partnership	Description	Pros	Challenges
Research and Development Partnerships	Collaborations between industries, governments, and research institutions to develop and test innovative circular water technologies.	• Access to cutting-edge research and technology. • Accelerated development and implementation. • Shared costs and resources.	• Intellectual property and technology sharing challenges. • Balancing commercial interests with research goals.
Capacity- Building Programs	Collaborative efforts to provide training, education, and skills development to industries, governments, and communities in circular water management practices.	• Enhanced knowledge and expertise. • Increased adoption of circular practices. • Improved long-term sustainability.	• Resource and funding limitations for training programs. • Ensuring effective knowledge transfer and application.

prerequisites for the effective translation of circular economy principles and sustainability strategies into actionable policies and measures.

These collaborative partnerships play a crucial role in advancing circular water management by harnessing the collective strengths and resources of industries, governments, and communities. However, they also face various challenges that need to be addressed to ensure successful implementation of circular water initiatives.

1.7 INNOVATION ECOSYSTEMS FOR CIRCULAR WATER SOLUTIONS

1.7.1 Exploration of Innovation Ecosystems Fostering Circular Economy Innovations

The dynamic landscape of circular water solutions flourishes within innovation ecosystems, characterized by a nexus of actors, institutions, and processes that collectively propel the genesis, development, and diffusion of circular economy innovations (Yarnold et al., 2019). Such ecosystems transcend disciplinary boundaries, engendering a fertile milieu where diverse stakeholders synergistically collaborate to address water remediation challenges. Central to these ecosystems is the interconnectedness of academia, industry, governmental agencies, non-governmental organizations, and civil society. These stakeholders collaboratively incubate ideas, share knowledge, pool resources, and co-create breakthrough solutions, thus accelerating the transition toward circular water management paradigms (Liu et al., 2020).

1.7.2 Role of Technology, Research, and Development in Advancing Circular Water Remediation

Innovation ecosystems pivot around the indispensable triad of technology, research, and development, serving as the engine that drives the evolution of circular water solutions. Technological innovation, manifested in cutting-edge tools and methodologies, underpins the design, optimization, and deployment of circular water treatment technologies. Advanced sensor technologies, predictive modeling, and artificial intelligence not only enhance monitoring and process control but also drive resource recovery, minimize wastage, and optimize energy utilization within water remediation (Rizwan et al., 2018).

Concomitantly, research and development (R&D) catalyzes the evolution of circular water solutions through iterative experimentation, scientific inquiry, and interdisciplinary exploration. R&D endeavors deal with the elucidation of novel materials, treatment techniques, and process intensification strategies, often resulting in paradigm-shifting advances. The confluence of academic inquiry and practical application nurtures a cycle of continuous improvement, infusing innovation ecosystems with a repertoire of evidence-based and scalable circular water strategies (Sandefur et al., 2016).

1.7.3 Innovative Technologies Contributing to Circular Water Solutions

The symbiotic interplay between technology, research, and development spawns an array of innovative technologies designed to revolutionize circular water solutions. Membrane bioreactors (MBRs), for instance, epitomize the integration of biological treatment processes with advanced membrane filtration, engendering efficient pollutant removal, resource recovery, and water reuse (Sutherland et al., 2021; Zahmatkesh et al., 2022a). Additionally, microalga-based systems harness the natural powers of microorganisms to sequester pollutants, bioaccumulate valuable compounds, and synthesize biomass, thereby encapsulating a sustainable nexus of water treatment and resource generation (Shehata et al., 2022).

Furthermore, the proliferation of decentralized treatment systems, leveraging modular and compact designs, empowers localized water remediation efforts while minimizing infrastructure costs and environmental footprints (Zabed et al., 2020). The innovative coupling of anaerobic digestion and microbial fuel cells (MFCs) introduces an example of energy recovery from wastewater, wherein microbial metabolism is harnessed to concurrently treat water and generate bioelectricity (Thompson et al., 2020).

Collectively, these technologies epitomize the potential of innovation ecosystems to usher in transformative shifts in circular water solutions, reiterating the imperative of continuous technological advancement and cross-disciplinary collaboration in the pursuit of sustainable water management paradigms. Table 1.4 shows some innovative technologies contributing to circular water solutions.

1.8 POLICY INSTRUMENTS FOR CIRCULAR WATER MANAGEMENT

1.8.1 Policy Instruments Promoting the Circular Economy in Water Remediation

The transformative potential of circular economy principles in water remediation is buoyed by a suite of policy instruments designed to incentivize sustainable practices and encourage holistic resource management. These policy tools, which traverse legislative, regulatory, and market-based domains, act as pivotal drivers in shaping the landscape of circular water management. By aligning regulatory frameworks with circular principles, policy instruments exert a catalytic force, propelling industries, governments, and communities toward the adoption of circular water strategies (Thompson et al., 2020).

1.8.2 Extended Producer Responsibility (EPR) and Its Application in Water Treatment

A cornerstone of circular water management lies in the conception of extended producer responsibility (EPR), a policy instrument mandating manufacturers to bear accountability for the entire lifecycle of their products, including post-consumer

TABLE 1.4
Innovative Technologies Contributing to Circular Water Solutions

Innovative Technologies	Merits	Limitations
1. Membrane bioreactors (MBRs)	• Efficient removal of pollutants and pathogens. • Produces high-quality treated water for reuse. • Compact design. • Reduced sludge production.	• High initial and operational costs compared to conventional systems. • Requires regular maintenance. • Energy-intensive process.
2. Advanced oxidation processes (AOPs)	• Effective removal of organic and micro-pollutants. • Suitable for various contaminants. • Breaks down contaminants into non-toxic end-products.	• High energy consumption. • Generation of harmful by-products or residuals. • Complex to design and operate. • Costly implementation.
3. Alga-based treatment systems (algal ponds and photobioreactors)	• Natural nutrient removal by algae. • Production of biomass for various applications. • Capture carbon dioxide and releases oxygen. • Efficient removal of nutrients, metals, and other contaminants.	• Seasonal variations algal growth. • Need for monitoring and control of algal species. • Potential for nutrient imbalance. • High land and space requirement.
4. Nutrient recovery and removal technologies (struvite precipitation, electrocoagulation)	• Convert nutrients into valuable products (e.g., struvite). • Reduces nutrient pollution. • Prevents eutrophication and harmful algal blooms. • Energy efficient.	• High capital and operational costs for certain technologies. • Specific conditions required for nutrient recovery.
5. Decentralized wastewater treatment (green infrastructure, onsite systems)	• Reduced infrastructure and energy requirements. • Local water reuse potential. • Enhanced storm water management. • Nutrient cycling and ecosystem services. • Aesthetic and recreational benefits.	• Maintenance and monitoring of individual systems. • May not be suitable for dense urban areas. • Sizing and design challenges.

stages (Schroeder et al., 2019). EPR is particularly relevant within water treatment realms, wherein the integration of sustainable materials, modular designs, and effective end-of-life management achieves resource conservation and waste mitigation. Water treatment equipment and technologies can be engineered with a modularity that facilitates component replacement, refurbishment, or repurposing, aligning with the principles of EPR and emboldening circularity (Veleva et al., 2017).

1.8.3 Eco-design Regulations and Their Influence on Circular Water Remediation Technologies

Eco-design regulations, encapsulating the cradle-to-grave perspective, herald a proactive stance in shaping circular water remediation technologies. These regulations stipulate design criteria that promote durability, reparability, and recyclability, fostering the integration of circular principles at the nascent stages of technological innovation (Chai et al., 2021; Zahmatkesh et al., 2022b). For instance, eco-design directives can inspire the development of water treatment systems characterized by minimal resource consumption, streamlined maintenance, and harmonious integration into the circular water infrastructure. Through the alignment of technological design with regulatory imperatives, eco-design principles harmoniously intersect with circular economy ambitions.

1.8.4 Green Public Procurement and Its Impact on Circular Water Management

The potency of public procurement as an instrument of change is harnessed through green public procurement (GPP), wherein public authorities opt for products and services that showcase superior environmental performance, encompassing circular economy attributes (Cheng et al., 2019). In the context of water management, GPP materializes as a conduit for fostering the adoption of circular water technologies by leveraging the influential market demand of public entities. The integration of circular water strategies, such as resource recovery, reuse, and modular design, can resonate within GPP frameworks, bolstering the market presence of circular water solutions and stimulating industrial innovation (Goswai et al., 2021).

In summary, policy instruments form an indelible pillar of circular water management, harmonizing regulatory frameworks, market dynamics, and societal imperatives to forge a symbiotic nexus between sustainability and circularity.

1.9 HOLISTIC APPROACH TO THE CIRCULAR ECONOMY AND SUSTAINABILITY

In the quest for resilient and sustainable water management, a shift toward holistic circular economy strategies is needed, underpinned by an integration of economic, environmental, and social considerations. This section analyzes the imperative of adopting a comprehensive approach that traverses the lifecycle of water remediation solutions, fostering a harmonious convergence of circularity and sustainability objectives.

1.9.1 Holistic Approach to Circular Water Management

The intricacies and interdependencies inherent in water management systems necessitate a departure from fragmented strategies to embrace a holistic framework. Circular water management, an intricate amalgamation of circular economy principles and sustainability strategies, provides water treatment processes with multifaceted advantages, while concurrently acknowledging the various challenges facing water systems (Liyanaarachchi et al., 2021; Zahmatkesh et al., 2023). The pivotal paradigm shift toward circularity requires a holistic perspective that transcends siloed approaches, advocating for the synthesis of ecological, economic, and social dimensions.

1.9.2 Integration of Economic, Environmental, and Social Considerations

At the heart of a holistic circular water management ethos resides the meticulous integration of economic, environmental, and social considerations, forming an intricate tapestry of interrelated facets. This entails an astute assessment of economic viability, wherein circular strategies are scrutinized through the lens of cost-effectiveness, resource optimization, and long-term financial prudence (Mohsenpour et al., 2021). Concurrently, the environmental relevance of circular water solutions is paramount, highlighting the reduction of resource depletion, energy consumption, and waste generation. Moreover, the social dimension underscores the significance of equitable access to water resources, community engagement, and the propagation of societal well-being (Rizwan et al., 2018). By weaving these dimensions into the fabric of circular water management, a nexus of holistic sustainability unfolds, increasing the resilience and power of water treatment systems.

1.9.3 Importance of Considering the Entire Lifecycle of Water Remediation Solutions

A cardinal principle in the pursuit of holistic circular water management is the recognition of the entire lifecycle of water remediation solutions, spanning conception, design, implementation, operation, and eventual closure. This lifecycle perspective instills a cyclical rhythm, where end-of-life considerations resonate prominently as initial design choices. The infusion of circularity, predicated on concepts such as cradle-to-cradle design, underlines the imperative of materials and technologies that can be continuously rejuvenated, repurposed, or recycled, thereby extending the longevity of water treatment solutions (Shehata et al., 2022; Noorani et al., 2023). Moreover, this approach propels the exploration of regenerative processes, harnessing the innate propensity of natural systems toward restoration and replenishment, echoing the fundamental tenets of the circular economy.

In conclusion, a holistic approach to circular economy and sustainability involves circular water management with a network of interconnected dimensions, fostering a complex interaction between economic, environmental, and social spheres. By embracing the entire lifecycle of water remediation solutions, this approach weaves a resilient narrative that reverberates across the nexus of water systems and sustainable futures.

1.10 CASE STUDIES AND PRACTICAL APPLICATIONS

This section offers a panoramic view of the tangible fusion of the circular economy and sustainability strategies within real-world contexts of water remediation. By scrutinizing case studies and practical applications, we reveal the combination of operational success and transformative potential that emerges when circularity and sustainability converge.

1.10.1 Presentation of Real-world Case Studies

1.10.1.1 Case Study 1: Urban Wastewater Reclamation and Reuse

In the bustling urban landscape of Barcelona, Spain, a pioneering initiative has harnessed the symbiosis of circular economy and sustainability principles to revolutionize urban wastewater management. By meticulously designing a closed-loop system that recovers, treats, and reuses wastewater, this project exemplifies the quintessence of circular water management (Liu et al., 2020). The treatment process incorporates resource recovery technologies, capturing valuable nutrients and energy from wastewater. Such resource optimization significantly diminishes resource depletion and waste generation, simultaneously mitigating water scarcity and pollution burden.

1.10.1.2 Case Study 2: Circular Stormwater Management

The city of Rotterdam, in the Netherlands, has ingeniously embraced the circular ethos to curtail the environmental ramifications of stormwater runoff. Integrating circular economy principles, sustainable drainage systems (SuDS) have been strategically deployed to curtail flooding while enhancing ecosystem services (Sutherland et al., 2021). The multifunctionality of SuDS utilizes green spaces that promote biodiversity, carbon sequestration, and recreational opportunities, thereby intertwining social and environmental dimensions in a harmonious continuum.

1.10.2 Symbiosis of the Circular Economy and Sustainability

These case studies collectively illuminate the pivotal crossroads where circular economy and sustainability intertwine, amplifying the impact of water remediation initiatives. Their successful narratives underscore the potential of a symbiotic relationship between economic, environmental, and social dimensions, cultivating landscapes of resilience and adaptability.

1.10.3 Revealing the Path Forward

The practical applications showcased in these case studies have far-reaching implications. These tangible manifestations transcend theoretical paradigms, encapsulating the very essence of integration of the circular economy and sustainability strategies. While these case studies shine as beacons of progress, they also encourage us to explore uncharted frontiers, foster transformative innovations, and forge unbreakable alliances.

1.11 CONCLUSION AND FUTURE DIRECTIONS

The journey through the integration of circular economy and sustainability within the realm of water remediation has illuminated a path teeming with innovation, promise, and transformative potential. This section encapsulates the essence of our exploration, reflecting on the milestones achieved and paving the way for an inspiring future, brimming with possibilities.

The culmination of our investigation identifies the interplay of circular economy and sustainability as a combination of harmonious dynamics, capable of orchestrating an effective cycle of water management. The significance of adopting circular principles in water remediation becomes resoundingly clear, with closed-loop systems, resource recovery, and innovative product design demonstrating the art of harnessing resources, optimizing efficiencies, and mitigating environmental burdens. The holistic approach of the sustainability strategy encapsulates the essence of economic viability, environmental stewardship, and social equity.

This synthesis evokes a vision of a transformed water landscape, combining resilience and regenerative practices. Circular economy and sustainability, when combined together intricately, offer the potential to reshape the contours of water management, ushering in an era where resource scarcity is transcended, waste becomes a relic of the past, and communities thrive in the embrace of a rejuvenated natural environment.

The future of circular water management will involve uncharted realms of innovation and exploration, involving further refinement of closed-loop systems, ingenious resource recovery methodologies, and the ascent of circular design principles to new pinnacles. Collaborative platforms, the fusion of technology and research, and the harmonization of policy instruments hold promise to drive circular solutions toward unprecedented heights.

The odyssey of integrating circular economy and sustainability in water remediation has unfurled as a tapestry woven with scientific ingenuity, practical wisdom, and transformative zeal. It is an invitation, a call to action for researchers, policymakers, practitioners, and visionaries alike, to forge a future where the waters of circularity and sustainability flow ceaselessly, quenching the thirst of generations to come.

ACKNOWLEDGMENTS

We would like to express our sincere gratitude to Dr. Chingakham Chinglenthoiba (John) for his invaluable assistance in proofreading and providing constructive feedback on our manuscript. His expertise and meticulous attention to detail greatly enhanced the quality and clarity of our work. We are truly grateful for his time and dedication in helping us improve this research.

DECLARATION OF COMPETING INTEREST

The authors declare that they have no known competing financial interests or personal relationships that could have appeared to influence the work reported in this paper.

DECLARATION OF GENERATIVE AI AND AI-ASSISTED TECHNOLOGIES IN THE WRITING PROCESS

Statement: During the preparation of this work, the authors used ChatGPT as AI-assisted technology in order to improve the readability and language of this chapter. After using this tool/service, the authors reviewed and edited the content as necessary and take full responsibility for the content of the publication.

REFERENCES

Aditya, L., Mahlia, T.M.I., Nguyen, L.N., Vu, H.P., Nghiem, L.D., 2022. Microalgae-bacteria consortium for wastewater treatment and biomass production. *Sci. Total Environ.* 838, 155871.

Alamerew, Y.A., Brissaud, D., 2019. Circular economy assessment tool for end of life product recovery strategies. *Jnl. Remanufactur.* 9, 169–185. https://doi.org/10.1007/s13243-018-0064-8.

Alamerew, Y.A., Kambanou, M.L., Sakao, T., Brissaud, D., 2020. A multi-criteria evaluation method of product-level circularity strategies. *Sustainability* 12 (12), 5129. https://doi.org/10.3390/su12125129.

Aydin, R., Badurdeen, F., 2019. Sustainable product line design considering a multilifecycle approach. *Resour. Conserv. Recycl.* 149, 727–737. https://doi.org/10.1016/j.resconrec.2019.06.014.

Bolognesi, S., Bernardi, G., Callegari, A., Dondi, D., Capodaglio, A.G., 2021. Biochar production from sewage sludge and microalgae mixtures: Properties, sustainability and possible role in circular economy. *Biomass Convers. Biorefinery* 11 (2), 289–299.

Chai, W.S., Tan, W.G., Halimatul Munawaroh, H.S., Gupta, V.K., Ho, S.-H., Show, P.L., 2021. Multifaceted roles of microalgae in the application of wastewater biotreatment: A review. *Environ. Pollut.* 269, 116236.

Cheng, D.L., Ngo, H.H., Guo, W.S., Chang, S.W., Nguyen, D.D., Kumar, S.M., 2019. Microalgae biomass from swine wastewater and its conversion to bioenergy. *Bioresour. Technol.* 275, 109–122.

Chia, W.Y., Chia, S.R., Khoo, K.S., Chew, K.W., Show, P.L., 2021. Sustainable membrane technology for resource recovery from wastewater: Forward osmosis and pressure retarded osmosis. *J. Water Proc. Eng.* 39, 101758.

Cinà, P., Bacci, G., Arancio, W., Gallo, G., Fani, R., Puglia, A.M., Di Trapani, D., Mannina, G., 2019. Assessment and characterization of the bacterial community structure in advanced activated sludge systems. *Bioresour. Technol.* 282, 254–261. https://doi.org/10.1016/j.biortech.2019.03.018.

Cipolletta, G., Ozbayram, E.G., Eusebi, A.L., Akyol, C., Malamis, S., Mino, E., Fatone, F., 2021. Policy and legislative barriers to close water-related loops in innovative small water and wastewater systems in Europe: A critical analysis. *J. Clean. Prod.* 288, 125604. https://doi.org/10.1016/j.jclepro.2020.125604.

de Oliveira, C.T., Dantas, T.E.T., Soares, S.R., 2021. Nano and micro level circular economy indicators: Assisting decision-makers in circularity assessments. *Sustain. Prod. Consum.* 26, 455–468. https://doi.org/10.1016/j.spc.2020.11.024.

ECA (Economic Consulting Associates), 2019. *From Waste to Resource: Why and How Should We Plan and Invest in Wastewater? – Policy, Institutional and Regulatory Incentives*. Technical background paper prepared for the World Bank.

Fatimah, Y.A., Govindan, K., Murniningsih, R., Setiawan, A., 2020. Industry 4.0 based sustainable circular economy approach for smart waste management system to achieve sustainable development goals: A case study of Indonesia. *J. Clean. Prod.* 269, 122263 https://doi.org/10.1016/j.jclepro.2020.122263.

Ferrans, L., Avellòan, T., Muller, A., Hettiarachchi, H., Dornack, C., Caucci, S., 2020. Selecting sustainable sewage sludge reuse options through a systematic assessment framework: Methodology and case study in Latin America. *J. Clean. Prod.* 242, 118389. https://doi.org/10.1016/j.jclepro.2019.118389.

Ferreira, A., Ribeiro, B., Ferreira, A.F., Tavares, M.L., Vladic, J., Vidović, S., Cvetkovic, D., Melkonyan, L., Avetisova, G., Goginyan, V., 2019. Scenedesmus obliquus microalga-based biorefinery–from brewery effluent to bioactive compounds, biofuels and biofertilizers–aiming at a circular bioeconomy. *Biofuels Bioprod. Biorefin.* 13 (5), 1169–1186.

Ganora, D., Hospido, A., Husemann, J., Krampe, J., Loderer, C., Longo, S., et al., 2019. Opportunities to improve energy use in urban wastewater treatment: A European scale analysis. *Environ. Res. Lett.*, 14, 4.

Gherghel, A., Teodosiu, C., De Gisi, S., 2019. A review on wastewater sludge valorisation and its challenges in the context of circular economy. *J. Clean. Prod.*, 228, 244–263. https://doi.org/10.1016/j.jclepro.2019.04.240.

Ghisellini, P., Cialani, C., Ulgiati, S., 2016. A review on circular economy: The expected transition to a balanced interplay of environmental and economic systems. *J. Clean. Prod.* 114, 11–32. https://doi.org/10.1016/j.jclepro.2015.09.007.

Goswami, R.K., Mehariya, S., Verma, P., Lavecchia, R., Zuorro, A., 2021. Microalgae-based biorefineries for sustainable resource recovery from wastewater. *J. Water Process Eng.* 40, 101747.

Hapuwatte, B.M., Seevers, K.D., Jawahir, I.S., 2022. Metrics-based dynamic product sustainability performance evaluation for advancing the circular economy. *J. Manuf. Syst.* 64, 275–287. https://doi.org/10.1016/j.jmsy.2022.06.013.

Hart, J., Adams, K., Giesekam, J., Densley Tingley, D., Pomponi, F., 2019. Barriers and drivers in a circular economy: The case of the built environment. *Procedia Cirp* 80, 619–624. https://doi.org/10.1016/j.procir.2018.12.015.

IWA, 2018. *Wastewater Report 2018: The Reuse Opportunity*, The International Water Association, London.

Kehrein, P., van Loosdrecht, M., Osseweijer, P., Garfí, M., Dewulf, J., Posada, J., 2020. A critical review of resource recovery from municipal wastewater treatment plants–market supply potentials, technologies and bottlenecks. *Environ. Sci. Water Res. Technol.* 6, 877. https://doi.org/10.1039/C9EW00905A.

Kristensen, H.S., Mosgaard, M.A., 2020. A review of micro level indicators for a circular economy – Moving away from the three dimensions of sustainability? *J. Clean. Prod.* 243, 18531. https://doi.org/10.1016/j.jclepro.2019.118531.

Liu, J., Pemberton, B., Lewis, J., Scales, P.J., Martin, G.J.O., 2020. Wastewater treatment using filamentous algae - A review. *Bioresour. Technol.* 298, 122556.

Liu, R., Wei, T., Zhao, Y., Wang, Y., 2018. Presentation and perspective of appealing green facilities for eco-cyclic water management. *Chemical Engineering Journal*, 337, 671–683. https://doi.org/10.1016/j.cej.2017.12.127.

Liyanaarachchi, V.C., Premaratne, M., Ariyadasa, T.U., Nimarshana, P.H.V., Malik, A., 2021. Two-stage cultivation of microalgae for production of high-value compounds and biofuels: A review. *Algal Res.* 57, 102353.

Mannina, G., Badalucco, L., Barbara, L., Cosenza, A., Di Trapani, D., Gallo, G., Laudicina, V.A., Marino, G., Muscarella, S.M., Presti, D., et al., 2021. Enhancing a transition to a circular economy in the water sector: The EU project Wider Uptake. *Water*, 13, 946. https://doi.org/10.3390/ w13070946.

Mannina, G., Presti, D., Montiel-Jarillo, G., Carrera, J., Suárez-Ojeda, M.E., 2020. Recovery of polyhydroxyalkanoates (PHAs) from wastewater: A review. *Bioresour. Technol.*, 297, 122478. https://doi.org/10.1016/j.biortech.2019.122478.

Maryam, B., Büyükgüngör, H., 2019. Wastewater reclamation and reuse trends in Turkey: Opportunities and Challenges. *J. Water Proc. Eng.*, 30, 100501. https://doi.org/10.1016 /j.jwpe.2017.10.001.

Mesa, J., Esparragoza, I., Maury, H., 2018. Developing a set of sustainability indicators for product families based on the circular economy model. *J. Clean. Prod.* 196, 1429–1442. https://doi.org/10.1016/J.JCLEPRO.2018.06.131.

Mohsenpour, S.F., Hennige, S., Willoughby, N., Adeloye, A., Gutierrez, T., 2021. Integrating micro-algae into wastewater treatment: A review. *Sci. Total Environ.* 752.

Morseletto, P., 2020. Targets for a circular economy. *Resour. Conserv. Recycl.* 153, 104553. https://doi.org/10.1016/j.resconrec.2019.104553.

Mukherjee, M., Jensen, O., 2020. Making water reuse safe: A comparative analysis of the development of regulation and technology uptake in the US and Australia. *Saf. Sci.*, 21, 5–14. https://doi.org/10.1016/j.ssci.2019.08.039.

Naidoo, D., Nhamo, L., Lottering, S., Mpandeli, S., Liphadzi, S., Modi, A.T., Mabhaudhi, T., 2021. Transitional pathways towards achieving a circular economy in the water, energy, and food sectors. *Sustainability* 13 (17), 9978. https://doi.org/10.3390/su13179978.

Noorani, K.R.P.M., Flora, G., Surendarnath, S., Stephy, M.G., Amesho, K.T.T., Chinglenthoiba, C., Thajuddin, N., 2023. Recent advances in remediation strategies for mitigating the impacts of emerging pollutants and ensuring environmental sustainability. *J. Environ. Manag.*, 351, 119674. https://doi.org/10.1016/j.jenvman.2023.119674.

Panchal, R., Singh, A., Diwan, H., 2021. Does circular economy performance lead to sustainable development? A systematic literature review. *J. Environ. Manag.* 293, 112811. https://doi.org/10.1016/j.jenvman.2021.112811.

Papa, M., Foladori, P., Guglielmi, L., Bertanza, G. 2017. How far are we from closing the loop of sewage resource recovery? A real picture of municipal wastewater treatment plants in Italy. *J. Environ. Manag.*, 198, 9–15. https://doi.org/10.1016/j.jenvman.2017.04.061.

Pradel, M., Aissani, L., Villot, J., Baudez, J.-C., Laforest, V., 2016. From waste to added value product: Towards a paradigm shift in life cycle assessment applied to wastewater sludge: A review. *J. Clean. Prod.* 131, 60–75. https://doi.org/10.1016/j.jclepro.2016.05.076.

Priyadarshini, P., Abhilash, P.C., 2020. Circular economy practices within energy and waste management sectors of India: A meta-analysis. *Bioresour. Technol.* 304, 123018 https://doi.org/10.1016/j.biortech.2020.123018.

PUB, 2016. *Our Water, Our Future*, Singapore's National Water Agency (PUB), Singapore.

Rizwan, M., Mujtaba, G., Memon, S.A., Lee, K., Rashid, N., 2018. Exploring the potential of microalgae for new biotechnology applications and beyond: A review. *Renew. Sust. Energ. Rev.* 92, 394–404.

Rocchi, L., Paolotti, L., Cortina, C., Fagioli, F.F., Boggia, A., 2021. Measuring circularity: An application of modified Material Circularity Indicator to agricultural systems. *Agricul. Food Econ.* 9 (1), 1–13. https://doi.org/10.1186/s40100-021-00182-8.

Salminen, J., Määttä, K., Haimi, H., Maidell, M., Karjalainen, A., Noro, K., Koskiaho, J., Tikkanen, S., Pohjola, J., 2022. Water-smart circular economy – Conceptualisation, transitional policy instruments and stakeholder perception. *J. Clean. Prod.*, 334, 130065. https://doi.org/10.1016/j.jclepro.2021.130065.

Sandefur, H.N., Asgharpour, M., Mariott, J., Gottberg, E., Vaden, J., Matlock, M., Hestekin, J., 2016. Recovery of nutrients from swine wastewater using ultrafiltration: Applications for microalgae cultivation in photobioreactors. *Ecol. Eng.* 94, 75–81.

Sartal, A., Ozcelik, N., Rodriguez, M., 2020. Bringing the circular economy closer to small and medium enterprises: Improving water circularity without damaging plant productivity. *J. Clean. Prod.* 256, 120363. https://doi.org/10.1016/j. jclepro.2020.120363.

Schroeder, P., Anggraeni, K., Weber, U., 2019. The relevance of circular economy practices to the sustainable development goals. *J. Ind. Ecol.* 23, 77–95. https://doi.org/10.1111/jiec.12732.

Shehata, N., Obaideen, K., Sayed, E.T., Abdelkareem, M.A., Mahmoud, M.S., El-Salamony, A.- H.R., Mahmoud, H.M., Olabi, A.G., 2022. Role of refuse-derived fuel in circular economy and sustainable development goals. *Process Saf. Environ. Prot.* 163, 558–573.

Sutherland, D.L., McCauley, J., Labeeuw, L., Ray, P., Kuzhiumparambil, U., Hall, C., Doblin, M., Nguyen, L.N., Ralph, P.J., 2021. How microalgal biotechnology can assist with the UN sustainable development goals for natural resource management. *Curr. Res. Environ. Sustain.* 3, 100050.

Thompson, T.M., Young, B.R., Baroutian, S., 2020. Efficiency of hydrothermal pretreatment on the anaerobic digestion of pelagic Sargassum for biogas and fertiliser recovery. *Fuel* 279, 279.

Valentino, F., Morgan-Sagastume, F., Campanari, S., Villano, M., Werker, A., Majone, M., 2017. Carbon recovery from wastewater through bioconversion into biodegradable polymers. *New Biotechnol.*, 37, 9–23. https://doi.org/10.1016/j.nbt.2016.05.007.

Veleva, V., Bodkin, G., Todorova, S., 2017. The need for better measurement and employee engagement to advance a circular economy: Lessons from Biogen's "zero waste" journey. *J. Clean. Prod.* 154, 517–529. https://doi.org/10.1016/J.JCLEPRO.2017.03.177.

Wang, Q., Wei, W., Gong, Y., Yu, Q., Li, Q., Sun, J., Yuan, Z., 2017. Technologies for reducing sludge production in wastewater treatment plants: State of the art. *Sci. Total Environ.*, 587–588, 510–521. https://doi.org/10.1016/j.scitotenv.2017.02.203.

Werker, A., Bengtsson, S., Korving, L., Hjort, M., Anterrieu, S., Alexandersson, T., Johansson, P., et al., 2020. Consistent production of high quality PHA using activated sludge harvested from full scale municipal wastewater treatment–PHARIO. *Water Sci. Technol.*, 78, 2256–2269. https://doi.org/10.2166/wst.2018.502.

World Bank. 2020. *Shifting Paradigms for Smarter Wastewater Interventions in Latin America and the Caribbean*, World Bank, Washington, DC. https://www.worldbank.org/en/topic/water/publication/wastewater-initiative.

Yarnold, J., Karan, H., Oey, M., Hankamer, B., 2019. Microalgal aquafeeds as part of a circular bioeconomy. *Trends Plant Sci.* 24 (10), 959–970.

Zabed, H.M., Akter, S., Yun, J., Zhang, G., Zhang, Y., Qi, X., 2020. Biogas from microalgae: Technologies, challenges and opportunities. *Renew. Sust. Energ. Rev.* 117, 109503.

Zahmatkesh, S., Amesho, K.T.T., Sillanpaa, M., 2022a. A critical review on diverse technologies for advanced wastewater treatment during SARS-CoV-2 pandemic: What do we know? *Journal of Hazardous Materials Advances*, 7, 100121, https://doi.org/10.1016/j.hazadv.2022.100121

Zahmatkesh, S., Klemeš, J.J., Bokhari, A., Rezakhani, Y., Wang, C., Sillanpää, M., Amesho, K.T.T., Ahmed, W.S.A., 2022b. Reducing COD from weak wastewater through microalgae, sludge and activated carbon roles: A novel application of fuzzy logic-based simulation in MATLAB. *Comput. Chem. Eng.*, 166, 107944. https://doi.org/10.1016/j.compchemeng.2022.107944

Zahmatkesh, S., Klemeš, J.J., Bokhari, A.., Wang, C., Sillanpaa, M., Amesho, K.T.T., Vithanage, M., 2023. Various advanced wastewater treatment methods to remove microplastics and prevent transmission of SARS-CoV-2 to airborne microplastics. *Int. J. Environ. Sci. Technol.* 20, 2229–2246 (2023). https://doi.org/10.1007/s13762-022-04654-2

2 Circular Economy in Developing Countries
Present Status, Pathways, Benefits, and Challenges

Yash Aryan
Environmental Science and Engineering Department,
Indian Institute of Technology Bombay, Mumbai, India

Nikhilesh Nagare
Environmental Science and Engineering Department,
Indian Institute of Technology Bombay, Mumbai, India

Anil Kumar Dikshit
Environmental Science and Engineering Department,
Indian Institute of Technology Bombay, Mumbai, India

Amar Mohan Shinde
Environmental Science and Engineering Department,
Indian Institute of Technology Bombay, Mumbai, India

2.1 INTRODUCTION

The demand for sustainable development has grown as a result of the increased anxiety and ambiguity surrounding the effects of the world's most concerning problems, including climate change, resource scarcity, energy supply, and food insecurity. According to a report by the United Nations Environment Programme (UNEP), natural resource exploitation has increased threefold in the past 40 years (UNEP, 2016). The growing population, increasing resource demand, climate change, and the scarcity of natural resources have led to a shift from a linear model to a circular loop. The linear model includes resource extraction, manufacturing, use, and discarding of the product as waste, whereas the circular model promotes the reuse of the resources and their maintenance in the production loop, to be used as a resource and not discarded as waste. About $2 trillion extra can be generated for the world economy by 2050 if the resources are used wisely (Panel, 2017). There is both descriptive and

DOI: 10.1201/9781003432869-2

linguistic meaning to the term "circular economy" and the origins of the phrase "circular economy" are debatable (Murray et al., 2017). Greyson (2007) asserted that Kenneth Boulding (1966) originated the term circular economy (CE) when he wrote "Man must find his place in a cyclical ecological system which is capable of continuous reproduction of material form even though it cannot escape having inputs of energy". A few researchers have stated that the concept of CE was originated and first used in China (Yuan et al., 2006; Liu et al., 2009). Another study claimed that the term "circular economy" was first mentioned in the Western literature in the 1980s by Stahel and Reday-Mulvey (1976), who first introduced the closed-loop economy. The circular economy (CE) replaces the linear economy and is defined as a way of managing resource circularity, efficiency, and optimization that promotes the use of waste as a resource to generate value instead of discarding it as a waste by-product (Pearce and Turner, 1989). Linear economy and CE are presented in Figure 2.1 and Figure 2.2, respectively. A circular economy improves resource usage by minimizing natural resource extraction and waste generation, resulting in sustainable development.

Kirchherr et al. (2017) analysed 114 definitions of CE and provided their own CE definition. According to them, CE is based on business models that operate at micro, meso, and macro levels, replacing the end-of-life concept with the reduction, reuse, recycling, and recovery of any product aimed at promoting sustainable development, including environmental, economic, and social aspects. The three principles of CE

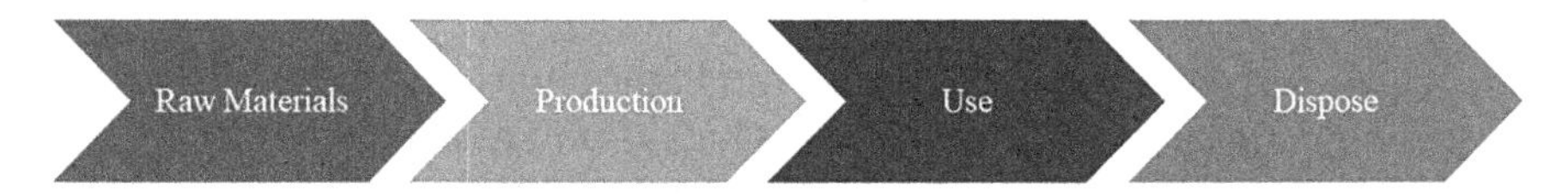

FIGURE 2.1 Liner Economy

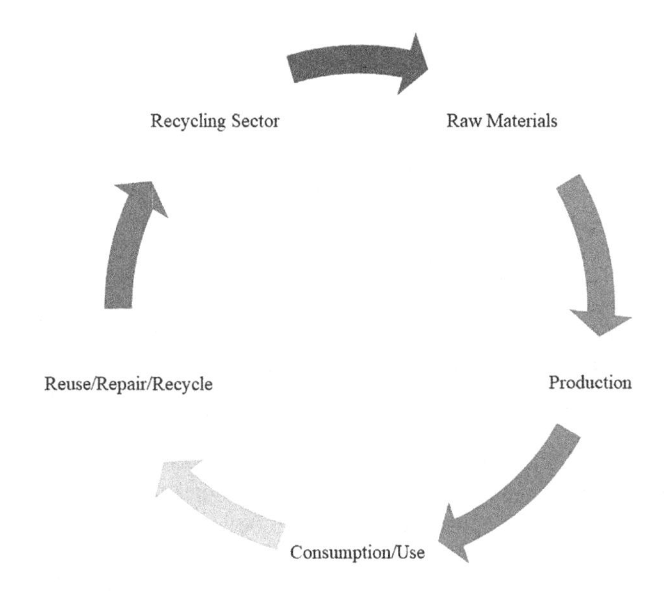

FIGURE 2.2 Circular Economy

are based on the 3R idea, which stands for "Reduce, Reuse, Recycle" (MacArthur, 2013). The first concept of the circular economy is the "appropriate design" approach, which underlines the importance of the design stage in putting forward options for avoiding waste discharge. The second principle advocates reclassifying materials and dividing them into "nutrients" and "technical" materials, while the third concept is primarily focused on "renewability". According to this idea, the circular economy's primary energy source should be renewable energy (Halog and Anieke, 2021). Several researchers have mentioned "R" frameworks, such as 6R(s) and 9R(s), which are beyond the 3R concept (Sihvonen and Ritola, 2015; Van Buren et al., 2016; Potting et al., 2017). The 9R framework, proposed by Potting et al. (2017), included refuse, rethink, reduce, reuse, repair, refurbish, remanufacture, repurpose, recycle, and recover. There is a close relationship between CE and Sustainable Development Goals (SDGs), and, to meet the SDG(s), many of these nations must adopt the circular economy. Several of the SDGs' aims can be accomplished with the aid of CE practices and related business models. Twenty-one of the SDGs' targets are directly or indirectly achieved as a result of CE practices. SDG 6, 7, 8, 12, and 15 have the strongest connections to the CE practices in terms of their targets. However, CE practices will not tackle all the SDGs' concerns because at least 35 of the targets have little or no influence on CE practices (Ghosh, 2020a).

2.2 GLOBAL OVERVIEW OF THE CIRCULAR ECONOMY?

The circular economy has been highlighted as a large economic opportunity worth up to $4.5 trillion p.a., with the potential to promote the development of new companies, create jobs, reduce greenhouse gas emissions, and increase the efficient use of natural resources (WBCSD, 2017). Developed countries like the USA, the UK, France, Denmark, Germany, Sweden, and others have performed extensive research on the concept of circular economy and discovered innovations and solutions toward a sustainable future (Halog and Anieke, 2021). According to EU objectives, municipal and package waste recycling rates should reach 65% and 70%, respectively, by 2035, with landfilling restricted to a maximum of 10% of municipal waste (EU, 2021). Zisopoulos et al. (2022), in their study, found that the median circularity rate of the EU countries was 6.9% with the Netherlands having the highest circularity rate of 27.1%. The Netherlands is one of the few European countries that aims to be entirely circular by 2050 (MIE, 2016). Among EU countries, Denmark, Austria, France, Czech Republic, Belgium, Italy, France, Netherlands, Germany, Estonia, Finland, Ireland, Lithuania, Portugal, Luxembourg, Spain, Slovenia, and Sweden are performing well in the transition toward a circular economy and implementing various initiatives (Marino and Pariso, 2020). On the other hand, some EU countries, such as Bulgaria, Croatia, Cyprus, Greece, Hungary, Malta, Latvia, Poland, Romania, and Slovakia, are having a slow transition from the linear to the circular economy and have a weak position toward CE implementation (Marino and Pariso, 2020). Marino and Pariso (2020) reported that the top three countries with the highest municipal waste recycling rates were Germany, Denmark, and Austria while the bottom three were Romania, Malta, and Slovakia. In terms of reusing material, the Netherlands was at the top while Greece was at the bottom. Zisopoulos et al. (2022)

found that, despite the fact that European countries have established a highly organized section of their material and energy flow networks that favors efficiency over resilience, none of them has attained near-maximum robustness.

Guerra and Leite (2021) found that the CE in the construction industry in the USA is lagging behind countries in Europe and Asia, possibly due to extensive resource and land availability. Australia is taking several measures to promote CE for sustainable development. The circular economy is actively practiced by numerous sectors in South Australia, including the construction and demolition (C&D) sector, water utilities, electronics, and composting activities. South Australia is recycling more than 90% of C&D waste and several measures of circularity are being performed, such as recycling of water resources, aquifer recharge, container deposit legislation, and renewable energy developments (Levitzke, 2020). The resource recovery increased from 2 million tons in 2003–2004 to 4.4 million tons in 2016–2017, and ~84% of waste is diverted to resource recovery in South Australia (Levitzke, 2020). Daskal and Ayalon (2020) concluded that Israel has poor CE implementation despite having strict regulations, economic penalties, and financial incentives, and needs to develop suitable infrastructure to promote CE in the country.

The concept and implementation of CE practices have not been restricted to developed countries but also appeal to rapidly industrializing countries like China and India. Unlike Europe, China's efforts to implement CE are not limited to the private sector but also involve the national government, which follows a top-down approach (Zhu et al., 2019). Chinese leadership has promoted "Green Development" and "Beautiful China", presenting the idea that a clean environment results in bringing profit to the country (Kuhn, 2016; Pesce et al., 2020). In China, the circular economy is being advanced through the promotion of waste material recirculation through the establishment of targets, the implementation of policies and regulations, and the implementation of financial measures. The CE implementation in China places a strong emphasis on replacing traditional industrial culture with cutting-edge technology and procedures that considerably boost production efficiency and profitability (Geng et al., 2012). China is among the top nations and has contributed significantly to CE-related research, possibly because China adopted the CE concept in its national development strategy in 2002 and was one of the earliest countries to adopt CE (Ngan et al., 2019). India is the world's fastest-growing economy and is anticipated to be the world's third-largest economy by 2030, contributing approximately 8.5% of global GDP (Kalaari, 2022; Ghosh, 2020b). Traditionally, India has been a thrifty society that incorporates circularity into daily life. Under a medium-growth scenario, India's material needs are anticipated to be just over 15 billion tons by 2030 and slightly more than 25 billion tons by 2050. The Indian government has developed a supportive regulatory framework to promote the growth of CE in India, such as "Zero Defect Zero Effect" and made revisions to various waste management rules (Ngan et al., 2019). The initiatives to implement CE in India have been taken at both the government and industry levels and many industries in India, such as the Mahindra group, Green Vortex, and India Glycols Ltd., are adopting initiatives for the transition from a linear to a circular economy (Ngan et al, 2019). Researchers in India and other developing countries, like Brazil, Kenya, and Ghana, are working in the field of CE and have published several papers related to CE (e.g., Halog and Anieke, 2021). In order to capture at least 60% of the electronic trash produced in Pune, one of India's largest metropolitan areas, the World Institute of Sustainable Energy has created an action plan (Fiksel et al., 2021). The Government of India formed 11 committees to achieve the transition from a linear to a circular economy which are led by ministers and officials from the Ministry of Environment, Forest, and Climate Change (MoEFCC) and NITI Ayog (PIB, 2021a). Malaysia is actively pursuing the adoption of green technology. Energy, environment, economy, and social issues are among the four core policy goals (Isa et al., 2021). The Eleventh Plan of Malaysia proposed a comprehensive strategy for CE implementation and aims to achieve 22% recycling of municipal solid waste (Agamuthu and Mehran, 2020; Ngan et al., 2019). The Malaysian Environment Department is encouraging industrial ecology and proposes to construct industrial ecology for two forms of trash, namely bleached earth and abandoned vehicles (Agamuthu and Mehran, 2020). Bangladesh also has huge potential to implement CE like other developing countries. Fileted Ltd. is the first factory in Bangladesh to adopt "Circular Fashion" and the 3R (reduce, reuse, recycle) concept (Ahmed et al., 2022). The Mymensingh-based Simco Spinning and Textiles Limited can make 15 tons of yarn each day from cotton clips left over after sewing clothes (Ahmed et al., 2022). The government of Bangladesh installed the CETP (Central Effluent Treatment Plant) to treat harmful effluent from tannery industries, which is also capable of generating an energy of 5 MW/hour (Ahmed et al., 2022). The CE implementation in Bangladesh is facing many challenges and is getting hampered by many reasons, such as lack of technical expertise, shortage of funds, lack of public awareness, ineffective policy, lack of law implementation, and an absence of research-based knowledge (Ahmed et al., 2022). There is major scope for improvements in CE implementation in both developed and developing countries and a need to be more effective in the transition from a linear to a circular economy.

2.3 NEED FOR AND BENEFITS OF THE CIRCULAR ECONOMY IN DEVELOPING COUNTRIES

The world population surpassed the 8 billion mark in November 2023 and the number of people on the Earth has grown by one-third, or 2.1 billion, over the previous 25 years. By 2050, it is anticipated that there will be just under 10 billion people on the planet (UNCTAD, 2022a). The majority of population growth during the past 25 years has occurred in developing nations, primarily in Africa (an additional 700 million people) and Asia and Oceania (1.2 billion more people) (UNCTAD, 2022a). According to the UN, the proportion of the world's population residing in developing nations has risen over time, from 66% in 1950 to 83% today, and is expected to reach 86% by 2050 (UNCTAD, 2022a). The population growth in developing nations emphasizes the significance of addressing issues that affect these countries, such as hunger, access to clean water, sanitation, and health services, as well as connecting people to sustainable, inexpensive sources of electricity and energy. The implementation of CE in various sectors, such as food, will address the hunger issues, CE implementation in the wastewater sector will address the access to clean water problem and CE implementation in the energy sector will provide affordable and sustainable energy to all.

The whole world is experiencing the adverse effects of climate change but developing nations like India, Pakistan, South Africa are more vulnerable to climate change (Wijaya, 2014). According to the report from the Asian Development Bank (ADB) Institute, the Pacific and Asia will be crucial to preventing the worst effects of climate change. Developing countries like India are greatly concerned regarding climate change and have aggressively determined their Nationally Determined Contributions (NDCs) to limit global warming to within 1.5–2 degrees Celsius above pre-industrial levels. India has recently revised its NDCs to achieve its commitment of net zero by 2070. India aims to fulfill 50% of its total energy demand from non-fossil-fuel-based power sources by the year 2030 and to reduce the emission intensity of its Gross Domestic Product (GDP) by 45% compared with 2005 levels by 2030 (UNFCCC, 2022). South Africa also updated its NDC and aims to reach net zero emissions by 2050 (UNDP, 2021). Bangladesh has targeted to reduce its greenhouse gas (GHG) emissions by 12 million tons (5%) by 2030 from that of the 2011 level by itself and proposed further reductions of 24 million tons (10%) with international support (UNFCCC, 2021). The implementation of CE in developing countries will shift their economy from linear to circular and will play a key role in achieving the committed NDCs. The targets of NDCs will be fulfilled only if these developing countries properly implement the CE initiatives throughout their sectors. A circular economy, if implemented properly, has huge potential for revenue generation, especially in developing countries. According to the Ministry of Housing and Urban Affairs (MoHUA), Government of India, CE implementation in the waste management sector in India could generate revenue of about INR (Indian rupees) 13,880 crores (138.8 billion) per annum (MoHUA, 2021). In another report, it has been estimated that if India adopts CE it could save $624 million annually and GHG emissions could be reduced by 44% by 2050 (UNCTAD 2022b). The benefits of CE are not only restricted to limiting GHG emissions but extend to many areas as shown in Figure 2.3. The CE implementation provides benefits in terms of all three pillars of sustainability, i.e., economic, environmental, and social.

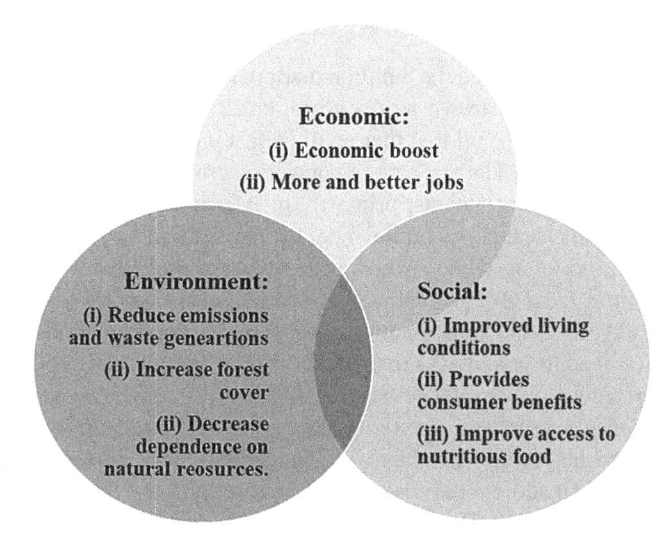

FIGURE 2.3 Benefits of Circular Economy

2.4 POLICIES AND RECOMMENDATIONS TO PROMOTE A CIRCULAR ECONOMY IN DEVELOPING COUNTRIES

This section discusses the important policies, regulations, and acts of various developing countries. The majority of developing countries, except for a few such as China and India, lack legislative frameworks and policies related to CE implementation. In China, the most important laws to promote CE are the "Circular Economy Promotion Law", "Clean Production Promotion Law", and "Control of Environmental Pollution by Solid Waste Law" (Zhu et al., 2019). The Circular Economy Promotion Law was passed by the National People's Congress and went into effect in January 2009. The Circular Economy Promotion Law aims to promote the circular economy, increase resource utilization, safeguard and enhance the environment, and achieve continuous development. The Cleaner Production Promotion Law was issued in 2002 and aims to encourage cleaner manufacturing, increase resource utilization, reduce and prevent pollution, safeguard the environment and human health, and encourage the economy and society's sustainable growth (Peng and Liu, 2016). In 1995, China announced and adopted a law on the Control of Environmental Pollution by Solid Waste; this law was revised in 2004 and 2020 (Guo et al., 2021). The revised 2020 law consists of many changes but the most important is that all solid waste is prohibited from importation into China with effect from 1 January 2021. The revision also included the establishment of an extended period of producer responsibility (EPR) for lead storage batteries, automotive traction batteries, and electrical and electronic products. Recently, the National Development and Reform Commission (NDRC) and Ministry of Finance, China have introduced a policy for automobile sectors to encourage the use of old products in re-manufacturing (Mac Arthur Foundation, 2022). Recently, China had introduced the Fourteenth Five-Year Plan (2021–2025) which aimed to promote resource conservation and recycling (Bleischwitz et al., 2022). The major commitments in this plan are (i) enhancing productivity of resources by 20% compared to 2020 levels, (ii) decreasing energy and water use by 13.5% and 16%, respectively, to that of 2020, (iii) consuming 320 million tons of scrap steel and 60 million tons of paper waste, (iv) generating 20 million tons of recycled non-ferrous metals, and (v) achieving 86% utilization rate for crop stalks and 60% for solid and construction waste (Bleischwitz et al., 2022). There are several other policies and programmes in China which are directly or indirectly linked to promotion of CE, such as "Policy and Regulation of Mineral Resources Exploitation", "Comprehensive Utilization of Resources", and "Utilization Technology for Wastes and Tailings" (Zhu et al., 2019).

As in China, the CE has been promoted by legislations and regulations for a quite long time in India. Although there is no official law in India governing the circular economy directly, the concept and its implementation criteria are incorporated in numerous rules and regulations (Ghosh, 2020b). The foundation of India's environmental plan is comprised of the action plans following the 1972 Stockholm Conference and the 42nd amendment to Articles 48 Part IV and 51A(g) of the Indian Constitution (Ghosh, 2020b). There are several rules and policies in India which aim at shifting from a linear to a circular economy. The policies enabling resource efficiency during product design are "National Design Policy (2007)", "Science, Technology, and Innovation Policy (2013)", and "Bureau of Indian Standard Acts (2016)". The policies implementing resource conservation, eco-friendly disposal, and circulation of resources are "Solid Waste Management Rules (2016)", "E-Waste Management Rules (2016)", "Plastic Waste Management Rules (2016)", "Batteries (Management and Handling) Rules (2010)", and "C&D Waste Management Rules (2016)". The policies focused on efficient manufacturing are "National Manufacturing Policy (2011)" and "National Policy on Electronics (2012)". Recently, MoEFCC, Government of India, drafted National Resource Efficiency Policy (2019) which aims to implement resource efficiency across all resources. MoEFCC, Government of India has recently revised E-Waste Management Rules (2016) which came into effect from 1 April 2023. These amendments/additions to E-Waste Management Rules show India's commitment to adopting CE, which are as follows: (i) producers have been given annual electronic waste (e-waste) recycling targets; (ii) management of solar PV modules, panels, and cells has been introduced; (iii) EPR certificates have been introduced; and (iv) provision has been introduced for the constitution of the Steering Committee to monitor the implementation of these rules (PIB, 2022a). Similarly, MoEFCC, Government of India also amended the Plastic Waste Management Rules (2016) in 2021. These amendments prohibit the use of single-use plastic by 2022, while the thickness of plastic carrier bags has been increased from 75 microns to 120 microns with effect from September 2021. EPR has been provided with legal force in the recent amendments of 2021 (PIB, 2021b). MoEFCC, Government of India also notified revised Battery Waste Management Rules in 2022. The new regulations require the Producer to recycle, refurbish, or otherwise reuse unwanted batteries rather than dumping them in landfills or incinerators (PIB, 2022b). MoHUA, Government of India mandated the cement industries to replace 15% of its fuel intake with refuse-derived fuel (RDF) within three years from the date of amendment of these rules and 25% by 2025. (CPHEEO, 2018). Our Honourable Prime Minister of India introduced the concept "Lifestyle for the Environment (LiFE)" at the Conference of the Parties (COP) 26 in Glasgow on 1 November 2021. India is the first country to include LiFE in its NDC aims to promote and spread a healthy and sustainable way of life based on its traditions and conservation and moderation ideals, especially through a mass LiFE movement, as a means to addressing climate change (NITI Aayog, 2022). The States and Union Territories (Uts) in India are also taking initiatives; recently, Puducherry became the first UT to adopt the policy of green budgeting, on 5 June 2023 (TERI, 2023). A green budget aims to act as a planning and assessment tool and promote a system-wide approach by helping to institutionalize and integrate environmental sustainability into various government programs (TERI, 2023). The Government of India has been researching the opportunities for CE implementation in various sectors such as Municipal Solid Waste (MSW), C&D Waste, Municipal Wastewater, Automotive, Agriculture, Mining, Plastic Waste, and others (MoHUA, 2021; TERI, 2021; FICCI, 2018).

South Africa first introduced the concept of a circular economy in the Science, Technology, and Innovation (STI) White Paper in 2019 (Hoosain et al., 2023). The South African government has adopted the National Waste Management Plan 2020.

The National Waste Management Policy 2020 aims to promote the waste hierarchy and the goals of the circular economy, resulting in both socioeconomic gains and reduced environmental impacts (Hoosain et al., 2023). There are many individual state plans in South Africa, such as those of Eastern Cape, Gauteng, KwaZulu – Natal, Limpopo, and Western Cape which aim at enhancing skills, improving livelihoods, and promoting economic growth and sustainability (Hoosain et al., 2023). The Carbon Tax Act has also become a law in South Africa, which aims to reduce industrial greenhouse gas emissions significantly (Hoosain et al., 2023). In Malaysia, the CE concept is still in its initial stage, as a legal framework is lacking (Agamuthu and Mehran, 2020). There are a few acts and regulations, such as the Environmental Quality Act 1974, the Environmental Quality (Scheduled Waste) Regulation 2005, and the Solid, Waste and Public Cleansing Management Act 2007 to promote resource circulation in Malaysia. Malaysian Eleventh Plan aims to achieve 22% of municipal solid waste recycling (Agamuthu and Mehran, 2020). The Department of Environment (DoE), Malaysian Government has issued guidelines for co-processing of waste during cement production but robust legal framework and top-down and bottom-up approaches are required to properly implement CE in Malaysia (Agamuthu and Mehran, 2020). Still today, developing country governments face tremendous challenges in the proper administration, policymaking, policy execution, and technology constraints in implementing CE (Ding et al., 2019).

2.5 PRESENT SCENARIO OF CIRCULAR ECONOMY IN DEVELOPING COUNTRIES

This section discusses the present status and practices of CE being implemented in various developing countries like China, India, and Bangladesh, among others. At present, various ministries have been engaged in CE policymaking in China. In China, the basic CE regulations are developed by the NDRC, while resource recovery, product development, and EPR are looked after by the Ministry of Industry and Information Technology (MIIT), while the development of standards and Eco-Industrial Parks (EIPs) comes under the Ministry of Environment and Ecology (MoEE) (Bleischwitz et al., 2022). The proper implementation of CE requires funds and MoF and NDRC allocated a special fund for CE implementation. More than 10 billion yuan have been invested by the state, which is thought to have stimulated an additional 100 billion yuan in social investment (Bleischwitz et al., 2022). Bleischwitz et al. (2022) reported the significant success of decoupling GDP from resource consumption in China in the past 20 years, and relative resource productivity in China was almost double to that of the European Union. Several pilot projects are being undertaken by NDRC to bring innovation in CE; for this, about 178 units have been selected in important industries such as coal, iron and steel, electric power, non-ferrous metals, chemical industry, building materials, and light industry (Bleischwitz et al., 2022). Through its latest municipal waste separation schemes (e.g., Shanghai Master Plan 2017–2035), the city of Shanghai is performing exceptionally well in attempts to move toward a more circular system at the city level. Shanghai's potential for converting trash into energy and lowering overall energy consumption and CO_2 emissions has been estimated to be 6.6% and 4.9%, respectively (Xiao et al., 2020; Dong et al., 2018). China has been setting up EIP(s) since 2021 in collaboration with UNEP to implement CE in industrial systems across the country (Bleischwitz et al., 2022). China had positive growth toward vehicle remanufacturing and "Swap the Old for Remanufacturing" policy from NDRC, and MoF, Chinese Government, played a key role, with automobile refurbishment and recycling rates increased by more than 15% by 2019 in comparison with 2016 (MacArthur Foundation, 2022). There are several companies in China which are involved in CE, such as GEM Co. Ltd., a Shenzhen-based company involved in material recycling such as batteries, electronics, automobiles, and wastewater. Jingdong, also known as JD.com, Inc., is a Chinese e-commerce platform which purchases second-hand luxury items and then sells them to new consumers, thus promoting reuse of several items. The restaurants in Suzhou City, China, are mandated to send their food waste to Jiangsu Clean Environmental Technology for conversion to biogas, biofuel, and compost.

In India, many towns are implementing circularity in municipal solid waste management and transforming their waste into resources. The private companies are also involved in various CE initiatives in India which promote circularity and help to achieve sustainability. The CE initiatives from various sectors in India have been presented in Table 2.1. The CE initiatives in Bangladesh are in the initial stage, restricted to a few sectors such as renewable energy, e-waste, and ship breaking. Bangladesh has set up its largest solar power plant in Teknaf which has a power generation capacity of 28 MW (Ahmed et al., 2022). A furniture business in Bangladesh's Gazipur utilizes a biomass-powered boiler, which results in annual savings of Taka 5.25 million. A few initiatives have been taken in the e-waste sector, such as extracting lead from used lead acid batteries, saving about US $4.73 million. Dhaka is recycling 15% of the total e-waste generated per day. Rahimafrooz Batteries Ltd., the largest lead-acid battery manufacturer in Bangladesh, has established a smelting plant for battery recycling in an eco-friendly way. In Bangladesh, about 2 million tons of scrap ships are recycled every year (Ahmed et al., 2022). Farmers are utilizing wastewater to irrigate the land in some parts of Bangladesh and using compost as an organic fertilizer. Like Bangladesh, the CE initiatives in South Africa are also limited at present. The city of Cape Town is funding the Western Cape Industrial Symbiosis Programme (WISP) organization, involved in promoting circularity among the industries. Major companies in South Africa, like SAB, COKE, AB-InBev, and PETCO, have adopted circularity and recycling returnable glass and plastic bottles (Hoosain et al., 2023). Chile is a developing country in South America performing better than some of the developing countries in the South Asian region. Santiago, the capital of Chile, has recently introduced a community composting program. The private sector in Chile is also involved in promoting circularity and has launched the "Chile Circular" platform, bringing businesses, academia, and society together for better implementation and opportunities for the circular economy (Circular Lab, 2023). Another CE initiative in Chile, known as "Moda Circular Chile", is aimed at promoting circularity in the fashion sector, in which clothing is being reused and recycled (Circular Lab, 2023). Nestlé in Chile has implemented an

TABLE 2.1
Circular Economy Initiatives from Various Sectors in India

Company or Town	Sector	Initiatives	Benefits
Panaji, Goa	MSW management	99% source segregation and 80% waste processing.	Nearly INR 9 lakhs (approx. USD 11,000) revenue is being generated annually. Provides job opportunities to many.
Bobbili, Andhra Pradesh	MSW management	100% biodegradable waste processing into biogas and compost.	Nearly INR 6 lakhs (approx. USD 7,180) earned from selling compost. No dumpsite needed in the future.
Alappuzha, Kerala	MSW management	84% of MSW are being processed.	Compost is provided to farmers at zero cost. Self-help group (SHG) members are earning Rs 300–500 (USD 3.60 to USD 6.00) per day by selling recyclables. Decrease in water-borne diseases reported.
Indore, Madhya Pradesh	MSW management	100% source segregation and 100% waste processing. India's cleanest city since 2017.	Provides employment to more than 8,000 women. Negligible waste reaching landfill.
Mysuru, Karnataka	MSW management	70% of MSW collected is being processed in centralized composting plant and decentralized waste management plant.	Provides employment to many and improved environmental and health conditions.
Vengurla, Maharashtra	MSW management	There is no landfill in the town as 100% MSW is being processed.	Total revenue generated monthly is ~ INR 1,70,000 (~ USD 2035). Compost is sold to farmers at very cheap price of INR 10/kg. Dumpsite now transformed into parks and people living near dumpsite have improved health conditions.
Bhopal, Madhya Pradesh	MSW management	100% of MSW is being processed, recyclables are sent to recycling plant, non-recyclables for RDF and organic waste to compost and biogas plant. Awareness programs, like e-waste clinic, Bartan Bank, and Gobar se Gamle, were conducted.	Negligible waste reaching landfill. Integration of informal sector into formal sector waste management system.

(Continued)

TABLE 2.1 (CONTINUED)
Circular Economy Initiatives from Various Sectors in India

Company or Town	Sector	Initiatives	Benefits
Dhenkanal, Odisha	MSW management	100% of MSW is being processed.	Revenue generation of ~ INR 1.3–1.8 million (USD 15,560 to USD 21,545) monthly. Improvement in city's aesthetic. Provides employment to many in the form of SHGs.
Jamshedpur, Jharkhand	MSW and e-waste management	95% source segregation and 100% waste processing. Plastic waste is used for road construction. Concept of eco-bricks introduced in schools and residential societies. A total of 230 tons of e-waste has been collected and processed up to 2021.	Provided better opportunities to about 200 rag-pickers and 1,200 other people. City now spends less money on MSW management. Total revenue generated INR 26 million (USD 3,11,203). Integration of informal sector into formal sector. Air pollution has been reduced by eliminating burning of e-waste.
Surat, Gujarat	MSW management	100% MSW collected, 100% source segregation, and 100% waste processing.	Total revenue generation is ~ INR 1,030 million (USD 1,23,28,412) annually. Employment to about 15,000 people.
Bicholim, Goa	MSW management	100% plastic waste is being processed.	No new landfill needed. Provides employment to women.
Gangtok, Sikkim	MSW management	90% MSW collection, 80% source segregation, and 63% waste processing.	Revenue of INR 5.3 million (USD 63,437) generated per month. Compost is sold to tea gardens at very low price of INR 7–8/kg. Complete ban on the use of single-use plastic.
Kumbakonam, Tamil Nadu	MSW management	100% MSW collection, 70% source segregation, and 60% waste processing. Plastic waste used for road construction and sent to cement industries. Coconut shells are sent to factories to produce mosquito repellents.	Negligible waste is reaching landfill and dumpsites, using bio-mining technology.
Gurugram, Haryana	C&D waste management	100% (1,200 tons C&D waste per day + 300 tons per day legacy waste) of C&D waste is being collected and processed.	About 0.35 million tons of C&D waste has been processed and revenue of more than INR 57 million (USD 6,82,252) has been generated to date.

(Continued)

TABLE 2.1 (CONTINUED)
Circular Economy Initiatives from Various Sectors in India

Company or Town	Sector	Initiatives	Benefits
Delhi, India	C&D waste management	C&D waste is being converted into cement bricks, pavement blocks, and kerbstones. More than 4.5 million tons of C&D waste have been transformed up to 2021.	Prevented debris reaching the eastern and western banks of the Yamuna River.
Ambuja Cement	Industry CE initiatives	Producing alternate fuel for cement kiln from municipal, agricultural, and industrial wastes.	Achieved thermal substitution rate of 5.4% in 2019. About 94,570 tons of plastic waste processed in kilns. Compared to 1990 levels, about 31% reduction in CO_2 emissions per tonne of cement manufactured.
Mahindra Sanyo	Industry CE initiatives	Metal waste, such as slag, is being recovered and reused. Zero-discharge of processed water.	Significant reduction in waste reaching to landfill. About 1,242 tons of metallic scrap were sold for reuse in 2018. About 9,267 tons of different slag were reused in 2018 instead of sending to landfill.
SABIC	Industry CE initiatives	CO_2 waste from one plant is purified and used as feedstock in another plant for production of urea and methanol.	Achieved reduction in manufacturing expenses.
JSW Steel	Industry CE initiatives	Established zero-liquid discharge facilities. Iron was recovered and recycled from processing sludge and fine dust.	100% water was utilised and about 3.77 million tons of material were recycled in 2022.
Mahindra Group	Industry CE initiatives	Bio-CNG plant and car shredding facilities.	About 8–10 tons of organic waste from Mahindra world city are treated, resulting in the generation of 400 kg purified CNG per day and 4 tons of organic fertilizer per day. Reductions in car scrappage reaching landfill.

(Continued)

TABLE 2.1 (CONTINUED)
Circular Economy Initiatives from Various Sectors in India

Company or Town	Sector	Initiatives	Benefits
Tata Steel	Industry CE initiatives	Producing bio-fertilizer from recycled slag waste. Recycling of ferrous scrap and utilizing solid waste at various plants. Installed India's first CO_2 capture facility from blast furnace gas plant and deployed India's first biofuel-powered ship for transportation of raw materials.	Improved the crop yield per acre by 20%. Revenue of ~ INR 4.6 billion (USD 5,50,58,928) was generated from ferrous recycling in 2022. Tata steel utilized about 97% to 100% of solid waste at various plants.
IFFCO	Industry CE initiatives	Transforming highly polluted ash into granulated ash to be used in producing flyash bricks and in cement manufacturing.	Prevented dumping of highly polluted flyash into ponds and lakes.

(*Sources:* CSE, 2021; FICCI, 2018; Ambuja Cement, 2019; Mahindra Sanyo, 2018; JSW, 2022; Tata Steel, 2022.)

initiative called "Cero Residuos", meaning Zero Waste, which targets achieving no waste reaching landfills. A steel manufacturer in Chile, the Gerdau AZA company, has introduced a closed-loop system for steel production, utilizing all the waste generated during the manufacturing processes (Circular Lab, 2023). Malaysia has been implementing circularity in various sectors, such as palm oil production and waste management. The domestic recycling rate in Malaysia increased from 5% in 2010 to 10.5% in 2012 by adopting the Tenth Malaysian Plan (Agamuthu and Mehran, 2020). Waste, such as fly ash and bottom ash generated from industries and coal-fired power plants, is being reused by the cement industries in Malaysia (Agamuthu and Mehran, 2020). Several other scheduled wastes, such as gypsum, metal sludge, glue, petroleum by-products, and mineral sludge are being reused back by the producer companies according to EPR regulations (Agamuthu and Mehran, 2020). At present, palm oil biomass pellets are being used to produce electricity in Malaysia. About 23% of agro-industrial by-products and waste are reused in the palm oil production process (Bejarano et al., 2022).

2.6 PATHWAYS FOR CIRCULAR ECONOMY IN DEVELOPING COUNTRIES

This section discusses in brief the different pathways for implementing circularity in various sectors of developing countries. The pathways are limited not only to opportunities related to technologies but are also related to regulations and policies.

2.6.1 MSW Sector

According to the World Bank Report, about 2.01 billion tons of municipal solid waste (MSW) is being generated globally and at least 33% of MSW is not being managed scientifically (Maalouf and Agamuthu, 2023). The majority of developing countries are still struggling to properly manage their solid waste and are mainly disposing of their waste in open landfills (Diaz, 2017). Waste generation in low-income countries is anticipated to increase more than threefold by 2050. There has been tremendous scope for implementation of CE in various sectors of developing countries. The MSW stream contains dry waste, which mainly constitutes recyclables (plastic, paper, metal, glass, aluminum, rubber) and wet waste, which is biodegradable (food waste, kitchen waste). Most of the technologies for implementing CE, such as pyrolysis, gasification, refuse-derived fuel (RDF), use of plastic waste in road construction, bio-methanation, and composting, are known to the world but still the circularity in developing countries is not fully implemented. The first requirement for implementing circularity in MSW is proper source segregation. In developing countries, source segregation is not being performed properly mainly because of the lack of awareness among citizens, and in many cases, even after being aware of the benefits of source segregation, citizens are careless toward the environment. The government must make the citizens aware of the need to segregate their waste and, if needed, it should impose penalties. Stakeholder involvement is essential for successful implementation of a proper waste management system. There must be incentives for industries to be involved in recycling or research, and recycled products should be provided with incentives to encourage the use of recycled products. There is also a need to spend on research and development related to the technologies to develop new advanced technologies for the recycling or transformation of MSW. At present, recycled products are inferior to virgin products and are less preferred by the user. The informal sectors play an important role, especially in the recycling of dry wastes and this sector needs to be mainstreamed or merged with the formal sector. Often, the baled polyethylene terephthalate (PET) waste is kept at the material recovery facilities (MRFs) but is not purchased or sent to the recycling industry for further processing. There is a need for a strong and efficient recycling network to ensure that all the collected plastic waste has been recycled or reused. Policies, such as EPR, need to be enforced strictly and EPR should also be introduced for other recyclables, such as paper, rubber, glass, metals, and others. In the case of wet waste, composting and bio-methanation are the two most used technologies (MoHUA, 2021). Bio-methanation is more techno-commercially viable than the traditional composting plant for capacities greater than 50 thermal design power (MoHUA, 2021). There is a need for developing countries to establish new bio-methanation plants and to replace old composting plants. The other pathways to implement circularity for dealing with wet waste is to provide more refined biogas which can replace compressed natural gas (CNG) and liquified petroleum gas (LPG) as fuel in automobiles and for cooking, respectively. Food waste can be reused by establishing an efficient route between potential food donors and hunger relief charities. Landfilling of recyclables and wet waste must be strictly banned and digitization of the entire solid waste management system will contribute to achieving improved circularity.

2.6.2 C&D Waste Sector

Construction and Demolition (C&D) waste is one of the world's major waste flows, with numerous studies estimating that C&D waste accounts for 30–40% of total solid waste (Islam et al., 2019). India generates ~12 million tons of C&D waste (20–25% of total solid waste) annually, while, in China, 30–40% of total waste is C&D waste (MoHUA, 2021; Huang et al., 2018). The developed countries have properly implemented CE in management of C&D waste and have very high recycling rates, unlike developing countries which have very low recycling rates. Developed countries have recycling rates of 86% in the UK, Italy 75%, the USA 70%, Australia 67%, and France 47%, whereas developing countries like China and India have recycling rates of 5% and 1%, respectively (Tezeswi et al., 2022; Huang et al., 2018; CSE, 2020). There are several measures which can be adopted to implement effective circularity in C&D waste management. Firstly, developing countries must accurately estimate and characterize the generation of C&D waste. After proper estimation and characterization, Urban Local Bodies (ULBs) must make arrangements for proper collection and transportation of C&D waste, which is a challenge for developing countries like China and India. The collection points and waste processing plants must be properly located to minimize the transport distances involved. There should be proper guidance and standards for reusing C&D waste which will encourage the construction industry to reuse the recycled C&D waste. Developing countries are still using immature technologies to process their C&D waste. Mostly wastes consisting of bricks, stones, and tiles are crushed to produce Recycled Aggregates (RA), while wastes consisting of concrete rubbles are crushed to produce Recycled Concrete Aggregates (RCA). These recycled aggregates are of inferior quality when compared with virgin materials and contractors hesitate to use the recycled materials. Developed countries like Japan and Germany are using advanced recycling technologies, such as the carbonization of C&D waste and concrete and cement separation (Huang et al., 2018). The USA is technologically capable of producing recycled concrete that satisfies a wide range of strength requirements (Huang et al., 2018). The developing countries need to upgrade their C&D waste recycling technologies to improve the circularity of C&D waste. Circular economy implementation for C&D waste in developing countries can also be improved by introducing "Design for Deconstruction", "Design for Reuse", and "Design for Longevity", which will promote the reuse of construction material and structures, reducing the C&D waste generation. Circularity can also be improved by introducing the concept of Life Cycle Assessment (LCA), Material Flow Analysis (MFA), and Environmental Product Declarations (EPDs) for the construction sector of developing countries.

2.6.3 E-waste Sector

Over the past few years, e-waste has emerged as a major global environmental concern, attracting the attention of numerous nations. Globally, about 44.7 million tons of e-waste was generated in 2016 which increased to 50 million tons in 2018 and 53.6 million tons in 2019 (Gollakota et al., 2020; Sengupta et al., 2022). E-waste generation per capita in Oceania is about 17.3 kg/capita, Europe 16.6 kg/capita, America 11.6 kg/capita, Asia 4.2 kg/capita, and Africa 1.9 kg/capita, whereas, in terms of overall e-waste generation, Asia is at the top, generating about 18.2 million tons of e-waste in 2016, followed by Europe (12.3 million tons), America (11.3 million tons), Africa (2.2 million tons), and Oceania (0.7 million tons) (Gollakota et al., 2020). E-waste generation in Asia increased from 18.2 million tons in 2016 to 24.9 million tons in 2019 and it is expected that globally e-waste generation will exceed 74 million tons by 2030 (Andeobu et al., 2021). In emerging nations, the rate of e-wate production is rapidly rising at a rate of 3 to 5% annually (Rasheed et al., 2022). The e-waste amounts generated by developing countries in 2019 were China (10.1 million tons), India (3.2 million tons), Indonesia (1.6 million tons), Bangladesh (2.8 million tons), and Pakistan (0.4 million tons) (Rasheed et al., 2022; Masud et al., 2019; Andeobu et al., 2021). The e-waste recycling rates in some of the developed nations are Germany (52%), the USA (15%), and Japan (22%), while the values in developing nations include China (15%) and India (0.93%) (Sengupta et al., 2022). The majority of e-waste generated in developing countries, such as India, Bangladesh, Pakistan, and Indonesia, is dumped into landfills or the open sea (Masud et al., 2019). There are various alternatives to the introduction of circularity in e-waste management: (i) refurbish the discarded product for reuse; (ii) recover valuable metals such as gold and silver from e-waste; (iii) generate significant amounts (40 MJ/kg) of energy electronic chips and plastic in e-waste using the Catalytic De-polymerization Process Technology (Masud et al., 2019); (iv) pyrolysis of e-waste will transform waste into combustible fuel; or (v) companies may start providing e-waste specific managment service. The pathways for implementing CE in e-waste management or promoting sustainable e-waste management in developing countries require several reforms at various levels. Firstly, there should be proper data inventorization on the e-waste generated; secondly, there is the utmost need to merge the large informal sector into the formal sector to improve the collection rate and to reduce the bad side effects of the improper handling of e-waste. There should be a proper take-back system which must be accessible and less time consuming than the current one. Take-back collection points should use a Global Positioning System (GPS) to share information about their location and preferred visit time to the consumers. There must be stringent rules and regulations for e-waste management and the departments concerned must ensure that these rules are followed. There should be awareness campaigns for e-waste to make citizens aware of the take-back system and the recycling of e-waste. The infrastructure of advanced e-waste recycling technologies needs to be established in greater numbers to improve the recycling rate. The companies should provide support in establishing and operating the e-waste recycling industries and must improve after-sales service of the electronic items so that the consumer gets them repaired instead of discarding them. There should be more research on eco-friendly technologies such as bio-sorption, bioleaching, and bioremediation which have the ability to recover metals from e-waste. LCA and MFA can help in improving the circularity of electronic items.

2.6.4 Wastewater Sector

The freshwater supplies available on Earth are insufficient to meet increasing human activity-related water demands. Additionally, it is predicted that climate change will make it more challenging to acquire usable water due to extreme weather events and a decrease in rainfall quantity and regularity (Mannina et al., 2022). It is predicted that 17 countries, home to one-quarter of the world's population, will face extremely high drought stress worldwide (Hofste et al., 2019). Developing countries like India and Pakistan are under high water stress while China and South Africa are under moderate water stress (Hofste et al., 2019). About 80% of wastewater is directly discharged into the environment without any treatment and ~2 billion people worldwide consume feces-contaminated water (Kakwani and Kalbar, 2020). The quantity and quality issues facing potable water supplies due to the linear approach of extracting water and discharging wastewater have induced nations to implement a circular approach to the water and wastewater sector. Many developed nations, such as the USA, Spain, and Japan, are reusing their significant volumes of wastewater for other purposes and some developing nations, such as China and Mexico, are also efficient in reusing their wastewater (Guerra-Rodríguez et al., 2020). There are several pathways to implementing circularity in the wastewater sector: (i) using treated wastewater for agriculture and landscape irrigation, (ii) using treated wastewater for industrial and construction activities, (iii) reusing wastewater for domestic purposes, such as flushing, car washing, or watering garden plants, (iv) recovery of energy from wastewater using microbial fuel cells, or (v) recovery of nutrients from wastewater. The developing nations need to introduce policies on wastewater and sludge reuse and recycling. Water losses from infrastructure leakage and faulty equipment must be minimized. The government must control groundwater extraction and provide freshwater at a high price to discourage overuse of freshwater. The developing nations need to shift from conventional wastewater treatment technologies (e.g., the activated sludge process) toward nature-based treatment technologies, such as wetland and phytoremediation. The circular economy in the wastewater sector cannot be implemented properly without public acceptance. Consumers are hesitant in reusing wastewater for any useful purposes and there is a need to increase awareness and train the public on the importance of freshwater and the safety of reusing the treated wastewater. Farmers are also hesitant to use treated wastewater for irrigation because they think consumers will not favor crops that are watered with reclaimed water (Frijns et al., 2016).

2.6.5 Agriculture Sector

Feeding a population that is always expanding is one of the greatest difficulties facing humanity (Pandey and Dwivedi, 2020). Research indicates that food production needs to be increased by 5.1 billion tons (70%) by 2050 (Velasco-Muñoz et al., 2022). The agriculture sector accounts for 24% of GHG emissions globally and more than 90% of environmental impacts related to land and water (Velasco-Muñoz et al., 2021; Barros et al., 2020). The implementation of circularity will reduce food wastage along with decreasing the impacts on the environment by the agriculture sector. A study on potato farming (TERI, 2018) discovered that adopting various sustainable farming practices grounded in CE principles can cut GHG emissions by 55% and biodiversity loss by 15%, while also lowering agricultural costs by reducing the use of machinery, fertilizers, and pesticides. This section will mainly focus on the agriculture sector in India as India is the world's second-largest producer of agricultural goods, contributing close to 7.68% of the overall agricultural output (TERI, 2018). There are various pathways to implementing CE in the agriculture sector of any developing country. Farming fields can be installed with drip or sprinkler irrigation systems, resulting in saving freshwater of up to 80%. The use of treated wastewater for irrigation purposes can also improve circularity. The shelf life of food can be extended to reduce food wastage. Crop residues can be properly utilized in the form of livestock feed or can be converted into compost, biochar, or biofuel. Crop residues such as rice husk can be used to generate electricity. The development of cold chain infrastructure can promote sustainable agriculture by reducing food loss by 76% and carbon emissions by 16% (TERI, 2018). Red seaweed derivatives can be used as organic solutions for crop enhancement. Used coffee grounds, having a high calorific value, can be utilized as a fuel in boilers. Awareness and training campaigns for farmers and the implementation of lifecycle assessments can play a key role in implementing circularity in the agricultural sector.

2.7 TOOLS TO ANALYSE A CIRCULAR ECONOMY

There are several tools by which to assess circularity for better implementation of a circular economy. The important tools are as follows:

1. Material Passport. A material passport is a document in digital form that lists all of the materials used in a product or a construction throughout its lifecycle to support supply chain managers in strategizing circularity decisions. Passports contain a set of data indicating specific properties of materials in products, allowing for the identification of their value for recovery, recycling, and reuse.
2. Material Circularity Indicator (MCI). MCI was developed by the Ellen MacArthur Foundation and Granta Design. MCI helps the companies to improve product design and material procurement by measuring the circularity of material flows. An MCI value between 0 and 1 is achieved, and larger values denote greater circularity. The inputs required by the tool are details about the production process, utility during use, the end-of-life scenario, and recycling efficiency.
3. Environmental Product Declarations (EPDs). EPDs are third-party validated records that offer information on a product's environmental impact throughout its lifecycle. Manufacturers, governments, and consumers can utilize them to make effective decisions about product sustainability. EPDs play an important role in the circular economy because they provide transparency regarding the environmental implications of items throughout their lifecycle.
4. Lifecycle Costing (LCC). The circular economy is not limited to environmental aspects but economic and social aspects are equally important. LCC is a tool to calculate an asset's total cost throughout its lifecycle, which includes original capital expenditure, maintenance costs, operating costs, and the asset's residual value at the end of its life. LCC has been derived from ISO 15686 (Hoosain et al., 2023).
5. Lifecycle Assessment (LCA). The LCA tool is widely used to identify the environmental impacts of any product, system, or activity throughout its lifecycle. LCA is performed according to the guidelines and framework described in ISO 14040 and ISO 14044 (Aryan et al., 2023). LCA helps in assessing and comparing the sustainability impacts of various CE strategies to identify the optimum CE strategy. Social LCA also analyzes the circular economy initiatives with respect to social and sociological aspects of the products for the entire lifetime.

2.8 CHALLENGES IN ADOPTING THE CIRCULAR ECONOMY IN DEVELOPING COUNTRIES

There are several barriers and challenges involved in implementing circularity in developing countries. Most of the developing countries lack effective policies and regulations for the implementation of CE. Those countries which have the policy and regulations are not able to implement them strictly due to mismanagement, lack of willingness, and lack of a proper supply chain. Developing countries do not have a strong economy and are low in technological advances. The lack of financial capability and the high cost of investment makes CE implementation financially challenging for a developing economy. Proper and effective CE implementation requires research and innovation in technology which is not available in the majority of developing countries. There is a limited market for recycled products and society is unaware of the various uses of recycled products. The recycled products are often more expensive than the products made from virgin raw materials which makes them less attractive to the consumer. There is a lack of awareness among the consumers regarding the adverse effects of the linear approach. Consumers are hesitant to use recycled products as they have a perception that recycled products are always inferior to virgin products. Stakeholders are also resistant to investing in the implementation of the circular economy. Both the consumer and the supplier lack the involvement and support to promote circularity. The adoption of circularity-related initiatives appears to be motivated by economic rather than environmental concerns, with a clear preference for practices that yield a financial return in the short term (Masi et al., 2018).

2.9 CONCLUSIONS

The transition from the linear to the circular approach is necessary for developing countries to move toward a sustainable future. CE is one of the important development frameworks for developing countries to address the natural resource scarcity and excess waste concerns while preventing additional development that would have an impact on the cost to the environment and resources for the next generation. The present study discussed in detail the circular economy in developing countries, along with various other aspects, such as the concept of CE, the current status in developed and developing nations, policies to promote circularity, benefits of and pathways to implementing CE, tools involved in the analysis of CE implementation and challenges to implementing circularity in developing countries. Developing nations like China and India have policies and regulations to implement circularity and are performing circularity but there is scope for improvement in CE implementation, whereas developing nations like Pakistan and Bangladesh lack the effective policies and regulations to support CE implementation. The circular economy provides benefits covering all the aspects of sustainability viz., economic, environmental, and social. There is a need for effective policies, a proper supply chain, investment, stakeholder involvement, technological advancement, public awareness, and a protective attitude toward the environment to promote and successfully implement circularity. Sustainability-promoting tools such as Material Circularity Indicators, Environmental Product Declarations, and Life Cycle Assessment should help to identify effective strategies for circularity. Every country, especially emerging nations, should pursue policies that aid in decoupling growth, thereby sustaining economic growth while promoting inclusive development.

ACKNOWLEDGMENTS

The first author would like to acknowledge the Indian Institute of Technology, Bombay, for providing research facilities.

COMPETING INTERESTS

The authors have no relevant financial or non-financial interests to disclose.

REFERENCES

Agamuthu, P. and Mehran, S.B., 2020. Circular economy in Malaysia. *Circular Economy: Global Perspective*, pp. 241–268.

Ahmed, Z., Mahmud, S. and Acet, H., 2022. Circular economy model for developing countries: Evidence from Bangladesh. *Heliyon*, 8, p. e09530.

Ambuja Cement, 2019. *Sustainable Development Report 2019.* Retrieved from: https://www.ambujacement.com/Upload/PDF/Ambuja-SD-Rep_11820_WEB.pdf.

Andeobu, L., Wibowo, S. and Grandhi, S., 2021. A systematic review of e-waste generation and environmental management of Asia Pacific countries. *International Journal of Environmental Research and Public Health*, 18(17), p. 9051.

Aryan, Y., Dikshit, A.K. and Shinde, A.M., 2023. A critical review of the life cycle assessment studies on road pavements and road infrastructures. *Journal of Environmental Management*, 336, p. 117697.

Azizuddin, M., Shamsuzzoha, A. and Piya, S., 2021. Influence of circular economy phenomenon to fulfil global sustainable development goal: Perspective from Bangladesh. *Sustainability*, 13(20), p. 11455.

Barros, M.V., Salvador, R., de Francisco, A.C. and Piekarski, C.M., 2020. Mapping of research lines on circular economy practices in agriculture: From waste to energy. *Renewable and Sustainable Energy Reviews*, 131, p. 109958.

Bejarano, P.A.C., Rodriguez-Miranda, J.P., Maldonado-Astudillo, R.I., Maldonado-Astudillo, Y.I. and Salazar, R., 2022. Circular economy indicators for the assessment of waste and by-products from the palm oil sector. *Processes*, 10(5), p. 903.

Bleischwitz, R., Yang, M., Huang, B., Xiaozhen, X.U., Zhou, J., McDowall, W., Andrews-Speed, P., Liu, Z. and Yong, G., 2022. The circular economy in China: Achievements, challenges and potential implications for decarbonisation. *Resources, Conservation and Recycling*, 183, p. 106350.

Boulding, K.E., 1966. The economics of coming spaceship earth. In H. Jarret (Ed.), *Environmental Quality in a Growing Economy.* Baltimore, MD: John Hopkins University Press.

Circular Lab, 2023. Chile is leading the way as an example of a regional circular economy. Retrieved from: https://www.circularinnovationlab.com/post/chile-is-leading-the-way-as-an-example-of-a-regional-circular-economy#:~:text=Chile%20has%20been%20implementing%20policies,waste%20and%20maximise%20resource%20efficiency.

CPHEEO, 2018. Central public health and environmental engineering organisation. Guidelines on usage of refuse derived fuel in various industries. Retrieved from: http://cpheeo.gov.in/upload/5bda791e5afb3SBMRDFBook.pdf.

CSE, 2020. Centre for science and environment. India recycles only 1% of its construction and demolition waste. Retrieved from: https://www.downtoearth.org.in/news/waste/india-recycles-only-1-of-its-construction-and-demolition-waste-cse-73027.

Daskal, S. and Ayalon, O., 2020. Circular economy—Situation in Israel. *Circular Economy: Global Perspective*, pp. 187–200.

Diaz, L.F., 2017. Waste management in developing countries and the circular economy. *Waste Management & Research*, 35(1), pp. 1–2.

Ding, X., Zhou, C., Zhong, W. and Tang, P., 2019. Addressing uncertainty of environmental governance in environmentally sensitive areas in developing countries: A precise-strike and spatial-targeting adaptive governance framework. *Sustainability*, 11(16), p. 4510.

Dong, H., Geng, Y., Yu, X. and Li, J., 2018. Uncovering energy saving and carbon reduction potential from recycling wastes: A case of Shanghai in China. *Journal of Cleaner Production*, 205, pp. 27–35.

EU, 2021. European Commission. First circular economy action plan. Retrieved from: https://ec.europa.eu/environment/circular-economy/first_circular_economy_action_plan.html.

FICCI, 2018. Federation of Indian chambers of commerce & industry. Accelerating India's circular economy shift. Retrieved from: https://www.ficcices.in/pdf/FICCI-Accenture_Circular%20Economy%20Report_OptVer.pdf.

Fiksel, J., Sanjay, P. and Raman, K., 2021. Steps toward a resilient circular economy in India. *Clean Technologies and Environmental Policy*, 23, pp. 203–218.

Frijns, J., Smith, H.M., Brouwer, S., Garnett, K., Elelman, R. and Jeffrey, P., 2016. How governance regimes shape the implementation of water reuse schemes. *Water*, 8(12), p. 605.

Geng, Y., Fu, J., Sarkis, J. and Xue, B., 2012. Towards a national circular economy indicator system in China: An evaluation and critical analysis. *Journal of cleaner production*, 23(1), pp. 216–224.

Ghosh, S.K., 2020a. Introduction to circular economy and summary analysis of chapters. In: Ghosh, S. (eds) *Circular Economy: Global Perspective*, pp. 1–23. Singapore.

Ghosh, S.K., 2020b. Circular economy in India. In: Ghosh, S. (eds) *Circular Economy: Global Perspective*, pp. 157–185. Singapore.

Gollakota, A.R., Gautam, S. and Shu, C.M., 2020. Inconsistencies of e-waste management in developing nations–Facts and plausible solutions. *Journal of Environmental Management*, 261, p. 110234.

Greyson, J., 2007. An economic instrument for zero waste, economic growth and sustainability. *Journal of Cleaner Production*, 15(13–14), pp. 1382–1390.

Guerra, B.C. and Leite, F., 2021. Circular economy in the construction industry: An overview of United States stakeholders' awareness, major challenges, and enablers. *Resources, Conservation and Recycling*, 170, p. 105617.

Guerra-Rodríguez, S., Oulego, P., Rodríguez, E., Singh, D.N. and Rodríguez-Chueca, J., 2020. Towards the implementation of circular economy in the wastewater sector: Challenges and opportunities. *Water*, 12(5), p. 1431.

Guo, W., Xi, B., Huang, C., Li, J., Tang, Z., Li, W., Ma, C. and Wu, W., 2021. Solid waste management in China: Policy and driving factors in 2004–2019. *Resources, Conservation and Recycling*, 173, p. 105727.

Halog, A. and Anieke, S., 2021. A review of circular economy studies in developed countries and its potential adoption in developing countries. *Circular Economy and Sustainability*, 1, pp. 209–230.

Hofste, R.W., Reig, P. and Schleifer, L., 2019. 17 countries, home to one-quarter of the world's population, face extremely high water stress. Retrieved from: https://www.wri.org/insights/17-countries-home-one-quarter-worlds-population-face-extremely-high-water-stress?trk=public_post_comment-text.

Hoosain, M.S., Paul, B.S., Doorsamy, W. and Ramakrishna, S., 2023. Comparing South Africa's sustainability and circular economic roadmap to the rest of the world. *Materials Circular Economy*, 5(1), p. 2.

Huang, B., Wang, X., Kua, H., Geng, Y., Bleischwitz, R. and Ren, J., 2018. Construction and demolition waste management in China through the 3R principle. *Resources, Conservation and Recycling*, 129, pp. 36–44.

Isa, N.M., Sivapathy, A. and Kamarruddin, N.N.A., 2021. Malaysia on the way to sustainable development: Circular economy and green technologies. In B. S. Sergi and A. R. Jaaffar (Eds.), *Modeling Economic Growth in Contemporary Malaysia* (pp. 91–115). Emerald Publishing Limited.

Islam, R., Nazifa, T.H., Yuniarto, A., Uddin, A.S., Salmiati, S. and Shahid, S., 2019. An empirical study of construction and demolition waste generation and implication of recycling. *Waste Management*, 95, pp. 10–21.

JSW, 2022. Integrated report 2021–2022. Retrieved from: https://www.jsw.in/sites/default/files/assets/industry/steel/IR/Financial%20Performance/Annual%20Reports%20_%20STEEL/JSW%20Steel%20Integrated%20Report%202021-22.pdf.

Kakwani, N.S. and Kalbar, P.P., 2020. Review of circular economy in urban water sector: Challenges and opportunities in India. *Journal of Environmental Management*, 271, p. 111010.

Kalaari, 2022. Circular economy a Kalaari Capital report. Retrieved from: https://www.kalaari.com/wp-content/uploads/2022/04/Circular-Economy-Report-2022.pdf

Kirchherr, J., Reike, D. and Hekkert, M., 2017. Conceptualizing the circular economy: An analysis of 114 definitions. *Resources, Conservation and Recycling*, 127, pp. 221–232.

Kuhn, B., 2016. Sustainable development discourses in China. *Journal of Sustainable Development*, 9(6), pp. 158–167

Levitzke, P.V., 2020. The development of a circular economy in Australia. In S. Ghosh (Ed.),*Circular Economy: Global Perspective*, pp. 25–42. Springer, Singapore. https://doi.org/10.1007/978-981-15-1052-6_2

Liu, Q., Li, H.M., Zuo, X.L., Zhang, F.F. and Wang, L., 2009. A survey and analysis on public awareness and performance for promoting circular economy in China: A case study from Tianjin. *Journal of Cleaner Production*, 17(2), pp. 265–270.

Maalouf, A. and Agamuthu, P., 2023. Waste management evolution in the last five decades in developing countries–A review. *Waste Management & Research*, p. 0734242X231160099.

MacArthur Foundation, 2022. Ellen MacArthur Foundation. Advancing vehicle remanufacturing in China: The role of policy. Retrieved from: https://emf.thirdlight.com/file/24/6vfmew76vosTAS-6v8h26Xv_O5R/Case%20Studies%20%20Advancing%20vehicle%20remanufacturing%20in%20China.pdf.

MacArthur, E., 2013. Towards the circular economy. *Journal of Industrial Ecology*, 2(1), pp. 23–44.

Mahindra Sanyo, 2018. Sustainability report 2018. Retrieved from: http://www.mssspl-india.com/images/pdf/mahindra-sanyo-sustainability-report-2017-2018.pdf.

Mannina, G., Gulhan, H. and Ni, B.J., 2022. Water reuse from wastewater treatment: The transition towards Circular Economy in the water sector. *Bioresource Technology*, 363, p. 127951.

Marino, A. and Pariso, P., 2020. Comparing European countries' performances in the transition towards the circular economy. *Science of the Total Environment*, 729, p. 138142.

Masi, D., Kumar, V., Garza-Reyes, J.A. and Godsell, J., 2018. Towards a more circular economy: Exploring the awareness, practices, and barriers from a focal firm perspective. *Production Planning & Control*, 29(6), pp. 539–550.

Masud, M.H., Akram, W., Ahmed, A., Ananno, A.A., Mourshed, M., Hasan, M. and Joardder, M.U.H., 2019. Towards the effective E-waste management in Bangladesh: A review. *Environmental Science and Pollution Research*, 26(2), pp. 1250–1276.

MIE, 2016. Ministry of infrastructure and the environment. A circular economy in the Netherlands by 2050. Retrieved from: https://www.government.nl/documents/leaflets/2016/09 /22/a-circular-economy-in-the-netherlands-by-2050.

MoHUA, 2021. Ministry of housing and urban affairs, Government of India. Circular economy in municipal solid and liquid waste management. Retrieved from: https://mohua.gov.in/pdf/627b8318adf18Circular-Economy-in-waste-management-FINAL.pdf.

Murray, A., Skene, K. and Haynes, K., 2017. The circular economy: An interdisciplinary exploration of the concept and application in a global context. *Journal of Business Ethics*, 140, pp. 369–380.

Ngan, S.L., How, B.S., Teng, S.Y., Promentilla, M.A.B., Yatim, P., Er, A.C. and Lam, H.L., 2019. Prioritization of sustainability indicators for promoting the circular economy: The case of developing countries. *Renewable and Sustainable Energy Reviews*, 111, pp. 314–331.

NITI Aayog, 2022. Lifestyle for environment. Retrieved from: https://www.niti.gov.in/sites/default/files/2022-10/Brochure-10-pages-op-2-print-file-20102022.pdf.

Pandey, S. and Dwivedi, N., 2020. Utilisation and management of agriculture and food processing waste. In P. Mishra, R. R. Mishra, and C. O. Adetunji (Eds.), *Innovations in Food Technology: Current Perspectives and Future Goals*, pp. 269–288. Singapore: Springer. https://doi.org/10.1007/978-981-15-6121-4_19

Panel, I.R., 2017. Smarter use of resources can add $2 trillion annually to global economy. Retrieved from: https://www.unenvironment.org/news-and-stories/press-release/smarter-use-resources-can-add-2-trillion annually-global-economy.

Pearce, D.W. and Turner, R.K., 1989. *Economics of Natural Resources and the Environment. Land Economics*, 67(2), pp. 272–276. Johns Hopkins University Press.

Peng, H. and Liu, Y., 2016. A comprehensive analysis of cleaner production policies in China. *Journal of Cleaner Production*, 135, pp. 1138–1149.

Pesce, M., Tamai, I., Guo, D., Critto, A., Brombal, D., Wang, X., Cheng, H. and Marcomini, A., 2020. Circular economy in China: Translating principles into practice. *Sustainability*, 12(3), p. 832.

PIB, 2021a. Press information Bureau, Government of India. Retrieved from: https://pib.gov.in/PressReleasePage.aspx?PRID=1705772.

PIB, 2021b. Press information Bureau, Government of India. Retrieved from: https://pib.gov.in/PressReleaseIframePage.aspx?PRID=1745433.

PIB, 2022a. Press information Bureau, Government of India. Retrieved from: https://pib.gov.in/PressReleasePage.aspx?PRID=1881761.

PIB, 2022b. Press information Bureau, Government of India. Retrieved from: https://pib.gov.in/PressReleasePage.aspx?PRID=1854433.

Potting, J., Hekkert, M.P., Worrell, E. and Hanemaaijer, A., 2017. Circular economy: Measuring innovation in the product chain. *Planbureau voor de Leefomgeving*, 2544. Retrieved from: http://www.pbl.nl/sites/default/files/cms/publicaties/pbl-2016-circular-economy-measuring-innovation-in-product-chains-2544.pdf.

Rasheed, R., Rizwan, A., Javed, H., Sharif, F., Yasar, A., Tabinda, A.B., Mahfooz, Y., Ahmed, S.R. and Su, Y., 2022. Analysis of environmental sustainability of e-waste in developing countries—A case study from Pakistan. *Environmental Science and Pollution Research*, 29(24), pp. 36721–36739.

Sengupta, D., Ilankoon, I.M.S.K., Kang, K.D. and Chong, M.N., 2022. Circular economy and household e-waste management in India: Integration of formal and informal sectors. *Minerals Engineering*, 184, p. 107661.

Sihvonen, S. and Ritola, T., 2015. Conceptualizing ReX for aggregating end-of-life strategies in product development. *Procedia Cirp*, 29, pp. 639–644.

Stahel, W. and Reday-Mulvey, G., 1976. Jobs for tomorrow: The potential for substituting manpower for energy, Report to the Commission of the European Communities (now European Commission), Brussels.

Tata Steel, 2022. Integrated report 2021–2022. Retrieved from: https://www.tatasteel.com/media/16654/tata-steel-ir-2021-22-1.pdf.

TERI, 2018. The energy and resource institute. Circular Economy: A business imperative for India. Retrieved from: https://wsds.teriin.org/2018/files/teri-yesbank-circular-economy-report.pdf.

TERI, 2021. The energy and resources institute. Circular economy for plastics in India: A roadmap. TERI, New Delhi. Retrieved from: https://www.teriin.org/sites/default/files/2021-12/Circular-Economy-Plastics-India-Roadmap.pdf.

TERI, 2023. The energy and resource institute. Puducherry sets a green milestone. Retrieved from: https://www.teriin.org/press-release/puducherry-sets-green-milestone-unveils-pioneering-green-budget-report-world#:~:text=The%20green%20budget%20of%20the,crores%20in%20FY%202023%2D24.

Tezeswi, T.P. and MVN, S.K., 2022. Implementing construction waste management in India: An extended theory of planned behaviour approach. *Environmental Technology & Innovation*, 27, p. 102401.

UNCTAD, 2022a. United Nations Conference on trade and development. Retrieved from: https://unctad.org/data-visualization/now-8-billion-and-counting-where-worlds-population-has-grown-most-and-why.

UNCTAD, 2022b. United Nations Conference on trade and development. Circular economy principles could help India realise $624bn. Retrieved from: https://unctad.org/news/circular-economy-principles-could-help-india-realise-624bn.

UNDP, 2021. United Nations Development Programme. Retrieved from: https://climatepromise.undp.org/what-we-do/where-we-work/south-africa.

UNEP, 2016. Worldwide extraction of materials triples in four decades, intensifying climate change and air pollution. Retrieved from: https://www.unenvironment.org/news-and-stories/press-release/worldwide-extraction-materials-triples-four-decades-intensifying.

UNFCCC, 2021. United Nations framework convention on climate change. Retrieved from: https://unfccc.int/sites/default/files/NDC/2022-06/NDC_submission_20210826revised.pdf.

UNFCCC, 2022. United Nations framework convention on climate change. Retrieved from: https://unfccc.int/sites/default/files/NDC/202208/India%20Updated%20First%20Nationally%20Determined%20Contrib.pdf.

Van Buren, N., Demmers, M., Van der Heijden, R. and Witlox, F., 2016. Towards a circular economy: The role of Dutch logistics industries and governments. *Sustainability*, 8(7), p. 647.

Velasco-Muñoz, J.F., Aznar-Sánchez, J.A., López-Felices, B. and Román-Sánchez, I.M., 2022. Circular economy in agriculture. An analysis of the state of research based on the life cycle. *Sustainable Production and Consumption*, 34, pp. 257–270.

Velasco-Muñoz, J.F., Mendoza, J.M.F., Aznar-Sánchez, J.A. and Gallego-Schmid, A., 2021. Circular economy implementation in the agricultural sector: Definition, strategies and indicators. *Resources, Conservation and Recycling*, 170, p. 105618.

WBCSD, 2017. World Business Council for Sustainable Development. CEO guide to circular economy. Retrieved from: https://docs.wbcsd.org/2017/06/CEO_Guide_to_CE.pdf.

Wijaya, A.S., 2014, March. Climate change, global warming and global inequity in developed and developing countries (Analytical perspective, Issue, Problem and Solution). *In IOP Conference Series: Earth and Environmental Science* (Vol. 19, No. 1, p. 012008). IOP Publishing.

Xiao, S., Dong, H., Geng, Y., Tian, X., Liu, C. and Li, H., 2020. Policy impacts on municipal solid waste management in Shanghai: A system dynamics model analysis. *Journal of Cleaner Production*, 262, p. 121366.

Yuan, Z., Bi, J. and Moriguichi, Y., 2006. The circular economy: A new development strategy in China. *Journal of Industrial Ecology*, 10(1–2), pp. 4–8.

Zhu, J., Fan, C., Shi, H. and Shi, L., 2019. Efforts for a circular economy in China: A comprehensive review of policies. *Journal of Industrial Ecology*, 23(1), pp. 110–118.

Zisopoulos, F.K., Schraven, D.F. and de Jong, M., 2022. How robust is the circular economy in Europe? An ascendency analysis with Eurostat data between 2010 and 2018. *Resources, Conservation and Recycling*, 178, p. 106032.

3 Water Contamination and the Circular Economy

Nirmala Ganesan
Associate Professor, Vels Institute of Science, Technology & Advanced Studies (VISTAS), School of Bioengineering, Chennai, India

3.1 INTRODUCTION

Water is the main constituent of planet Earth, as well as being the primary component of the body fluids of all known living organisms. Whether it is used for drinking, domestic use, food production, or recreational purposes, safe and readily available water is of prime importance to public health. The economy of a country cannot flourish without a constant uninterrupted water supply, proper sanitation, and excellent management of water resources. Potable water can be defined as water that is suitable for purposes of human consumption, i.e., drinking or cooking. Consequently, it should be devoid of any unpleasant odor, color, or taste, and be available at a reasonable temperature range[1,2] Safe water means water that does not contain any toxins, carcinogens, pathogenic micro-organisms, or other health hazards. An adequate supply of safe drinking water is one of the major prerequisites for leading a healthy life; waterborne diseases continue to be a major cause of deaths worldwide, particularly of children, in addition to posing a grave economic threat to many subsistent economies.[3–12]

Basically, our Earth is full of water. Water is one of our most precious resources, without which we cannot imagine life on Earth. Water exists as a vapor, liquid, or solid (in the form of ice) on our planet. Water accounts for 71% of our planet's surface, yet, still, more than half of the world's human population faces water scarcity due to the unavailability of clean freshwater for daily usage purposes. On our Earth, there is a water cycle that continuously recycles water and also produces it. The conversion of vapors into liquids, liquids into solids, solids into liquids, and liquids into vapors is continuously occurring around the globe. So, there is plenty of water on our planet but the question is, how much of the water is accessible to us? Approximately 97% of the water on Earth is salty, being too loaded with minerals for humans to drink or for its use in agriculture.[13–15] Of the remaining 3% of potentially usable freshwater, more than two-thirds is frozen in polar ice caps and glaciers. That leaves less than 1% available for agricultural, industrial, and domestic use. Freshwater extends across our

 DOI: 10.1201/9781003432869-3

planet in rivers, lakes, streams, underground aquifers, groundwater, and permafrost. Groundwater is very difficult to extract so humans have mostly settled close to surface water like rivers and lakes. About 90% of the world's human population lives less than 10 kilometers away from a freshwater source. Any pollution of the freshwater area decreases the quality of this water. Out of the world's total water supply, only 0.007% of the water is considered safe for consumption, according to the World Health Organization (WHO), and this tiny amount needs to be shared by more than 6.9 billion people on the planet. The available freshwater sources are being rapidly depleted by humans, despite being slowly replenished by rain and snowfall. This freshwater supply isn't distributed evenly around the globe.[16–17] Diverse climatic conditions and geography result in more rainfall and natural water sources being supplied to some areas, while other areas have geographic features that result in less precipitation, making the transport of water much more difficult and also extremely expensive. According to the Index Mundi data, renewable endogenous freshwater resources per capita are calculated using the World Bank's estimates and, according to this, Iceland was the richest and Kuwait the poorest country in terms of water per capita in 2014.[18–20]

3.2 WATER CONTAMINATION

Water contamination is the presence of elevated concentrations of substances in the water above the natural background level for a given region and consuming organism. Water contamination is a common problem all over the world. According to the US Environmental Protection Safe Drinking Water Act, the term "contaminant" refers to any physical, chemical, biological, or radiological matter or substance in water. "Contaminant", in the context of water, can be broadly defined as the presence of anything other than water molecules.[21–24]

3.3 EFFECTS OF WATERBORNE CONTAMINANTS ON HEALTH

On the basis of the health threats posed, contaminants fall into two groups.

3.3.1 Acute Effects

These occur within hours or days of the consumption of contaminants. People can suffer acute health effects from almost any contaminant if they are exposed to extraordinarily high levels (as in the case of a spill). The presence of microbial contaminants, such as bacteria and viruses in drinking water, may have adverse health effects. The majority of people can resist the effect of microbial contaminants without suffering any permanent effects. However, excessive and/or prolonged exposure to these contaminants may cause serious ailments and be fatal for people with weak immune systems.

3.3.2 Chronic Effects

These occur when people have consumed a contaminant at levels in excess of the US Environmental Protection Agency (EPA) safety threshold over the course of many years. Contaminants in drinking water, like disinfection by-products, pesticides, solvents, radionuclides, and minerals, are capable of producing chronic effects including kidney or liver problems, reproductive complications, and cancers.

3.4 TYPES OF CONTAMINANTS

Drinking water may be expected to contain at least traces of some pollutants. Some pollutants may be detrimental if consumed at particular concentrations in drinking water.[25–26] The presence of pollutants doesn't necessarily imply that the water is unsuitable for drinking. The following are general types of drinking water pollutants.

3.4.1 Organic Pollutants

Major anthropogenic organic pollutants include pesticides, domestic waste, and artificial waste.

3.4.2 Inorganic Pollutants

Inorganic pollutants include significant quantities of natural calcium or magnesium compounds. In addition, there are several inorganic ions (viz. fluoride, arsenic, lead, copper, chromium, mercury, antimony, and cyanide) that can pollute water resources.

3.4.3 Physical Pollutants

These are primarily recognized by the physical appearance of the polluted water. For example, the deposition or release of organic material into the water of lakes, rivers, and canals due to soil erosion results in cloudiness.

3.4.4 Chemical Pollutants

Chemical pollutants can be chemical elements or compounds. These pollutants may be naturally occurring or anthropogenic in nature. Examples include nitrogen compounds, bleach, salts, pesticides, metals, toxins produced by bacteria, and human, crop, or animal medicines or their residues. Exposure to high chemical doses can cause skin discoloration and nerve or organ damage, whereas long-term exposure to lower doses can cause cancers. The effects of some water contaminants have yet to be determined (Figure 3.1).

3.4.4.1 Pesticides

As defined by the WHO, pesticides are chemical compounds that control pests and parasites, including insects, rodents, fungi, and weeds. They have the potential to contaminate drinking water supplies when applied to farmland, gardens, and lawns as they can leach into groundwater or surface water systems that feed drinking water supplies. The maximum allowable concentration set by the WHO for pesticide residue limit in drinking water is 0.1 ppb.[31]

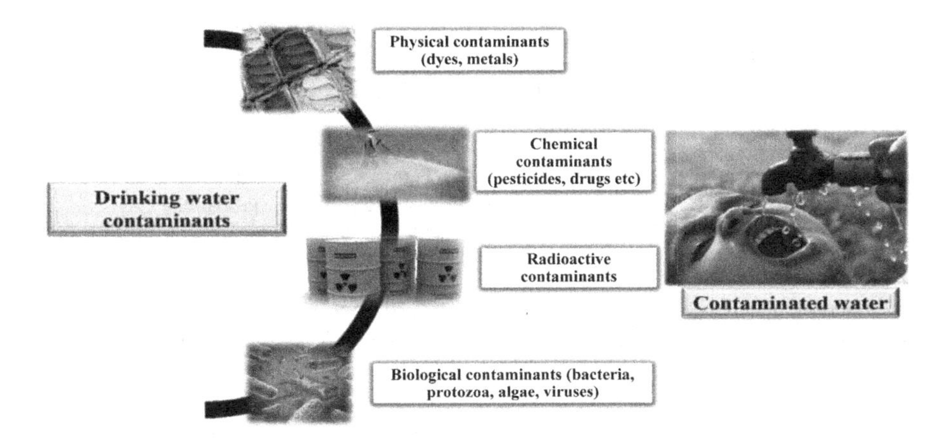

FIGURE 3.1 Types of water contaminants.

Whether these contaminants pose a health risk depends on the type of pesticide (and its site of action), the concentration, and the duration of exposure, e.g., carbamates and organophosphates are neurotoxic (e.g., carbofuran and trichlorfon), whereas others may act as irritants to the skin or eyes. Some may be carcinogenic (e.g., 1,2-dichloropropane, ethylene dibromide, and atrazine) or may affect the hormone or endocrine system of animals.[32]

3.4.4.2 Drugs

Pharmaceutical substances are widely used to treat health issues in humans and domestic animals. Due to incomplete metabolism, drugs are excreted through defecation and urination in the original form and can still remain at physiologically active concentrations in the environment. These pharmaceutical residues may enter water supplies and soil through the excretion of animals and human beings, surface water through the action of sewage treatment plants and agricultural runoff, and groundwater by leaching through the soil. According to researchers, about 90% of human medication consumed is metabolized by the human body while the remaining 10% enters the water supply through human waste.

3.4.5 Biological Pollutants

Biological pollutants are organisms present in water. They are also referred to as microbial pollutants and include bacteria, viruses, protozoa, and parasites.

3.4.6 Radiological Pollutants

Radiological pollutants are radioactive elements with a variable number of protons and neutrons in the nucleus, resulting in unstable nuclei that can emit ionizing radiation, e.g., cesium, plutonium, and uranium.[27–29]

TABLE 3.1
Seasonal (June–September) Rainfall Probability Forecast in Five Categories across the Nation

Category	Rainfall Range (% of LPA)	Forecast Probability (%)	Climatological Probability (%)
Deficient	<90	1	16
Below Normal	90–96	5	17
Normal	96–104	30	33
Above Normal	104–110	34	16
Excess	>110	30	17

LPA: long-period average

3.5 WATER SECURITY

Water scarcity (closely related to water stress or water crisis) is the presence of insufficient freshwater resources to meet water demand. Two types of water scarcity have been defined, namely physical or economic water scarcity. Physical water scarcity is where there is not enough water to meet all demands, including that needed for ecosystems to function effectively. Arid areas (for example, Central and West Asia, and North Africa) often suffer from physical water scarcity. Conversely, economic water scarcity results from either an inadequate human ability to meet water demand or from a lack of investment in infrastructure and technology to extract water from rivers, aquifers, and other sources. There is a severe lack of affordable water over much of sub-Saharan Africa. Some nations may still need to rely on water imports even if these accords are advantageous for everyone. Pakistan is a prime example; it gets 76% of its water from outside sources and has regular disputes with India, which gets 34% of its water from outside sources. There are a few rivers that are particularly popular for use as water supplies by neighboring nations. The Nile River's resources, for example, are shared by 11 countries. Situations such as these have given rise worldwide to over 3,800 unilateral, bilateral, or multilateral water declarations or conventions concerning water, and 286 treaties. Particularly, wildfires create problems for water security and sustainability resulting in increased stress on global water supplies. All the water conservation methods and water extraction technologies, such as desalination plants, underground water storage, dams, and cloud seeding, may be insufficient to meet the demands of a growing global population. As the water supply becomes limited and demand increases, the potential for water-related conflicts will probably increase and become more acute.[30]

3.6 THREATS TO WATER SECURITY

There are multiple factors which threaten water security at national, political, and social levels. There is a political and social power imbalance, with low political will, poor sustainability of services, gender inequality, poverty, inadequate financing, weak accountability, climate change, low institutional capacity, pollution, poor hygiene, poor sanitation, increasing demand for water, and physical challenges meaning that access to water is unequal. A water security action plan needs also to consider events related to extreme drought. As the supplies of irrigation water decrease, they also decrease vegetable production with related job and income losses.[33–35]

3.7 WATER UTILITY SECURITY

The US Environmental Protection Agency states that after 9/11, securing the country's water and sewage systems became an immediate and paramount concern. New security systems to identify and monitor pollutants and prevent security breaches are being developed, as are extensive efforts to evaluate and lessen vulnerabilities to possible terrorist attacks, as well as to prepare for and rehearse responses to crises and other disasters. One of the most important aspects of water security is the ability to detect pollution quickly and accurately. Guidelines and guidance materials have been developed by the EPA to guarantee that water utilities and suppliers are equipped with pollution warning systems. The security challenges that utilities frequently monitor revolve around fast detection, accuracy, and the ability to take fast action when there is a water problem. If contamination is detected early enough, it can be prevented from reaching consumers, and emergency water supplies can be put into effect. In cases where contamination might still reach consumers, fast and efficient communication systems are necessary.[36] All these factors also point to the need for organized and practiced emergency procedures and preparedness. According to the description by the WHO, 80% of all human diseases are directly or indirectly allied to the contamination of water. Water in its innate state is odorless and free from pathogens. The pH of water ranges from 6.5 to 8.5. This water is termed "potable water".[37]

3.8 CONTAMINATED DRINKING WATER AND ITS EFFECTS ON HUMAN HEALTH

Figure 3.2 shows the EPA-regulated drinking water contaminants. Chemical and biological contaminants in drinking water and tap water can cause the progression of non-infectious and infectious diseases.[38] Therefore, fast and sensitive detection techniques are crucial to ensuring a safe and clean water supply. Unsafe water supply affects human health, causing contagious diseases such as hepatitis, influenza, SARS, pneumonia, gastric ulcers, and pulmonary disease. Sodium, silica, chlorine, sulfur, and ammonia are just a few examples of the many non-biological pollutants found in the water supply.in addition to other dangerous materials, heavy metals such as cadmium (Cd), lead (Pb), arsenic (As), mercury (Hg), and nickel (Ni) can also be found in water supplies.[9] These non-biological contaminants are among regularly detected pollutants in metropolitan areas that cause an extensive array of human health effects.[39,40]

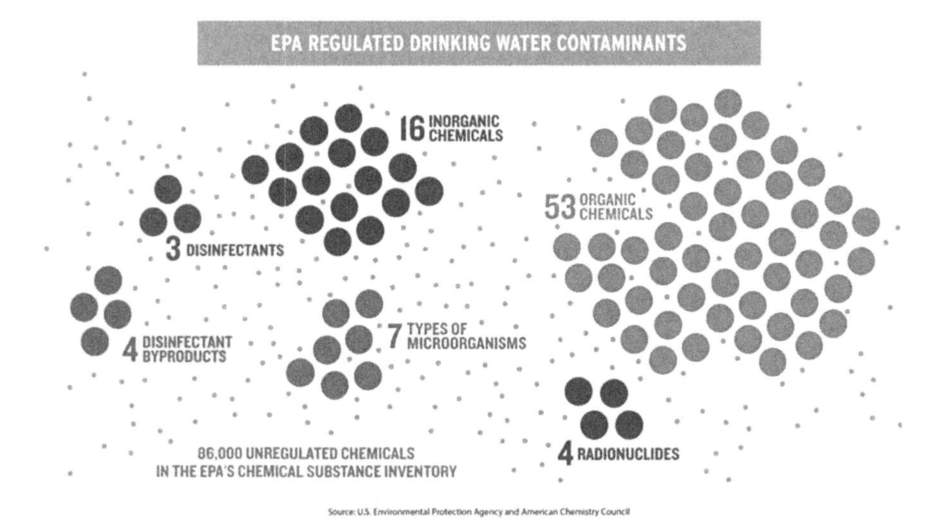

FIGURE 3.2 EPA-regulated drinking water contaminants.

The European Commission states that in order to maintain water quality, the European Union has implemented regulations outlined in the preamble of Directive 91/271/EEC. These regulations mandate precise treatment techniques that target pollutants including organic substances, phosphorus, and nitrogen. Other concerns on water quality include the presence of microbiological contaminants in tap and drinking water at the point of utilization. The presence of motile pathogens in water poses serious threats not merely to human beings but also to the entire water ecology. Pathogenic microbes can be categorized into bacteria (e.g., *Shigella*, *Salmonella typhi*, and *Vibrio cholera*), protozoa (e.g., *Cryptosporidium* and *Giardia lamblia*), and viruses (e.g., poliovirus). These types of microbes are occasionally detected in samples of drinking water, groundwater, and river.[41] According to the WHO, the most frequently found microbes in drinking water sources are *Escherichia coli*, *Pseudomonas*, *Legionella*, *Giardia*, and *Cryptosporidium*.[42]

3.9 COMBATING WATER POLLUTION

Recycling plastics when possible, properly disposing of household chemicals, not using pesticides, maintaining a vehicle that does not leak harmful substances, and making sustainable food and drink choices, with the goal of becoming vegan or vegetarian, are all ways an individual can help reduce water pollution.[43]

3.10 HEALTH EFFECTS

Depending on the pollutant and the overall exposure (time × concentration), the health impacts of drinking water pollutants may vary from moderate to severe, as seen in examples of documented or anticipated effects. There have been reports of human carcinogens in drinking water that have been linked to skin, kidney, bladder, liver, and liver cancers.[44]

In addition to the consequences on children's behavior and development, lead pollution may cause renal and cardiovascular issues. Contaminants discharged into groundwater during hydraulic fracturing, often known as fracking, can affect the reproductive and immune systems.

Safe drinking water is necessary for the well-being of humans. Water pollution is a serious ecological issue that can be caused by many contaminants. Human health can be compromised by drinking, entering, or washing in polluted water. If an individual experiences any negative effects of water pollution, they should talk with their doctor.[45]

3.11 APPROACHES TO ENHANCING WATER SECURITY

The development of flexible and adaptable trans-boundary water-sharing policies and planning depends on hydrologic, political, and socio-economic circumstances We need to identify interdependencies between water (both quantity and quality) and energy, agriculture, ecosystems, biodiversity, conservation, and climate change. We need to discuss and think about long-term planning and policy regarding these issues to ensure that competing users of water, especially in the energy and agriculture sectors, have adequate supplies to continue regional and national economic growth. The primary threats to water security are population growth, terrorism, climate change (sustained droughts), and industrialization.[46,47] These threats can be interdependent. So, we have to find the best procedures for developing comprehensive strategies by which to address them. We need to develop action plans to achieve stakeholder cooperation between the federal and private sectors for information sharing, vulnerability analysis, and risk assessment in order to improve water sustainability. India is experiencing challenges in the provision of drinking water, wastewater treatment, and management of water systems.[48] Climate change is resulting in extreme weather events, while pollution has exacerbated the water crisis, so that joint efforts are needed from the research community and policymakers to tackle the situation. The European Commission and the Government of India have co-funded seven projects to develop and deploy innovative solutions to tackle India's issue on water sanitation.[49]

3.12 CONCLUSION

Wastewater treatment plants can be an important part of the circular economy and sustainability owing to the integration of energy production and resource recovery during the production of clean water. Currently, the main drivers for the developing wastewater treatment industry are global nutrient needs and water and energy recovery from wastewater. Sustainable wastewater treatment systems should be designed to maximize the recovery of energy and resources (nutrients and water). Potable (drinking) water is essential for life. Since water is core for the existence of humans and other living things, its preservation and maintained availability cannot be overstressed. Although various wastewater treatment methods are being evaluated by industry and public water supply companies, untreated wastewater is still being discharged into waterbodies by some industries. Therefore, effective environmental protection policies must be implemented in order to ameliorate water pollution and ensure the availability of clean drinking water for human consumption. A rationalized use of water resources and more sustainable wastewater decontamination practices are needed. The steps taken together can help wastewater treatment plants move toward the circular economy and energy sustainability, pending the successful meeting of significant scientific and related challenges in the near future. Water cooperation could be improved by facilitating multi-level and multi-disciplinary dialog that promotes peace and development between nations. Water security could be improved by enhancing stakeholders' abilities to predict, avoid, and manage water conflicts.

CONFLICT OF INTERESTS

The authors declare that they have no conflict of interests with any of the commercial identities mentioned in the paper.

ACKNOWLEDMENTS

The author would like to thank Vels University for their support.

REFERENCES

1. Angelakis AN, Snyder SA (2015) Wastewater treatment and reuse: Past, present and future. *Water* 7:4887–4895.
2. BadeeNezhad A, Emamjomeh MM, Farzadkia M, JonidiJafari A, Sayadi M, DavoudianTalab AH (2017) Nitrite and nitrate concentrations in the drinking groundwater of Shiraz City, South- central Iran by statistical models. *Iran J Public Health* 46:1275–1284.
3. Badeenezhad A, Radfard M, Passalari H, Parseh I, Abbasi F, Rostami S (2019b) Factors affecting the nitrate concentration and its health risk assessment in drinking groundwater by application of Monte Carlo simulation and geographic information system. *Hum Ecol risk Assess*:1–14. https://doi.org/10.1080/ 10807039.2019.1655634.
4. Bierkens, Y Wada (2019) Non-renewable groundwater use and groundwater depletion a review. *Environ. Res. Lett.* 14:063002(1)–063002(43).
5. Borchardt MA, Bertz PD, Spencer S, Battigelli DA (2003) Incidence of enteric viruses in groundwater from household wells in Wisconsin. *Appl Environ Microbiol* 69:172–180.
6. Boule LA, et al. 2018. Developmental exposure to a mixture of 23 chemicals associated with unconventional oil and gas operations alters the immune system of mice. *ToxicolSci*; https://doi.org/10.1093/toxsci/kfy066.
7. Chang X, Xue Y, Li J, Zou L, Tang M (2019) Potential health impact of environmental micro- and Nano plastics pollution. *J Appl Toxicol*:3915–. https://doi.org/10.1002/jat.3915.
8. DiGiorgio CL, Gonzalez DA, Huitt CC (2002) Cryptosporidium and Gardia recoveries in natural waters by using environmental protection agency method 1623. *Appl Environ Microbiol* 68:5952–5955.

9. Diaz-Gonzalez M, Gutierrez-Capitan M, Niu P, Baldim A, Jimenez-Jorquera C, Fernandez- Sanchez C (2016) Electrochemical devices for the detection of priority pollutants listed in the EU water framework directive. *Trends Anal Chem* 77:186–202.
10. Dong W, Zhang Y, Quan X (2020) Health risk assessment of heavy metals and pesticides: A case study in the main drinking water source in Dalian, China. *Chemosphere* 242:125113. https://doi.org/10. 1016/J.Chemosphere.2019.125113.
11. EPA (2013) Revised total coliform rule and total coliform rule, drinking water requirements for states and public water systems. https://www.epa.gov/dwreginfo/revised-total -coliform-rule- and-total-coliform-rule.
12. Ghaderpoori M, et al. (2018) Comparison of bottled waters current brands in term of important chemical parameters (nitrate, fluoride, chloride, sulfate) effecting on health international. *J Pharm Res* 10:328–333.
13. He S, Wu J (2019) Relationships of groundwater quality and associated health risks with land use/land cover patterns: A case study in a loess area, northwest China. *Human Ecol Risk Assess: An Int J* 25:354–373.
14. Heydari M, Karimyan K, Darvishmotevalli M, Karami A, Vasseghian Y, Azizi N, Ghayebzadeh M, Moradi M (2018) Data for efficiency comparison of raw pumice and manganese-modified pumice for removal phenol from aqueous environments—Application of response surface methodology. *Data Brief* 20:1942–1954. https://doi.org /10.1016/J. Dib.2018.09.027.
15. Kumar M, Singh G, Chaminda T, Quan PV, Kuroda K (2014) Emerging water quality problems in Developing countries Sci. *World J*:1–2.
16. Mohammadi AA, Zarei A, Majidi S, Ghaderpoury A, Hashempour Y, Saghi MH, Alinejad A, Yousefi M, Hosseingholizadeh N, Ghaderpoori M (2019) Carcinogenic and non-carcinogenic health risk assessment of heavy metals in drinking water of Khorramabad, Iran. *Methodsx* 6:1642–1651.
17. Mishra N, Khare D, Gupta K, Shukla R (2014) Impact of land use change on groundwater—A review. *Adv Water Resour Protect* 2:28–41.
18. Mirzaei N, Ghaffari HR, Karimyan K, Moghadam FM, Javid A, Sharafi K (2015) Survey of effective parameters (water sources, seasonal variation and residual chlorine) on presence of thermotolerant coliforms bacteria in different drinking water resources. *Int J Pharm Technol* 7:9680–9689.
19. Malakootian M, Mobini M, Sharife I, Haghighifard A (2014) Evaluation of corrosion and scaling potential of wells drinking water and aqueducts in rural areas adjacent to Rafsanjan fault in during October to December 2013. *J RafsanjanUniv Med Sci* 13:293–304.
20. Krausfeldt LE, Steffen MM, McKay RM, Bullerjahn GS, Boyer GL, Wilhelm SW (2019) Insight into the molecular mechanisms for microcystin biodegradation in lake erie and lake taihu. *Front Microbiol* 10:2741. doi: 10.3389/fmicb.2019.02741.
21. NTP. (2016) *Report on Carcinogens*, Fourteenth ed; Research Triangle Park, NC: U.S. Department of Health and Human Services, Public Health Service. ntp.niehs.nih.gov/go/roc14.
22. Okoh AI, Sibanda T, Gusha SS (2010) Inadequately treated wastewater as a source of human enteric viruses in the environment. *Int. J. Environ. Res. Public Health* 7:2620–2637.
23. Patra S, Sahoo S, Mishra P, Mahapatra SC (2018) Impacts of urbanization on land use/cover changes and its probable implications on local climate and groundwater level. *J Urban Manag* 7:70–84.
24. Pesticides (NIEHS). www.niehs.nih.gov/health/topics/agents/pesticides.

25. Pirsaheb M, Khosravi T, Sharafi K, Babajani L, Rezaei M (2013) Measurement of heavy metals concentration in drinking water from source to consumption site in Kermanshah-Iran. *World ApplSci J* 21:416–423.
26. Pirsaheb M, Khosravi T, Sharafi K, Mouradi M (2016a) Comparing operational cost and performance evaluation of electrodialysis and reverse osmosis systems in nitrate removal from drinking water in Golshahr, Mashhad. *Desalin Water Treat* 57:5391–5397. https://doi.org/10.1080/19443994.2015.1004592.
27. Pirsaheb M, Moradi M, Ghaffari H, Sharafi K (2016b) Application of response surface methodology for efficiency analysis of strong nonselective ion exchange resin column (A 400 E) in nitrate removal from groundwater. *Int J Pharm Technol* 8:11023–11034.
28. Prabhakar A, Tiwari H (2015) Land use and land cover effect on groundwater storage. *Model Earth Syst Environ* 23:1–10.
29. Radfard M, Rahmatinia M, Tabatabaee H, Solimani H, Mahvi AH, Azhdarpoor A (2018a) Data on health risk assessment to the nitrate in drinking water of rural areas in the Khash city, Iran. *Data Brief* 21:1918–1923.
30. Salman SA, Shahid S, Mohsenipour M, Asgari H (2018) Impact of landuse on groundwater quality of Bangladesh. *Sustain Water ResourcManag* 4:1031–1036.
31. Simpi B, Hiremath SM, Murthy KNS, Chandrashekhar KN, Patel AN, Puttiah ET (2011) Analysis of water quality using physic-chemical parameters, Hosahalli tank in Shimoga district, Karnataka, India. *Global Journal of Science Frontier Research* 1(3):31–34.
32. Sapouckey SA, et al. (2018) Prenatal exposure to unconventional oil and gas operation chemical mixtures altered mammary gland development in adult female mice. *Endocrinology* 159(3):1277–1289.
33. Sharafi K, Nodehi RN, Yunesian M, HosseinMahvi A, Pirsaheb M, Nazmara S (2019) Human health risk assessment for some toxic metals in widely consumed rice brands (domestic and imported) in Tehran, Iran: Uncertainty and sensitivity analysis. *Food Chem* 277:145–155. https://doi.org/10.1016/J.Foodchem.2018.10.090.
34. Shyamala R, Shanthi M, Lalitha P (2008) Physicochemical analysis of bore well water samples of Telungupalayam area in Coimbatore district, Tamil Nadu, India. *E-J of Chem* 5(4):924.
35. Smallwood C (1998) *Guidelines for Drinking Water Quality* (Addendum to Vol. 2), second ed. Geneva: World Health Organization.
36. Susan DR (2004) Environmental mass spectroscopy: Emerging contaminants and current issues. *Anal Chem* 76:3337–3364.
37. Soleimani H, et al. (2020) Groundwater quality evaluation and risk assessment of nitrate using Monte Carlo simulation and sensitivity analysis in rural areas of Divandarreh county, Kurdistan Province, Iran. *International J Environ Anal Chem* 20:1–19.
38. Swedish National Agency for Food (Livsmedelsverket) (2015) *Rådomenskilddricksvattenförsörjning*. Uppsala: Swedish National Agency for Food (Livsmedelsverket).
39. The United Nations World Water Development Report (2017) The United Nations World Water development report wastewater: The unstrapped resource United Nations Educational Scientific Cultural Organization (UNESCO), Paris (2017). www.unesco.org/new/en/natural- sciences/environment/water/wwap/wwdr/2017-wastewater-the-untapped-resource/
40. United Nations Educational, Scientific and Culture Organization [UNESCO] (2009) *Water in a Changing World*. Paris and London: United Nations Educational, Scientific and Culture Organization (UNESCO).

41. World Health Organization (WHO) (2007) Legionella and The Prevention of Legionellosis. http://www.who.int/water_sanitation_health/emerging/legionella_rel/en/
42. World Health Organization (WHO) Water Sanitation and Health (WSH) (2011) *Guidelines for Drinking-Water Quality*, fourth ed. https://www.who.int/publications/i/item/9789241549950
43. WHO, World Health Organization (2011) *Guidelines for Drinking-water Quality.* Geneva: WHO.
44. World Health Organization [WHO] (2012) *Rapid Assessment of Drinking-Water Quality; a Handbook for Implementation.* Geneva: World Health Organization.
45. World Population Prospects. (2017) The 2017 revision United Nations Department of Economics and Social Affairs (UNDESA) (2017).
46. Yousefi M, Mohammadi AA, Yaseri M, Mahvi AH (2017) Epidemiology of drinking water fluoride and its contribution to fertility, infertility, and abortion: An ecological study in west Azerbaijan Province, Poldasht county, Iran. *Fluoride* 50:343–353.
47. Yousefi M, Ghoochani M, Mahvi AH (2018a) Health risk assessment to fluoride in drinking water of rural residents living in the Poldasht city, northwest of Iran. *Ecotoxicol Environ Saf* 148:426–430.
48. Yousefi M, Yaseri M, Nabizadeh R, Hooshmand E, Jalilzadeh M, Mahvi AH, Mohammadi AA (2018b) Association of hypertension, body mass index, and waist circumference with fluoride intake; water drinking in residents of fluoride endemic areas, Iran. *Biol Trace Elem Res* 185:282–288.
49. Yousefi M, Ghalehaskar S, Asghari FB, Ghaderpoury A, Dehghani MH, Ghaderpoori M, Mohammadi AA (2019) Distribution of fluoride contamination in drinking water resources and health risk assessment using geographic information system, northwest Iran. *RegulToxicol Pharmacol* 107:10440.

4 Nitrate and Phosphorus Removal from Industrial and Domestic Wastewater Using Agricultural Biomass and Agro-industrial Waste Composites

K. Sureka
Department of Soil Science and Agricultural Chemistry,
Faculty of Agriculture,
Annamalai University, Tamil Nadu, India

S. Sathiyamurthi
Department of Soil Science and Agricultural Chemistry,
Faculty of Agriculture,
Annamalai University, Tamil Nadu, India

S. Praveenkumar
Department of Soil Science and Agricultural Chemistry,
Faculty of Agriculture,
Annamalai University, Tamil Nadu, India

M. Sivasakthi
Department of Soil Science and Agricultural Chemistry,
Faculty of Agriculture,
Annamalai University, Tamil Nadu, India

M. Santhoshkumar
Department of Soil Science and Agricultural Chemistry,
Faculty of Agriculture,
Annamalai University, Tamil Nadu, India

DOI: 10.1201/9781003432869-4

4.1 INTRODUCTION

Water is an explicit indicator of life. Of the water available on the Earth's surface, only 3% of the surface freshwater contributed by the dams, rivers, lakes, and streams are available for consumption. Urbanization, climate change, rainfall scarcity, and industrialization have been among the most challenging problems for matching water supply with demand for the living beings in the world.

Wastewater is nothing but used water from domestic sources, industries, agriculture, and other sources, the composition of which is altered due to the addition of compounds (1). Because of the increased population, industrial, and agricultural activities, the production of wastewater has also increased. Wastewater can also be considered as a reliable freshwater resource after it is treated, but many countries like India have undervalued or not tapped into this resource (2). Disposal of untreated wastewater in streams, lakes, or rivers has resulted in contamination of freshwater bodies, which reduces the availability of clean water for consumption.

Major elements present in wastewater that cause contamination are nitrogen (as nitrate) and phosphorus (as phosphate). Nitrate and phosphate are the most growth-limiting nutrients required for all living organism to complete their physiological processes, but they are considered as pollutants when their concentration exceeds the threshold limit. Excess load of NO_3^- or PO_4^{3-} in waterbodies leads to eutrophication by accelerating the growth of algal clumps, bad odor, and discoloration. Excess levels of NO_3^- and NO_2^- in potable water cause methemoglobinemia and teratogenic effects in infants. To reduce this effect, several countries and organizations have set NO_3^- and PO_4^{3-} limits to lakes and rivers (3).

The permissible limit of PO_4^{3-} discharge to lakes and rivers is set to be <0.01 and 0.07 mg/L, respectively, by European countries (3–6). The US EPA has recommended 10 ppm as the maximum permissible level of NO_3^- in drinking water (7, 8). The World Health Organization (WHO) has established a maximum P discharge limit of 0.5–1.0 ppm as a recommendation. So, it is important to remove NO_3^- and PO_4^{3-} from wastewater.

Numerous studies have found that it is possible to remove NO_3^- and PO_4^{3-} from wastewater, using natural adsorbents like charcoal, ash, biochar, nanoparticles, coir pith from coconut shells, rice husk, and sugarcane bagasse can adsorb NO_3^- and PO_4^{3-} from wastewater. Thus, treating wastewater with agricultural biomass and agro-industrial waste composite could be an effective, eco-friendly, cheap method, without causing harm to the environment. Considerable research has been carried out using this approach to reduce the nitrate and phosphate levels. In this chapter, we will discuss the different methods, processes, and mechanisms involved in nitrate and phosphate removal by agricultural biomass and agro-industrial waste composites.

4.2 WASTEWATER PRODUCTION SCENARIO

Qadir and coworkers (9) reported that about 380 billion m^3 of wastewater are generated per year worldwide. Based on the projected population growth rate and urbanization, the daily wastewater generated is expected to increase by 24% (to 470 billion m^3) by the end of the Sustainable Development Goal (SDG) era in 2030 and by 51%

(to 574 billion m^3) by 2050. The largest volume of wastewater, about 42% (159 billion m^3), is generated in Asia, and, by 2030, it is expected to increase by 44%. Jones et al. (10) reported that annual wastewater generation was estimated to be 359 billion m^3, of which 63% (225 billion m^3) is collected and 52% (188 billion m^3) is treated globally.

According to the the Urban Wastewater Scenario, India generates nearly 72,368 million liters per day (MLD) of urban wastewater as of 2020–21, of which only 28% (20,236 MLD) is treated, with the remaining 72% being untreated and disposed of in rivers, lakes, or streams (9). According to the Central Pollution Board of India, wastewater is generated in India at a rate of 60 m^3/pa, where more than 63% of wastewater is disposed of untreated in the freshwater bodies (11).

4.3 SOURCES OF NO_3^- AND PO_4^{3-} POLLUTION IN WATER

In addition to being essential nutrients for all living organisms, nitrogen and phosphorus are the main pollutants of waterbodies. The major source of the NO_3^- and PO_4^{3-} pollution of surface water is when rainwater passes through industrial areas, agricultural land, urban areas, etc. The natural sources of nitrate and phosphate ions are mineral weathering processes (P only), decomposition of organic matters, and soil erosion. Anthropogenic activities like mining, agricultural activities, sewage, industrial effluent, and urban and domestic waste play a significant role in NO_3^- and PO_4^{3-} pollution of waterbodies. Wastewaters are a complex mixture of organic and inorganic components and sewage from point and non-point sources (12).

Solaraj et al. (13) revealed that maximum levels of phosphate of 0.43 mg/L were recorded, exceeding the permissible levels at certain sampling stations at Cauvery River in India. Therefore, agricultural runoff, sewage, and industrial effluents (IE) are probable water pollution sources. Rahman et al. (14) identified that about 42.94% of the pollution was from urban wastewater in the dry season and 41.80% of the total pollution was by soil erosion as non-point sources in the wet season. Gizaw and coworkers (15) reported that 3 million tons/year of fertilizer P and 50% of applied urea are lost to freshwater bodies. Luthra (16) reported that wastewater from detergents, sewage, IE, and domestic sources are some of the sources of nitrate and phosphate pollutants which have increased from the safety limit of 0.5 to 2 ppm and from 0.1 to 0.7 ppm, respectively, in the Yamuna River in India.

4.4 METHOD OF ANALYSIS

The well-known procedure for estimating nitrate ion concentration (spectrophotometric cadmium reduction) in water samples is given by the US Environmental Protection Agency (EPA). Drolc and Vrtovšek (17) used on-line UV spectroscopy and data analysis to determine concentrations of nitrate and nitrite nitrogen in various wastewaters. Based on matrix calibration, this method has an average recovery rate of 99.6%. Baezzat and Parsaeian (18) determined traces of NO_3^- in water samples using the spectrophotometric method after preconcentration on microcrystalline naphthalene. The method was based on the complex formation between 2,6-bis (4-methoxyphenyl)-4-phenyl pyrylium perchlorate (PPP) and nitrate, with extraction being conducted with microcrystalline naphthalene. The nitrate adsorption (15–135 μg/L) was detected at 328 nm with a detection limit of 10 μg/L. Kurniawati et al. (19) verified the use of the APHA 2012 Section 4500 NO_3-B spectrophotometric method to determine nitrate concentration in water samples. The method has level of linearity of 50 m/L with a 99% level of confidence and used standards over a nitrate concentration range of 10 to 50 mg/L. Pang and Christison (20) described a method for nitrate and nitrite estimation using capillary ion chromatography (IC) with UV detection. Good linearity was observed from 0.05 to 100 ppm with method detection limits (MDL) of 0.010 and 0.015 ppm for NO_3^- and NO_2^-, respectively.

Phosphate ions in water samples were analyzed by the well-known orthophosphate molybdenum blue methods published by Murphy and Riley (21). The method relies on maximum absorption at wavelengths of 880, 825, 774, and 660 nm, where the molybdenum blue complex has a high absorbance. However, it encountered some interference from other complex anions such as AsO_3^{3-} and AsO_4^{3-}, SiO_4^{4-}, VO_4^{3-}, and WO_4^{2-}, which are also potential environmental toxins (22). Akhter et al. (23) developed a revolutionary, affordable, and energy-efficient phosphate sensor for precision agriculture production. An Internet-of-Things (IoT)-enabled smart PO_4^{3-} detection system proved to be capable of determining concentrations of PO_4^{3-} varying from 10 μg/L to 40 mg/L. A machine-learning method was adopted to train the Arduino-based system to estimate the phosphate concentration in actual water samples. O'Grady et al. (24) developed a molybdenum blue-based fully integrated sensor for phosphate detection. The updated approach enables on-board storage of 54–600 g/L, with a maximum detection of 16 g/L. The system's detector is comprised of a pair of inexpensive optical transducers (LED and a photodiode).

Ibnul and Tripp et al. (25) established a simple, rapid, and solvent-free technique for converting phosphate in solution to a solid that can be quantified using visible spectroscopy. By neutralizing the anionic heteropolymolybdate ions with cetyltrimethylammonium bromide (CTAB), it is converted into a solid colloidal precipitate, which is subsequently trapped on a visible transparent membrane. The visible spectrum is then recorded across the membrane in transmission mode, and the PO_4^{3-} concentration is measured by the intensity of a band at 700 nm. The detection limit for PO_4^{3-} in water was thus reduced to 0.64 g/L.

4.5 NITRATE AND PHOSPHORUS REMOVAL BY BIOCHAR AND A BIOCHAR-BASED COMPOSITE

Biochar is the plant-based solid carbonic material produced by pyrolysis under an oxygen-limited environment. Applications of biochar on NO_3^- and PO_4^{3-} removal are presented in Tables 4.1 and 4.2. Li et al. (26) investigated NO_3^- and PO_4^{3-} removal from water using biochar (BC) made from wheat straw using low-oxygen pyrolysis at 450°C, activated with HCl, and coated with iron ($FeCl_3.6H_2O$). A Maximum adsorption capacities for NO_3^- and PO_4^{3-} were 2.47 at pH 3 and 16.58 mg/g at pH 6, respectively. Jung et al. (27) created various biochars, using oak wood, bamboo wood, maize residue, soybean stover, and peanut shell as a sustainable PO_4^{3-} adsorptive material from an aqueous solution. Among the various materials used, peanut shell biochar performed best, with the maximum phosphate adsorption rate of 61.3% (6.79 mg/g for PO_4^{3-} at 30°C). Qiu et al. (28) produced biochar from *Broussonetia*

TABLE 4.1
Phosphate Removal by Biochars and Biochar Composites

Adsorbent	Modification	Maximum Adsorption (mg/g)	Kinetics	Isotherm Studied	Dose of Adsorbent	ET	pH	Mechanism of Absorption Involved	Reference
oak sawdust	$LaCl_3$	NH_4^+ 10.1 mg/g; NO_3^- 100.0 mg/g, and 142.7 mg/g	PFO, PSO	Langmuir, Freundlich	0.1 g				Wang et al. (43)
Water hyacinth	Fe oxide	5.07	PFO, PSO	Langmuir, Freundlich, Langmuir-Freundlich	0.2 g		2	Electrostatic attraction and surface complexation via ligand exchange	Cai et al. (44)
Pineapple peel	$La(OH)_3$		PFO, PSO	Langmuir, Freundlich	0.025 g			Precipitation, electrostatic interaction, ligand exchange,inner-sphere complexation	Liao et al. (39)
Wheat straw	UiO-66 nanoparticle	30.7 to 69.3	PFO, PSO	Langmuir, Freundlich	1–20 mg/L		4.5 and 5.5	Coulombic interaction.	Qiu et al. (45)
Pine cone flakes	Mg/Fe-LDHs	17.46	PFO, PSO, intraparticle diffusion	Langmuir, Freundlich	5 g/L	60 min	2–4	Electrostatic attraction, ligand exchange, surface complex formation	Bolbol et al. (46)
Tobacco stalk	MgAl-LDHs	41.16	PFO, PSO,	Langmuir, Freundlich,	0.1 g	24 h		Electrostatic interaction, ligand exchange, ion exchange	He et al. (47)

(Continued)

TABLE 4.1 (CONTINUED)
Phosphate Removal by Biochars and Biochar Composites

Adsorbent	Modification	Maximum Adsorption (mg/g)	Kinetics	Isotherm Studied	Dose of Adsorbent	ET	pH	Mechanism of Absorption Involved	Reference
Sewage sludge and walnut shell		303.49	PFO, PSO, intraparticle diffusion model	Langmuir, Freundlich, Langmuir-Freundlich	0.1 g		4	Electrostatic attraction	Yin et al. (48)
Chinese cabbage and rape	Biochar/Mg-Al layered double oxides (LDOs)	50 mg/L	PSO	Langmuir, Freundlich	0.05 g	5 min	2–10	Memory effect, electrostatic attraction, surface complexation, and anion exchange	Zhang et al. (49)
Banana bract	ACM@BBAC	91.78	PFO, PSO, IPD	Langmuir, Freundlich, Dubinin-Radushkevich	100 mg			Electrostatic interaction and surface complexation.	Karthikeyan et al. (35)
Rice husk	BRH-C	10.72	PFO, PSO	Langmuir, Freundlich,		11.7 h	2–10	Electrostatic interactions	Ramola et al. (36)
Bamboo	Zr/CTAB/BAC		PFO, PSO	Langmuir, Freundlich, and Temkin	0.1 –1.0 g	180 min		Liquid-solid sorption interaction	Shao et al. (37)

Abbreviations: Pseudo first order (PFO); Pseudo second order (PSO); IPD: Intra-particle diffusion

TABLE 4.2
Nitrate Removal by Biochar and Biochar Composite

Name of Adsorbent	Modified	Maximum Adsorption (mg/g)	Kinetics	Isotherm Studied	Dose of Adsorbent	ET	pH	Mechanism of Absorption Involved	Reference
Oak sawdust	$LaCl_3$	NH_4^+ 10.1 mg/g, NO_3^- 100.0 mg/g, and 142.7 mg/g	PFO, PSO	Langmuir, Freundlich	0.1				Wang et al. (43)
Sugarcane bagasse	Epichlorohydrin (EPI) *N,N*-dimethylformamide (DMF) Pyridine Dimethylamine (DMA)	28.21	PFO, PSO, intraparticle diffusion, Avrami.	Langmuir, Freundlich, Sips, Dubinin-Raduskovich	2 g L^{-1}	60 min	4.64	Electrostatic attraction	Hafshejani et al. (51)
Wheat straw	Mg/Al-LDH	81.83	PFO, PSO	Langmuir, Freundlich	1.0 g		3.0	Ion exchange, electrostatic attraction, surface complex	Li et al. (52)
Eucalyptus wood		2.3–3.4 mg/L N and 0.2–4.1 mg/L N of nitrate and ammonia, respectively			10 and 50 mg/L of nitrate and ammonia, respectively				Hanandeh et al. (53)

(Continued)

TABLE 4.2 (CONTINUED)
Nitrate Removal by Biochar and Biochar Composite

Name of Adsorbent	Modified	Maximum Adsorption (mg/g)	Kinetics	Isotherm Studied	Dose of Adsorbent	ET	pH	Mechanism of Absorption Involved	Reference
Corn cob	Sulfuric acid (1 mol/L) and sodium hydroxide (2 mol/L)	13.20	PFO, PSO, Elovich	Langmuir, Freundlich	0.1 g/L	24 h	3	Electrostatic interaction	Hu et al. (54)
Banana straw	$MgCl_2$	31.15	PFO, PSO, intraparticle diffusion,	Langmuir, Freundlich	0.25 g	240 min		Electrostatic interactions	Jiang et al. (55)
Sewage sludge and walnut shell		22.85 mg/g	PFO, PSO, intraparticle diffusion model	Langmuir, Freundlich, Langmuir-Freundlich	0.1 g	36 h	2	Electrostatic repulsion	Yin et al. (48)
Woody waste of apple branches	Mg/Al-LDHs	156.84 mg/g.	PFO, PSO, intraparticle diffusion,Elovich	Langmuir, Freundlich, and Temkin		30 min	>9,	Surface physical sorption, intraparticle diffusion, electrostatic adsorption, ion exchange, and metal-bonded bridges	Wang et al. (56)

Abbreviations: Pseudo first order (PFO); Pseudo second order (PSO); IPD: Intra-particle diffusion

TABLE 4.3
Phosphate and Nitrate Removal by Agricultural Biomass and its Composites

Name of Adsorbent	Modification	Maximum Adsorption (mg/g)	Kinetics	Isotherm Studied	Dose of Adsorbent	ET	pH	Reference
Coconut shell fibers	Ammoniumquaternarysalt(2-hydroxypropyltrimethyl ammonium chloride)	200	PFO, PSO, IPD	Langmuir, Freundlich, SIPS, Redlich–Peterson, Temkin	0.1 g			De Lima et al. (65)
Rice husk	Epichlorohydrin		Morris–Weber, Lagergren, PFO, PSO	Langmuir, Freundlich, Dubinin-Radushkevich,	0.4 g	90 min	7	Katal et al. (66)
Sunflower seed husk		NO_3^- 38.8 to 68.4%		Langmuir, Freundlich	3.0 g	120 min	2	Moyo et al. (67)
Cotton stalk (CS) and wheat stalk (WS)	EPI Diethylenetriamine (DEA) Trimethylamine (TMA)	60.61 (AC-WS) 41.9 (AC-CS) 49.05 (AC-WS)						Xu et al. (68)
Sugarcane bagasse	Epichlorohydrin (EPI) *N,N*-dimethylformamide (DMF) Pyridine Dimethylamine (DMA)	21.3						Zhang et al. (2012) (69)

(Continued)

TABLE 4.3 (CONTINUED)
Phosphate and Nitrate Removal by Agricultural Biomass and its Composites

Name of Adsorbent	Modification	Maximum Adsorption (mg/g)	Kinetics	Isotherm Studied	Dose of Adsorbent	ET	pH	Reference
Soybean	Calcium chloride, hydrochloric acid, and calcination	Nitrate 34.4, Nitrite -37.2	PFO, PSO	Langmuir, Freundlich	1.0 g/L	Nitrate 24 h, nitrite 16 h,	3.0	Ogata et al. (70)
Euclayptus leaves	Cetyltrimethylammonium bromide (CTAB), iron oxide nanoparticles (IONP)	PO_4^{3-} 7.49	PFO, PSO	Langmuir, Freundlich	0.5 mg/L	90 min		Cao et al. (71)
Chinese reed	hydrous zirconium oxide nanoparticle	PO_4^{3-} 1043.5	PSO, IPD	Langmuir, Freundlich	1.0 g/L			Shang et al. (72)

Abbreviations: Pseudo first order (PFO); Pseudo second order (PSO); IPD: Intra-particle diffusion

TABLE 4.4
Nitrate and Phosphate Removal by Activated Carbon from Agricultural Waste and its Composites

Name of Adsorbent	Modification	Maximum Adsorption (mg/g)	Kinetics	Isotherm Studied	Dose of Adsorbent	ET	pH	Reference
Sugar beet bagasse	$ZnCl_2$	NO_3^- 41.2%	PFO, PSO, IPD	Langmuir, Freundlich, Temkin	0.1 g		3	Demiral et al. (79)
Sawdust		NO_3^- 1.68		Langmuir, Freundlich	1.0 g			Curran (80)
Rice straw			PFO, PSO, IPD		0.1 g	10 h	8	Hanafi and Azeema (81)
Coconut	$ZnCl_2$	NO_3^- 14.01	PFO, PSO, IPD	Langmuir, Freundlich, Temkin	2 g	12 h		Liu et al. (82)
Prosopis juliflora		PO_4^{3-} 13.55 mg/g NO_3^- -10.99 mg/g	PFO, PSO, IPD, liquid-film diffusion, Elovich, Bangham	Langmuir, Freundlich, Dubinin-Radushkevich, Temkin,Elovich	1 g/L		6.7±0.2	Manjunath, and Kumar (83)
Olive stone	$ZnCl_2$	NO_3^- 5.76	PFO, PSO, IPD	Langmuir, Freundlich,	2 g	90 min	4	Nabhan (84)

Abbreviations: Pseudo first order (PFO); Pseudo second order (PSO); IPD: Intra-particle diffusion

papyrifera leaves with the goal of preventing eutrophication by eliminating PO_4^{3-} ions from the water. Due to the presence of $CaCO_3$ in the *B. papyrifera* biochar, its sorption capacity increased as the pH value increased. Under alkaline circumstances, the hydrolysis products of $CaCO_3$ precipitated the phosphate to be adsorbed. Within 120 minutes, the adsorption equilibrium was reached. Vijayaraghavan and Balasubramanian (29) developed pine wood waste biochar by pyrolysis and investigated its capacity to absorb NO_3^- and PO_4^{3-} from single and binary solutions. The result showed that biochar produced at 600°C showed a maximum adsorption of nitrate (20.5 mg/g) and phosphate (4.20 mg/g) at pH 2 within 360 min due to its high C content (63.8%), pore volume (0.201 cm^3/g), surface area (204.2 m^2/g), and reduced acidic binding groups.

Hafshejani and coworkers (30) compared the performance of sugarcane bagasse biochar (SBBC) and vermicompost (VC) on NO_3^- removal and also determined the optimum conditions for the adsorption process. The results revealed the optimum pHs for NO_3^- adsorption by SBBC and VC were 4.64 and 3.78, respectively, with the optimum adsorbent dosage being 2 g/L. This difference between the two materials may be due to the specific surface area, carbon content, and anion exchange capacity (AEC) of SBBC being higher than those of VC.

Yin et al. (31) prepared and studied the potential of Al, Mg-soybean straw biochar (BC) (Mg/BC, Al/BC, and Mg-Al/BC) for ammonium, NO_3^- and PO_4^{3-} removal from eutrophic water. The maximum adsorption capacities exhibited by Mg-Al/BC, Al/BC, and Mg/BC for ammonium, nitrate, and phosphate were 0.70 mg/g, 40.63 mg/g, and 74.47 mg/g, respectively. The PO_4^{3-} adsorption capacities of Mg/BC, Al/BC, and Mg-Al/BC increased 43-, 26-, and 21-fold, respectively. Jena and coworkers (32) prepared Mg-modified waste corn BC as an adsorbent for phosphorus from source-separated urine and achieved 96% of phosphate recovery from synthetic urine. Xu et al. (33) tested different wood biochars modified in salt solutions of $MgCl_2$, $AlCl_3$, $CaCl_2$, or $FeCl_3$ to release PO_4^{3-} from human urine. Among the tested biochar, the highest PO_4^{3-} removal capacity of 118 mg P/g was exerted by 2.3 M $MgCl_2$ from the hydrolyzed urine.

Zhu and coworkers (34) examined the removal of phosphorus from water using a novel hierarchical porous nano-Fe_2O_3/Fe_3O_4-coated bamboo biochar-based material. The P adsorption capacity of HPA-Fe/C-B enhanced from 0.20 to 2.81 mg-P/g (contact time 12 h). Karthikeyan and coworkers (35) synthesized and examined magnetic activated charcoal from banana bract (ACM@BBAC) crosslinked with amines. The PO_4^{3-} and NO_3^- ions had the highest adsorption densities at 91.78 and 75.81 mg/g, respectively. The removal process is mediated through exchange of hydroxyl ions with Cl^- , PO_4^{3-}, and NO_3^- ions, and the positive Fe^{3+} ions in the adsorbent form surface complexes with the negatively charged PO_4^{3-} and NO_3^- via electrostatic interactions.

Ramola and co-workers (36) created a new rice husk biochar–calcite (BRH-C) composite (4.2:1) for phosphate removal from an aqueous solution. At low phosphate concentrations, the maximum removal and adsorption capabilities of optimized BRH-C were 87.3 % and 1.76 mg/g, respectively. At a PO_4^{3-} content of 95 mg/L, 0.24 g of BRH-C, a pH of 5.4, and a contact period of 11.75 hours, the maximum PO_4^{3-} adsorption (54.2 %) and 10.72 mg/g of adsorption potential were reached. Shao et al. (37) created a new adsorbent Zr/CTAB/BAC from $ZrOCl_2.8H_2O$ and CTAB as

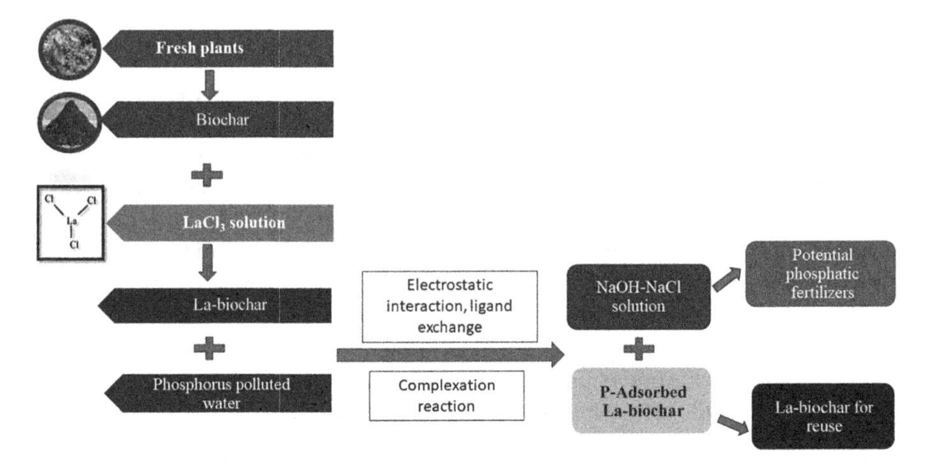

FIGURE 4.1 Phosphorus adsorption by La-biochar (40).

well as bamboo-based biochar. With a dosage of 1.0 g, the highest phosphate removal capacity of Zr/CTAB/BAC was 99.42%. Banu et al. (38) carried out an experiment to remove NO_3^- and PO_4^{3-} ions from eutrophic water using Zr-embedded chitosan-soya bean husk activated biochar composite beads (Zr-CS-SAC). The results showed that 0.1 g of adsorbent has an adsorption capacity of NO_3^- (90.09 mg/g) and PO_4^{3-} (131.29 mg/g) ions at 45 min and 30 min contact time for NO_3^- and PO_4^{3-} ions, respectively. $La(OH)_3$-modified pineapple biochar (Lax-MC) demonstrated good magnetic characteristics and adsorption capacity of up to 101.16 mg P/g. However, the removal efficiency decreased after three adsorption-desorption cycles. During the leaching performance investigation, the ions were shown to be extremely stable (39).

Xu and coworkers (40) synthesized lignocellulose biochar using discarded cattail plants and La and used it as a phosphate adsorption agent from wastewater. A maximum phosphorus adsorption capacity of 36.06 mg P/g, under a pH of 3 to 12 was obtained. It has a reusability of 92.3% desorption efficiency and retained 85% adsorption capacity after five recycles. The mechanism involved in the phosphorus adsorption by La-biochar is presented in Figure 4.1

Jiang and co-workers (41) prepared Zn-Al-layered double hydroxide-loaded banana straw biochar (ZnAl-LDO-BSB) as an efficient phosphorus removal agent from wastewater by a hydrothermal method and calcination at 500°C. According to the results of their study, the composite had a phosphorus adsorption capacity of 185.19 mg/g. Lee et al. (42) developed and examined the phosphorus adsorption characteristics of rice husk biochar with Mg/Al-calcined layered double hydroxides (RHBC/MgAl-CLDHs) via co-pyrolysis in an aqueous environment. Different molar concentrations of 1: 2.5 and 1: 5 affect the charge density and co-pyrolysis temperature of 300–700°C affects surface functionality and porosity for phosphorus adsorption. The results showed that RHB/MgAl-CLDHs (2:1/500) exhibited the highest phosphate removal of 97.6%.

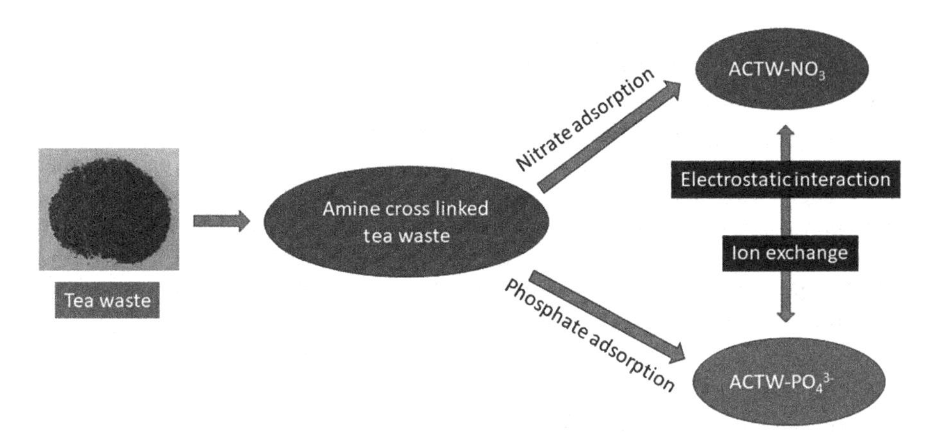

FIGURE 4.2 Mechanism involving removal of nitrate and phosphate using amine-cross-linked tea waste (60).

4.6 NITRATE AND PHOSPHATE REMOVAL BY AGRICULTURAL WASTE BIOMASS AND BIOMASS COMPOSITES

Hredya et al. (57) used lemon peel and activated charcoal for removal of NO_3^- from groundwater. At a layer thickness of 7.5 cm activated charcoal and 2.5 cm lemon peel, it has a maximum nitrate absorption efficiency of 82.5% with a concentration of the 12.03 mg/L. Xu et al. (58) developed the agriculture-waste-based anion exchanger using wheat straw (WS–AE) to evaluate the potential for removal of NO_3^- and PO_4^{3-}. The maximum absorption capacity of WS–AE for NO_3^- and PO_4^{3-} were 52.8 and 45.7 mg/g, respectively. The WS–AE with sorbed ions can be used at least three or four times by regenerating in both HCl and NaCl solutions through an ion-exchange mechanism without losing its sorption capacity.

Song et al. (59) synthesized magnetic amine-crosslinked biopolymer-based corn stalk (MAB-CS) and used it for NO_3^- removal. The NO_3^- adsorption capacity of MAB-CS was 102.04 mg/g at 318K under pH 6 to 9. Qiao et al. (60) prepared amine cross-linking tea waste and compared the ability of tea waste (TW) and amine cross-linked tea waste (ACTW) to remove the NO_3^- and PO_4^{3-} from an aqueous solution. Compared to unmodified-TW, ACTW has considerable adsorption capacities over a wide range of pH of 3–10. The maximum sorption capacity of ACTW for NO_3^- was 136.43 mg/g and for PO_4^{3-} ions was 98.72 mg/g. The mechanism involved in adsorption was electrostatic attraction as well as ion exchange. The mechanism involved in nitrate and phosphate adsorption by using amine cross-linked tea waste is presented in Figure 4.1.

Hu et al. (61) synthesized a nano-Zr(IV) oxide-corn straw anion exchanger (Zr@MCS) and evaluated the phosphorus adsorption capacity in municipal wastewater. The Zr@MCS with high Zr content (10.6 wt%) and large surface area exhibited a high phosphate adsorption capacity with a wide operating pH range (2–8). Nano-La(III) hydroxide-modified wheat straw (Ws-N-La) was also reported as a novel adsorbent with the ability of high PO_4^{3-} removal. This adsorbent was recommended for soil amendment after phosphate removal (62).

Qiu et al. (63) fabricated a new nanocomposite (HFO@St+) for enhanced phosphate adsorption under acidic conditions. The results showed that phosphorus adsorption affinity by biomass-based nanocomposite were greatly promoted after HFO impregnation, with a separation factor K value of 2.5–38 or 2.5–15 for near-neutral or acidic pHs, respectively. Alagha et al. (64) synthesized a date palm biochar MgAl-augmented double-layered hydroxide (biochar–MgAl–LDH) nanocomposite for removal of NO_3^- and PO_4^{3-} from wastewater. The maximal NO_3^- and PO_4^{3-} adsorption capacities from the monolayer, by the non-linear Langmuir model, were 177.97 mg/g and 28.06 mg/g at 308 and 298 K, respectively. Faheem et al. (50) also modified biochar to improve its phosphate removal capacity.

4.7 NITRATE AND PHOSPHATE REMOVAL THROUGH ACTIVATED CARBON FROM AGRO-INDUSTRIAL WASTE COMPOSITE

Kilpimaa et al. (73) employed carbon residue from wood gasification as a precursor for removing NO_3^- and PO_4^{3-} from water by physical activation. Compared with the adsorption capacity of unactivated carbon residue, the results demonstrated that the activated carbon residue was much more adsorbent. Its removal efficiency was somewhat greater than 40%. Shahmoradi et al. (74) explored the effectiveness in nitrate adsorption of activated carbon absorbents produced from pyrolysis of rice husk or sludge from paper waste industries. The maximum NO_3^- absorption occurred at pH 4 with absorption of 93.5 mg/g from rice husk and 79.5 mg/g from sludge from the paper industry. Najmi et al. (75) used *Glycyrrhiza glabra* for the preparation of activated carbon, with a surface area of 959.22 m^2/g to remove NO_3^- and PO_4^{3-} from wastewater. They reported that activated carbon prepared from residues had maximum removal capacities of NO_3^- and PO_4^{3-} of 4 g/L, and 6 g/L, respectively. Sow et al. (76) used activated carbon prepared from mango shells for removing NO_3^- and PO_4^{3-} from Bafing River. They reported that an average of 81.50% efficiency in the removal of NO_3^- and PO_4^{3-} ions from contaminated water was achieved.

Ping and coworkers (77) prepared $FeCl_3$-coated activated carbon using wheat straw to study the NO_3^- adsorption capacity from wastewater. The optimal conditions for maximal adsorption were maintained at a 527°C carbonization temperature with a 92.5% chemical weight ratio ($FeCl_3$/wheat straw). Plenty of pore space, with increases in specific surface area and pore volume, and more iron oxide, resulted in maximum adsorption of nitrate ions from water than wheat straw biochar. The maximum adsorption of NO_3^- was 14.68 mg/g. Hafshejani and Naseri (78) used activated sugarcane bagasse hydrocar to remove the PO_4^{3-} ions from the agricultural wastewater by adsorption. The phosphorus adsorption of 1 g/L was optimal at a pH of 6 with the optimal contact being for 180 min. The maximum phosphorus adsorption was 54.69 mg/g.

4.8 NITRATE AND PHOSPHATE REMOVAL USING PLANT ASH AND ASH COMPOSITES

Research into the efficacy of removing PO_4^{3-} with rice husk ash (RHA) found that the highest removal (up to 89%) occurred at a pH of 6 with a 2 g/L dosage for 120 min (85). Hamzah et al. (86) prepared and tested the ash from thermally heated rice husk ash at 450–700°C on phosphate adsorption from wastewater. A phosphate removal of 97% was obtained by heating the rice husk at 750°C and a pH value of 9.0. Li et al. (87) tested a dielectrophoresis-assisted adsorption method using plant ash as the adsorbent for highly efficient removal of nitrate and phosphate from water to control eutrophication. This method showed that nitrate and phosphate removal efficiency increased by 66.06% and 43.04%, respectively, and processing time decreased by 92%, compared with the solution-phase adsorption method alone. Phan et al. (88) prepared and used triamine-bearing activated rice husk ash material (TRI-ARHA) as an advanced functional material for nitrate adsorption capacity from aqueous solutions. The TRI-ARHA material had a higher nitrate adsorption capacity (>160 mg NO_3^-/g) throughout ten cycles of adsorption–desorption.

4.9 NITRATE AND PHOSPHATE REMOVAL USING AN AGRO-INDUSTRIAL BY-PRODUCT AND ITS COMPOSITE

Mokif et al. (89) used *Actinidia deliciosa* peel as an absorbent for removing nitrate from water. The kiwi fruit peel has a high surface area which results in better adsorption. The maximum removal of NO_3^- of 100 mg/L occurred at a contact time of 3 h, with 0.1 g absorbent. Nkansah et al. (90) studied PO_4^{3-} absorbance capacity of sawdust and peanut shell powder. The results showed that 78% and 39% of phosphate were removed by sawdust and peanut shell powder, respectively, at adsorbate concentrations of 10 mg/L (total P) for a period of 180 min at an adsorbent dosage of 0.4 g/L. Fotsing et al. (91) investigated the removal of NO_3^- from wastewater by adsorption at the surface of an amine-modified cocoa shell adsorbent. In this study, cocoa shell, a by-product of the cocoa industry, was used as the adsorbent after chemical modification with 3-aminopropyltriethoxysilane grafting through salinization followed by protonation. The result showed that more than 53 mg/g was adsorbed to the amine-modified cocoa shell.

4.10 NITRATE AND PHOSPHATE REMOVAL USING A BIOSORBENT AND ITS COMPOSITE

Biosorption is a natural physiochemical process that allows specific biomass to passively concentrate and bind pollutants onto its cellular structure. Dey et al. (92) used five locally available biosorbents namely banana peel, orange peel, coconut wire, maize corn, and tea waste for removal of NO_3^- by biosorption. The order of percentage removal (i.e. biosorbent of nitrate from contaminated water was as follows: orange peel >tea waste≈coconut wire >banana peel >maize. From the result, orange peel has the highest biosorbent activity compared with other wastes at pH 6, dosage of 1.9 g/L, contact time of 60 min, optimum temperature of 30°C and optimum agitation speed of 60 rpm.

Dey et al. (93) produced a biosorbent using *Mangifera indica* leaves for removal of PO_4^{3-} ions from contaminated water. Complete removal of phosphate was observed at pH 8.5 using 4.3 g of biosorbents from mango leaves for 120 minutes at 35°C. Ang et al. (94) used six types of biosorbent, *viz.*, rambutan skin (RS), passion fruit skin (PFS), longan seeds (LS), chia seeds (CS), papaya seeds (PS), and luffa (L), for removal of nitrate from the water. They maintained optimal conditions of pH (3.0), room temperature (30°C), initial nitrate concentration (50 mg/L), biosorbent particle size (425 μm), contact duration (4 h), shaking speed (120 rpm) and biosorbent dosage (0.1 g) equally for all six biosorbents. Among the biosorbents, the acid- and heat-treated passion fruit skin biosorbent recorded the best biosorption performance with a nitrate uptake of 5.179 mg/g.

Nguyen et al. (95) studied the practicability of using Okara (soybean milk by-products) loaded with iron for removing PO_4^{3-} from wastewater. The maximal adsorption of phosphate of 4.785 mg/g occurred at pH 3, with sorbent dose of 20 mg/L within 7 hours. Ren et al. (96) explored the potential use of amine cross-linked reed (ACR) for removing NO_3^- from the aqueous solution. The grass reed is used as the biosorbent and it is a plant waste with cellulose and hemicellulose concentration of >60%. The maximum adsorption capacity of ACR was found to be about 118.9 mg/g for nitrate by electrostatic attraction. Wang et al. (97) prepared and used amine-cross-linked nano-Fe_3O_4 (ACMCS) by *in-situ* coprecipitation of magnetic corn cob with Fe^{2+} and Fe^{3+} solutions and amine functionalization as the biosorbent for removal of phosphate from a liquid. The results showed that PO_4^{3-} adsorption onto ACMCS was inefficient at extreme pH, but it exhibited a high performance of Cr (VI) adsorption under strongly acid conditions.

4.11 CONCLUSION

Even though nitrogen and phosphorus are two essential nutrients that play a vital role in plant growth, they are treated as toxic pollutants in wastewater that affect humans directly or indirectly, while also causing several metabolic diseases and are recognized as a serious threat to mankind. The studies described in this review prove that these elements can be removed from wastewater by using agro-industrial composites and agricultural biomass (rice husk ash,sugarcane bagasse, coirpith, biochar, corn stalk, etc.). After analyzing the research studies conducted by various teams, it has been concluded that the removal of nitrate and phosphate from wastewater can be achieved by an individual without any major financial investment. This can be achieved through simple agricultural practices at a very notional cost. As a result, treating wastewater with a composite made of agro-industrial waste and agricultural biomass can be an efficient, affordable, and environmentally protective strategy.

REFERENCES

1. Ambulkar, A. and Nathanson, J.A., 2021. Wastewater treatment. *Encyclopedia Britannica.*
2. Weerasekara, P., 2017. The United Nations world water development report 2017 wastewater. *Future of Food: Journal on Food, Agriculture and Society*, *5*(2), pp.80–81.
3. Berkessa, Y.W., Mereta, S.T. and Feyisa, F.F., 2019. Simultaneous removal of nitrate and phosphate from wastewater using solid waste from factory. *Applied Water Science*, *9*, pp.1–10.

4. Guignard, M.S., Leitch, A.R., Acquisti, C., Eizaguirre, C., Elser, J.J., Hessen, D.O., Jeyasingh, P.D., Neiman, M., Richardson, A.E., Soltis, P.S. and Soltis, D.E., 2017. Impacts of nitrogen and phosphorus: from genomes to natural ecosystems and agriculture. *Frontiers in Ecology and Evolution*, *5*, p.70.
5. Ruzhitskaya, O. and Gogina, E., 2017. Methods for removing of phosphates from wastewater. In *MATEC Web of Conferences* (Vol. 106, p. 07006). EDP Sciences.
6. Kim, K., Kim, D., Kim, T., Kim, B.G., Ko, D., Lee, J., Han, Y., Jung, J.C. and Na, H.B., 2019. Synthesis of mesoporous lanthanum hydroxide with enhanced adsorption performance for phosphate removal. *RSC Advances*, *9*(27), pp.15257–15264.
7. Bhatnagar, A. and Sillanpää, M., 2011. A review of emerging adsorbents for nitrate removal from water. *Chemical Engineering Journal*, *168*(2), pp.493–504.
8. Nodeh, H.R., Sereshti, H., Afsharian, E.Z. and Nouri, N., 2017. Enhanced removal of phosphate and nitrate ions from aqueous media using nanosized lanthanum hydrous doped on magnetic graphene nanocomposite. *Journal of Environmental Management*, *197*, pp.265–274.
9. Qadir, M., Drechsel, P., Jiménez Cisneros, B., Kim, Y., Pramanik, A., Mehta, P. and Olaniyan, O., 2020, February. Global and regional potential of wastewater as a water, nutrient and energy source. In *Natural Resources Forum* (Vol. 44, No. 1, pp. 40–51). Oxford: Blackwell Publishing Ltd.
10. Jones, E.R., Van Vliet, M.T., Qadir, M. and Bierkens, M.F., 2021. Country-level and gridded estimates of wastewater production, collection, treatment and reuse. *Earth System Science Data*, *13*(2), pp.237–254.
11. Gowd, S.C., Ramakrishna, S. and Rajendran, K., 2022. Wastewater in India: An untapped and under-tapped resource for nutrient recovery towards attaining a sustainable circular economy. *Chemosphere*, *291*, p.132753.
12. Alrumman, S.A., El-kott, A.F. and Keshk, S.M., 2016. Water pollution: Source and treatment. *American Journal of Environmental Engineering*, *6*(3), pp.88–98.
13. Solaraj, G., Dhanakumar, S., Rutharvel Murthy, K. and Mohanraj, R., 2010. Water quality in select regions of Cauvery Delta River basin, southern India, with emphasis on monsoonal variation. *Environmental Monitoring and Assessment*, *166*, pp.435–444.
14. Rahman, K., Barua, S. and Imran, H.M., 2021. Assessment of water quality and apportionment of pollution sources of an urban lake using multivariate statistical analysis. *Cleaner Engineering and Technology*, *5*, p.100309.
15. Gizaw, A., Zewge, F., Kumar, A., Mekonnen, A. and Tesfaye, M., 2021. A comprehensive review on nitrate and phosphate removal and recovery from aqueous solutions by adsorption. *Journal of Water Supply: Research and Technology-Aqua*, *70*(7), pp.921–947.
16. Luthra, S., 2022. Ammonia, phosphate levels way beyond safety limits in Yamuna. https://www.livemint.com/news/india/ammonia-phosphate-levels-way-beyond-safety-limits-in-yamuna-11657558557090.html.
17. Drolc, A. and Vrtovšek, J., 2010. Nitrate and nitrite nitrogen determination in waste water using on-line UV spectrometric method. *Bioresource Technology*, *101*(11), pp.4228–4233.
18. Baezzat, M.R., Parsaeian, G. and Zare, M.A., 2011. Determination of traces of nitrate in water samples using spectrophotometric method after its preconcentration on microcrystalline naphthalene. *Química Nova*, *34*, pp.607–609.
19. Kurniawati, P., Gusrianti, R., Dwisiwi, B.B., Purbaningtias, T.E. and Wiyantoko, B., 2017, December. Verification of spectrophotometric method for nitrate analysis in water samples. In *AIP Conference Proceedings* (Vol. 1911, No. 1, p. 020012). AIP Publishing LLC.

20. Pang, F. and Christison, T., 2017. Determination of nitrite and nitrate in wastewater using capillary IC with UV detection. Thermo Fisher Scientific, Sunnyvale, CA.
21. Murphy, J.A.M.E.S. and Riley, J.P., 1962. A modified single solution method for the determination of phosphate in natural waters. *Analytica Chimica Acta*, *27*, pp.31–36.
22. Doku, G.N. and Haswell, S.J., 1999. Further studies into the development of a micro-FIA (μFIA) system based on electroosmotic flow for the determination of phosphate as orthophosphate. *Analytica Chimica Acta*, *382*(1–2), pp.1–13.
23. Akhter, F., Siddiquei, H.R., Alahi, M.E.E. and Mukhopadhyay, S.C., 2021. Design and development of an IoT-enabled portable phosphate detection system in water for smart agriculture. *Sensors and Actuators A: Physical*, *330*, p.112861.
24. O'Grady, J., Kent, N. and Regan, F., 2021. Design, build and demonstration of a fast, reliable portable phosphate field analyser. *Case Studies in Chemical and Environmental Engineering*, *4*, p.100168.
25. Ibnul, N.K. and Tripp, C.P., 2021. A solventless method for detecting trace level phosphate and arsenate in water using a transparent membrane and visible spectroscopy. *Talanta*, *225*, p.122023.
26. Li, J.H., Lv, G.H., Bai, W.B., Liu, Q., Zhang, Y.C. and Song, J.Q., 2016. Modification and use of biochar from wheat straw (Triticum aestivum L.) for nitrate and phosphate removal from water. *Desalination and Water Treatment*, *57*(10), pp.4681–4693.
27. Jung, K.W., Hwang, M.J., Ahn, K.H. and Ok, Y.S., 2015. Kinetic study on phosphate removal from aqueous solution by biochar derived from peanut shell as renewable adsorptive media. *International Journal of Environmental Science and Technology*, *12*, pp.3363–3372.
28. Qiu, G., Zhao, Y., Wang, H., Tan, X., Chen, F. and Hu, X., 2019. Biochar synthesized via pyrolysis of Broussonetia papyrifera leaves: Mechanisms and potential applications for phosphate removal. *Environmental Science and Pollution Research*, *26*, pp.6565–6575.
29. Vijayaraghavan, K. and Balasubramanian, R., 2021. Application of pinewood waste-derived biochar for the removal of nitrate and phosphate from single and binary solutions. *Chemosphere*, *278*, p.130361.
30. Divband Hafshejani, L., Hooshmand, A., Naseri, A., Soltani Mohammadi, A. and Abbasi, F., 2016. Compare of Biochar and Vermicompost sugarcane bagasse performance on nitrate removal from contaminated water and determine the optimum conditions for adsorption process. *Iranian Journal of Irrigation & Drainage*, *10*(1), pp.104–116.
31. Yin, Q., Wang, R. and Zhao, Z., 2018. Application of Mg–Al-modified biochar for simultaneous removal of ammonium, nitrate, and phosphate from eutrophic water. *Journal of Cleaner Production*, *176*, pp.230–240.
32. Jena, J., Das, T. and Sarkar, U., 2021. Explicating proficiency of waste biomass-derived biochar for reclaiming phosphate from source-separated urine and its application as a phosphate biofertilizer. *Journal of Environmental Chemical Engineering*, *9*(1), p.104648.
33. Xu, K., Zhang, C., Dou, X., Ma, W. and Wang, C., 2019. Optimizing the modification of wood waste biochar via metal oxides to remove and recover phosphate from human urine. *Environmental Geochemistry and Health*, *41*, pp.1767–1776.
34. Zhu, Z., Huang, C.P., Zhu, Y., Wei, W. and Qin, H., 2018. A hierarchical porous adsorbent of nano-α-Fe2O3/Fe3O4 on bamboo biochar (HPA-Fe/CB) for the removal of phosphate from water. *Journal of Water Process Engineering*, *25*, pp.96–104.
35. Karthikeyan, P., Vigneshwaran, S. and Meenakshi, S., 2020. Removal of phosphate and nitrate ions from water by amine crosslinked magnetic banana bract activated carbon and its physicochemical performance. *Environmental Nanotechnology, Monitoring & Management*, *13*, p.100294.

36. Ramola, S., Belwal, T., Li, C.J., Liu, Y.X., Wang, Y.Y., Yang, S.M. and Zhou, C.H., 2021. Preparation and application of novel rice husk biochar–calcite composites for phosphate removal from aqueous medium. *Journal of Cleaner Production, 299*, p.126802.
37. Shao, Y., Li, J., Fang, X., Yang, Z., Qu, Y., Yang, M., Tan, W., Li, G. and Wang, H., 2022. Chemical modification of bamboo activated carbon surface and its adsorption property of simultaneous removal of phosphate and nitrate. *Chemosphere, 287*, p.132118.
38. Banu, H.T., Karthikeyan, P. and Meenakshi, S., 2019. Zr4+ ions embedded chitosan-soya bean husk activated bio-char composite beads for the recovery of nitrate and phosphate ions from aqueous solution. *International Journal of Biological Macromolecules, 130*, pp.573–583.
39. Li, T., Su, X., Yu, X., Song, H., Zhu, Y. and Zhang, Y., 2018. La (OH) 3-modified magnetic pineapple biochar as novel adsorbents for efficient phosphate removal. *Bioresource Technology, 263*, pp.207–213.
40. Xu, Q., Chen, Z., Wu, Z., Xu, F., Yang, D., He, Q., Li, G. and Chen, Y., 2019. Novel lanthanum doped biochars derived from lignocellulosic wastes for efficient phosphate removal and regeneration. *Bioresource Technology, 289*, p.121600.
41. Jiang, Y.H., Li, A.Y., Deng, H., Ye, C.H. and Li, Y., 2019. Phosphate adsorption from wastewater using ZnAl-LDO-loaded modified banana straw biochar. *Environmental Science and Pollution Research, 26*, pp.18343–18353.
42. Lee, S.Y., Choi, J.W., Song, K.G., Choi, K., Lee, Y.J. and Jung, K.W., 2019. Adsorption and mechanistic study for phosphate removal by rice husk-derived biochar functionalized with Mg/Al-calcined layered double hydroxides via co-pyrolysis. *Composites Part B: Engineering, 176*, p.107209.
43. Wang, Z., Guo, H., Shen, F., Yang, G., Zhang, Y., Zeng, Y., Wang, L., Xiao, H. and Deng, S., 2015. Biochar produced from oak sawdust by Lanthanum (La)-involved pyrolysis for adsorption of ammonium (NH_4^+), nitrate (NO_3^-), and phosphate (PO43–). *Chemosphere, 119*, pp.646–653.
44. Cai, R., Wang, X., Ji, X., Peng, B., Tan, C. and Huang, X., 2017. Phosphate reclaim from simulated and real eutrophic water by magnetic biochar derived from water hyacinth. *Journal of Environmental Management, 187*, pp.212–219.
45. Qiu, H., Ye, M., Zeng, Q., Li, W., Fortner, J., Liu, L. and Yang, L., 2019. Fabrication of agricultural waste supported UiO-66 nanoparticles with high utilization in phosphate removal from water. *Chemical Engineering Journal, 360*, pp.621–630.
46. Bolbol, H., Fekri, M. and Hejazi-Mehrizi, M., 2019. Layered double hydroxide–loaded biochar as a sorbent for the removal of aquatic phosphorus: Behavior and mechanism insights. *Arabian Journal of Geosciences, 12*, pp.1–11.
47. He, H., Zhang, N., Chen, N., Lei, Z., Shimizu, K. and Zhang, Z., 2019. Efficient phosphate removal from wastewater by MgAl-LDHs modified hydrochar derived from tobacco stalk. *Bioresource Technology Reports, 8*, p.100348.
48. Yin, Q., Liu, M. and Ren, H., 2019. Biochar produced from the co-pyrolysis of sewage sludge and walnut shell for ammonium and phosphate adsorption from water. *Journal of Environmental Management, 249*, p.109410.
49. Zhang, Z., Yan, L., Yu, H., Yan, T. and Li, X., 2019. Adsorption of phosphate from aqueous solution by vegetable biochar/layered double oxides: Fast removal and mechanistic studies. *Bioresource Technology, 284*, pp.65–71.
50. Faheem, Du, J., Bao, J., Hassan, M.A., Irshad, S., Talib, M.A. and Zheng, H., 2020. Efficient capture of phosphate and cadmium using biochar with multifunctional amino and carboxylic moieties: kinetics and mechanism. *Water, Air, & Soil Pollution, 231*, pp.1–16.

51. Hafshejani, L.D., Hooshmand, A., Naseri, A.A., Mohammadi, A.S., Abbasi, F. and Bhatnagar, A., 2016. Removal of nitrate from aqueous solution by modified sugarcane bagasse biochar. *Ecological Engineering*, *95*, pp.101–111.
52. Li, J.H., Lv, G.H., Bai, W.B., Liu, Q., Zhang, Y.C. and Song, J.Q., 2016. Modification and use of biochar from wheat straw (Triticum aestivum L.) for nitrate and phosphate removal from water. *Desalination and Water Treatment*, *57*(10), pp.4681–4693.
53. El Hanandeh, A., Bhuvaneswaran, A. and de Rozari, P., 2017. Removal of nitrate, ammonia and phosphate from aqueous solutions in packed bed filter using biochar augmented sand media. In *MATEC Web of Conferences* (Vol. 120, p. 05004). EDP Sciences.
54. Hu, X., Xue, Y., Long, L. and Zhang, K., 2018. Characteristics and batch experiments of acid-and alkali-modified corncob biomass for nitrate removal from aqueous solution. *Environmental Science and Pollution Research*, *25*, pp.19932–19940.
55. Jiang, Y.H., Li, A.Y., Deng, H., Ye, C.H., Wu, Y.Q., Linmu, Y.D. and Hang, H.L., 2019. Characteristics of nitrogen and phosphorus adsorption by Mg-loaded biochar from different feedstocks. *Bioresource Technology*, *276*, pp.183–189.
56. Wang, T., Zhang, D., Fang, K., Zhu, W., Peng, Q. and Xie, Z., 2021. Enhanced nitrate removal by physical activation and Mg/Al layered double hydroxide modified biochar derived from wood waste: Adsorption characteristics and mechanisms. *Journal of Environmental Chemical Engineering*, *9*(4), p.105184.
57. Hredya, E.M., Samad, A.A. and Saud, S.J., 2019. Nitrate removal from groundwater using crushed lemon peel and activated charcoal. *International Journal of Scientific Research and Reviews*, *8*(3), pp.148–164.
58. Xu, X., Gao, B.Y., Yue, Q.Y. and Zhong, Q.Q., 2010. Preparation of agricultural by-product-based anion exchanger and its utilization for nitrate and phosphate removal. *Bioresource Technology*, *101*(22), pp.8558–8564.
59. Song, W., Gao, B., Xu, X., Wang, F., Xue, N., Sun, S., Song, W. and Jia, R., 2016. Adsorption of nitrate from aqueous solution by magnetic amine-crosslinked biopolymer-based corn stalk and its chemical regeneration property. *Journal of Hazardous Materials*, *304*, pp.280–290.
60. Qiao, H., Mei, L., Chen, G., Liu, H., Peng, C., Ke, F., Hou, R., Wan, X. and Cai, H., 2019. Adsorption of nitrate and phosphate from aqueous solution using amine cross-linked tea wastes. *Applied Surface Science*, *483*, pp.114–122.
61. Hu, Y., Du, Y., Nie, G., Zhu, T., Ding, Z., Wang, H., Zhang, L. and Xu, Y., 2020. Selective and efficient sequestration of phosphate from waters using reusable nano-Zr (IV) oxide impregnated agricultural residue anion exchanger. *Science of the Total Environment*, *700*, p.134999.
62. Qiu, H., Liang, C., Yu, J., Zhang, Q., Song, M. and Chen, F., 2017. Preferable phosphate sequestration by nano-La (III)(hydr) oxides modified wheat straw with excellent properties in regeneration. *Chemical Engineering Journal*, *315*, pp.345–354.
63. Qiu, H., Ni, W., Zhang, H., Chen, K. and Yu, J., 2020. Fabrication and evaluation of a regenerable HFO-doped agricultural waste for enhanced adsorption affinity towards phosphate. *Science of the Total Environment*, *703*, p.135493.
64. Alagha, O., Manzar, M.S., Zubair, M., Anil, I., Mu'azu, N.D. and Qureshi, A., 2020. Comparative adsorptive removal of phosphate and nitrate from wastewater using biochar-MgAl LDH nanocomposites: Coexisting anions effect and mechanistic studies. *Nanomaterials*, *10*(2), p.336.
65. de Lima, A.C.A., Nascimento, R.F., de Sousa, F.F., Josue Filho, M. and Oliveira, A.C., 2012. Modified coconut shell fibers: A green and economical sorbent for the removal of anions from aqueous solutions. *Chemical Engineering Journal*, *185*, pp.274–284.
66. Katal, R., Baei, M.S., Rahmati, H.T. and Esfandian, H., 2012. Kinetic, isotherm and thermodynamic study of nitrate adsorption from aqueous solution using modified rice husk. *Journal of Industrial and Engineering Chemistry*, *18*(1), pp.295–302.

67. Moyo, M., Maringe, A., Chigondo, F., Nyamunda, B.C., Sebata, E. and Shumba, M., 2012. Adsorptive removal of nitrate ions from aqueous solutions using acid treated sunflower seed husk (Helianthus annuus). *International Journal of Advances in Science and Technology*, *5*(6), pp.47–66.
68. Xu, X., Gao, Y., Gao, B., Tan, X., Zhao, Y.Q., Yue, Q. and Wang, Y., 2011. Characteristics of diethylenetriamine-crosslinked cotton stalk/wheat stalk and their biosorption capacities for phosphate. *Journal of Hazardous Materials*, *192*(3), pp.1690–1696.
69. Zhang, J., Shan, W., Ge, J., Shen, Z., Lei, Y. and Wang, W., 2012. Kinetic and equilibrium studies of liquid-phase adsorption of phosphate on modified sugarcane bagasse. *Journal of Environmental Engineering*, *138*(3), pp.252–258.
70. Ogata, F., Imai, D. and Kawasaki, N., 2015. Adsorption of nitrate and nitrite ions onto carbonaceous material produced from soybean in a binary solution system. *Journal of Environmental Chemical Engineering*, *3*(1), pp.155–161.
71. Cao, D., Jin, X., Gan, L., Wang, T. and Chen, Z., 2016. Removal of phosphate using iron oxide nanoparticles synthesized by eucalyptus leaf extract in the presence of CTAB surfactant. *Chemosphere*, *159*, pp.23–31.
72. Shang, Y., Xu, X., Qi, S., Zhao, Y., Ren, Z. and Gao, B., 2017. Preferable uptake of phosphate by hydrous zirconium oxide nanoparticles embedded in quaternary-ammonium Chinese reed. *Journal of Colloid and Interface Science*, *496*, pp.118–129.
73. Kilpimaa, S., Runtti, H., Kangas, T., Lassi, U. and Kuokkanen, T., 2015. Physical activation of carbon residue from biomass gasification: Novel sorbent for the removal of phosphates and nitrates from aqueous solution. *Journal of Industrial and Engineering Chemistry*, *21*, pp.1354–1364.
74. Shahmoradi, M.H., Zade, B.A., Torabian, A. and Salehi, M.S., 2015. Removal of nitrate from ground water using activated carbon prepared from rice husk and sludge of paper industry wastewater treatment. *ARPN Journal of Engineering and Applied Sciences*, *10*(17), pp.7856–7863.
75. Najmi, S., Hatamipour, M.S., Sadeh, P., Najafipour, I. and Mehranfar, F., 2020. Activated carbon produced from Glycyrrhiza glabra residue for the adsorption of nitrate and phosphate: Batch and fixed-bed column studies. *SN Applied Sciences*, *2*, pp.1–22.
76. Sow, M.M., Cisse, A., Sakho, A.M. and Kante, C., 2023. Removal of nitrate and phosphate ions from the Bafing river by an adsorbent obtained from the shells of mango cores. *Journal of Geoscience and Environment Protection*, *11*(1), pp.67–78.
77. Li, J., Wei, P., Li, B., Guo, J., Li, J., Yang, B. and Song, J., 2021. Nitrate removal from water by activated carbon derived from wheat straw with FeCl3 activation. *Journal of Ecology and Rural Environment*, *37*, pp.224–233.
78. Divband Hafshejani, L. and Naseri, A., 2020. Optimizing the removal of phosphate from agricultural drains water using activated sugarcane bagasse hydrocar. *Iranian Journal of Irrigation & Drainage*, *14*(5), pp.1853–1865.
79. Demiral, H. and Gündüzoğlu, G., 2010. Removal of nitrate from aqueous solutions by activated carbon prepared from sugar beet bagasse. *Bioresource Technology*, *101*(6), pp.1675–1680.
80. Curran, D.T., 2015. Phosphate removal and recovery from wastewater by natural materials for ecologically engineered wastewater treatment systems. Graduate College Dissertations and Theses.
81. Hanafi, H.A. and Azeema, S.M.A., 2016. Removal of nitrate and nitrite anions from wastewater using activated carbon derived from rice straw. *Journal of Environmental and Analytical Toxicology*, *6*(346), pp.2161–0525.
82. Liu, L., Ji, M. and Wang, F., 2018. Adsorption of nitrate onto ZnCl2-modified coconut granular activated carbon: Kinetics, characteristics, and adsorption dynamics. *Advances in Materials Science and Engineering*, *2018*, 1939032.

83. Manjunath, S.V. and Kumar, M., 2018. Evaluation of single-component and multi-component adsorption of metronidazole, phosphate and nitrate on activated carbon from Prosopıs julıflora. *Chemical Engineering Journal, 346*, pp.525–534.
84. Nabhan, A., 2019. Nitrate removal from water using activated carbon prepared from olive stone by microwave heating. M.Sc. Thesis, MEDRC Water Research, Muscat.
85. Mor, S., Chhoden, K. and Ravindra, K., 2016. Application of agro-waste rice husk ash for the removal of phosphate from the wastewater. *Journal of Cleaner Production, 129*, pp.673–680.
86. Hamzah, S., Razali, N.A., Yatim, N.I., Alias, M., Ali, A., Zaini, N.S. and Abuhabib, A.A., 2018. Characterisation and performance of thermally treated rice husk as efficient adsorbent for phosphate removal. *Journal of Water Supply: Research and Technology-Aqua, 67*(8), pp.766–778.
87. Li, J., Jin, Q., Liang, Y., Geng, J., Xia, J., Chen, H. and Yun, M., 2022. Highly efficient removal of nitrate and phosphate to control eutrophication by the dielectrophoresis-assisted adsorption method. *International Journal of Environmental Research and Public Health, 19*(3), p.1890.
88. Phan, P.T., Nguyen, T.T., Nguyen, N.H. and Padungthon, S., 2019. Triamine-bearing activated rice husk ash as an advanced functional material for nitrate removal from aqueous solution. *Water Science and Technology, 79*(5), pp.850–856.
89. Mokif, L.A., Abdulhusain, N.A. and AL-Mamoori, S.O.H., 2018. The possibility of using the kiwi peels as an adsorbent for removing nitrate from water. *Journal of University of Babylon for Engineering Sciences, 26*(2), pp.192–197.
90. Nkansah, M.A., Donkoh, M., Akoto, O. and Ephraim, J.H., 2019. Preliminary studies on the use of sawdust and peanut shell powder as adsorbents for phosphorus removal from water. *Emerging Science Journal, 3*(1), pp.33–40.
91. Fotsing, P.N., Woumfo, E.D., Mezghich, S., Mignot, M., Mofaddel, N., Le Derf, F. and Vieillard, J., 2020. Surface modification of biomaterials based on cocoa shell with improved nitrate and Cr (vi) removal. *RSC Advances, 10*(34), pp.20009–20019.
92. Dey, S., Uppala, P., Sambangi, A., Haripavan, N. and Veerendra, G.T.N., 2022. Recycling of solid waste biosorbents for removal of nitrates from contaminated water. *Cleaner and Circular Bioeconomy, 2*, p.100014.
93. Dey, S., Sreenivasulu, A., Veerendra, G.T.N., Manoj, A.P. and Haripavan, N., 2022. Synthesis and characterization of mango leaves biosorbents for removal of iron and phosphorous from contaminated water. *Applied Surface Science Advances, 11*, p.100292.
94. Ang, B.Y.H., Ong, Y.H. and Ng, Y.S., 2021. Investigation on the removal of nitrate from water using different types of biosorbents. In *IOP Conference Series: Earth and Environmental Science* (Vol. 646, No. 1, p. 012010). IOP Publishing.
95. Nguyen, T.A.H., Ngo, H.H., Guo, W.S., Zhang, J., Liang, S. and Tung, K.L., 2013. Feasibility of iron loaded 'okara' for biosorption of phosphorous in aqueous solutions. *Bioresource Technology, 150*, pp.42–49.
96. Ren, Z., Xu, X., Wang, X., Gao, B., Yue, Q., Song, W., Zhang, L. and Wang, H., 2016. FTIR, Raman, and XPS analysis during phosphate, nitrate and Cr (VI) removal by amine cross-linking biosorbent. *Journal of Colloid and Interface Science, 468*, pp.313–323.
97. Wang, H., Xu, X., Ren, Z. and Gao, B., 2016. Removal of phosphate and chromium (VI) from liquids by an amine-crosslinked nano-Fe_3O_4 biosorbent derived from corn straw. *RSC Advances, 6*(53), pp.47237–47248.

5 Urban Sludge Management and Circular Transformation in Water Remediation

Kassian T.T. Amesho
Institute of Environmental Engineering,
National Sun Yat-Sen University, Kaohsiung, Taiwan
Center for Emerging Contaminants Research,
National Sun Yat-Sen University, Kaohsiung, Taiwan
Tshwane School for Business and Society,
Faculty of Management of Sciences, Tshwane University of Technology, Pretoria, South Africa
The International University of Management,
Centre for Environmental Studies, Main Campus,
Dorado Park Ext 1, Windhoek, Namibia
Destinies Biomass Energy and Farming Pty Ltd,
P. O. Box 7387, Swakopmund, Namibia
Regent Business School, Durban, South Africa

Abner Kukeyinge Shopati
Namibia Business School (NBS), Faculty of Commerce, Management and Law, University of Namibia, Private Bag 13301, Main Campus, Windhoek, Namibia

Sumarlin Shangdiar
Center for Emerging Contaminants Research,
National Sun Yat-Sen University, Kaohsiung, Taiwan
Tshwane School for Business and Society,
Faculty of Management of Sciences, Tshwane University of Technology, Pretoria, South Africa

Timoteus Kadhila
Department of Civil and Environmental Engineering, Birla Institute of Technology, Mesra Ranchi, Jharkhand, India
SVCET Engineering College and Technology (A),
Chittoor, Andhra Pradesh, India

DOI: 10.1201/9781003432869-5

Sioni Iikela
The International University of Management,
Centre for Environmental Studies, Main Campus,
Dorado Park Ext 1, Windhoek, Namibia

E.I. Edoun
Tshwane School for Business and Society,
Faculty of Management of Sciences, Tshwane University
of Technology, Pretoria, South Africa

Subrata Chowdhury
School of Education, Department of Higher
Education and Lifelong Learning, University of
Namibia, Private Bag 13301, Windhoek, Namibia

Nastassia Thandiwe Sithole
Department of Chemical Engineering,
Faculty of Engineering and the Built Environment at the
University of Johannesburg, South Africa

5.1 INTRODUCTION

5.1.1 Background and Context

Urbanization has ushered in significant advances in modern living, infrastructure, and economic development. However, alongside these transformative changes, there emerges a consequential challenge – the management of urban sludge. Urban sludge, a by-product of wastewater treatment processes, presents a formidable environmental concern due to its composition, volume, and potential hazards (Wei et al., 2020). As cities expand and populations burgeon, the efficient and sustainable management of urban sludge becomes an imperative endeavor that reverberates across ecological, economic, and societal dimensions (Collivignarelli et al., 2019).

5.1.2 Significance of Urban Sludge Management in Water Remediation

The effective management of urban sludge is intrinsically linked to the broader sphere of water remediation, which encapsulates multifaceted strategies aimed at mitigating water pollution, safeguarding ecosystems, and ensuring the availability of clean water resources (Nguyen et al., 2022). Urban sludge, while a challenge in its own right, holds the promise of transformation when viewed through the lens of circular economy principles. Its optimal management not only addresses waste disposal concerns but also opens up avenues for resource recovery, contributing to a paradigm shift in water remediation practices (Smol, 2023).

In line with the Sustainable Development Goals (SDGs) set forth by the United Nations, sustainable urban sludge management aligns with several crucial objectives, including clean water and sanitation (SDG 6), sustainable cities and communities (SDG 11), responsible consumption and production (SDG 12), and life below water (SDG 14). The judicious handling of urban sludge through circular transformation represents a harmonious nexus between urban development and environmental conservation (Preisner et al., 2022).

5.1.3 Purpose and Scope of the Chapter

The purpose of this chapter is to delve into the intricate landscape of urban sludge management and its circular transformation into a cornerstone of modern water remediation practices. By examining the combination of circular economy principles with urban sludge management, this chapter seeks to unravel the manifold implications, challenges, opportunities, and innovations that arise in the pursuit of sustainable water remediation.

The scope of this chapter encompasses a comprehensive exploration of urban sludge management, its diverse sources, and its inherent characteristics. It investigates the fundamental tenets of circular economy principles and their application to the context of urban sludge, thereby illuminating the potential for resource recovery, waste reduction, and enhanced environmental stewardship. Drawing from real-world case studies and examples, the chapter elucidates the practical manifestation of circular transformation in different settings.

Moreover, this chapter addresses the challenges and barriers that permeate the landscape of circular urban sludge management, ranging from regulatory hurdles to technical complexities. By critically assessing these impediments, the chapter aims to facilitate a nuanced understanding of the hurdles that necessitate concerted efforts for resolution.

Ultimately, the chapter underscores the pivotal role of circular transformation in urban sludge management, advocating for its integration as an indispensable component of holistic water remediation strategies. As the world grapples with pressing water scarcity and environmental degradation, the insights gleaned from this exploration can chart a course toward sustainable, resilient, and ecologically harmonious urban futures.

5.2 URBAN SLUDGE GENERATION AND CHARACTERISTICS

5.2.1 Sources and Types of Urban Sludge

Urban sludge, an inevitable by-product of wastewater treatment processes, comprises a diverse and complex blend of organic and inorganic substances arising from an array of domestic, industrial, and commercial activities (Wang et al., 2020). This amalgamation necessitates a systematic approach to management to mitigate potential environmental hazards and public health risks. The primary contributors to urban sludge generation are municipal wastewater treatment plants, industrial effluent discharges, and combined sewer overflows. Each of these sources introduces a distinct composition to the sludge, leading to a heterogeneous mixture with varying properties.

Municipal wastewater treatment plants play a significant role in the generation of urban sludge. Through the treatment of raw sewage, two primary types of sludge are produced: primary sludge and secondary sludge. Primary sludge consists of solid particles that settle during the initial treatment stages, including screenings and sedimentation (Gherghel et al., 2019). On the other hand, secondary sludge is a result of biological processes where microorganisms consume organic matter in the wastewater, producing excess biomass (Havukainen et al., 2016). Moreover, the stabilization of sludge through biological processes leads to biosolids, which are organic-rich, nutrient-enhanced products suitable for safe and beneficial reuse in various applications (Werle and Sobek, 2019).

Industrial activities contribute substantially to the diversity of urban sludge types. Industries ranging from food processing to pharmaceutical production generate sludge characterized by the specific processes and materials employed. For example, the agro-food sector may generate sludge with high organic content and nutrients, while pharmaceutical and chemical industries might introduce hazardous substances and pollutants (Raheem et al., 2018; Lipinska, 2018). This variation underscores the importance of tailored management strategies for different industrial sectors.

5.2.2 Physical, Chemical, and Biological Characteristics

The physical, chemical, and biological attributes of urban sludge collectively define its nature and behavior, significantly influencing its handling, treatment, and disposal. Physically, sludge exhibits a wide range of characteristics, including density, viscosity, and particle size distribution (Lamastra et al., 2018; Rorat et al., 2019). These properties influence its behavior during processes like dewatering, transportation, and application. The chemical composition of sludge encompasses pH, organic matter content, nutrient concentrations, trace element levels, and the presence of contaminants such as heavy metals and persistent organic pollutants (POPs) (Pikaar et al., 2022; Domini et al., 2022). These characteristics dictate the feasibility of beneficial applications, potential environmental impacts, and management options.

Biologically, urban sludge hosts a diverse microbial community that plays a vital role in the degradation and transformation of the sludge. Microorganisms in sludge contribute to processes like organic matter decomposition, nutrient cycling, and pathogen reduction (Wei et al., 2019). However, the presence of harmful microorganisms, including pathogens and antibiotic-resistant bacteria, poses health risks and necessitates appropriate treatment. The interplay between these physical, chemical, and biological attributes underscores the complexity of urban sludge, emphasizing the need for comprehensive management strategies that consider these intricate factors (Singh et al., 2020).

5.2.3 Environmental and Health Implications

Effective urban sludge management is critical due to its potential environmental and health consequences, if mishandled. Improper management practices can lead to the release of noxious gases, such as hydrogen sulfide, resulting in air pollution and malodorous emissions. These emissions can adversely affect local air quality and contribute to nuisance issues for nearby communities. The untreated disposal of sludge can introduce contaminants into surface water bodies and groundwater, compromising aquatic ecosystems and posing risks to human health through waterborne transmission (Yoshida et al., 2018).

Moreover, the application of untreated sludge to the land can impact soil quality, agricultural productivity, and groundwater quality. Pathogens and pollutants present in sludge may infiltrate soil and water, leading to potential food chain contamination. The presence of heavy metals and persistent organic pollutants in sludge can accumulate in soil, potentially reaching levels that pose ecological and human health concerns (Khalili et al., 2017). Additionally, the potential for emerging contaminants, pharmaceuticals, and antibiotic-resistant bacteria in sludge requires rigorous management to mitigate associated risks.

5.3 CIRCULAR ECONOMY PRINCIPLES IN WATER REMEDIATION

5.3.1 Overview of Circular Economy Concepts

The concept of a circular economy has gained prominence as a sustainable approach to resource management, advocating for a departure from the traditional linear "take-make-dispose" model. At its core, the circular economy aims to decouple economic growth from resource consumption and environmental degradation. This is achieved through a series of interconnected principles that prioritize the reduction of waste, the prolongation of product lifecycles, and the extraction of maximum value from resources (Yoshida et al., 2018; Hao et al., 2019b).

Central to the circular economy is the principle of designing products and processes with longevity in mind. This involves the use of durable materials, efficient production techniques, and modular designs that facilitate repair and upgradability. As a result, products are more resilient to wear and tear, extending their useful life and minimizing the need for frequent replacements.

5.3.2 Circular Economy Applied to Water Remediation

The application of circular economy principles to water remediation aligns with the goal of optimizing resource utilization while simultaneously addressing environmental challenges. In the context of urban sludge management, this involves transforming the conventional approach of sludge disposal into a resource-recovery-focused strategy (Hao et al., 2019a; Chen et al., 2012). By adopting a circular mindset, urban sludge can transition from a waste product to a valuable resource that contributes to a closed-loop system.

Circular economy strategies can be integrated into various stages of the urban sludge management process. For instance, during sludge treatment, technologies that promote energy recovery, such as anaerobic digestion and thermal processes, not only reduce the volume of sludge but also generate valuable biogas and thermal energy. These energy outputs can be harnessed to power treatment facilities or even fed back into the grid, showcasing a symbiotic relationship between waste treatment and energy production (Hao et al., 2020).

5.3.3 Resource Recovery and Reuse in a Circular Economy Context

Resource recovery and reuse are pivotal components of circular economy practices within the domain of urban sludge management. One of the key objectives is the extraction of valuable resources from sludge that can be reintroduced into the production cycle (Ramm, 2021). Nutrients like nitrogen and phosphorus, which are present in significant quantities in sludge, can be reclaimed and repurposed as fertilizers for agriculture (Zhao et al., 2019). This not only conserves precious mineral resources but also mitigates the environmental impact of excessive nutrient discharges into water bodies, preventing issues like eutrophication (Munasinghe-Arachchige and Nirmalakhandan, 2020).

Furthermore, the incorporation of sludge-derived biosolids into soil can enhance soil structure, improve water retention, and promote plant growth. This aligns with circular economy principles by creating a closed-loop system where sludge, once regarded as waste, is upcycled to enhance agricultural productivity (Wan et al., 2017; Chojnacka et al., 2023).

This section, focusing on the integration of circular economy principles into urban sludge management and its alignment with resource recovery and reuse, establishes the foundation for the subsequent exploration of practical applications and case studies in the field. By strategically embracing circularity, the management of urban sludge can evolve from a linear waste disposal approach to a sustainable and value-driven solution that contributes to water remediation and resource optimization.

5.4 CIRCULAR TRANSFORMATION OF URBAN SLUDGE MANAGEMENT

Urban sludge, once perceived solely as a waste by-product, is now being redefined through circular transformation approaches that emphasize resource recovery, recycling, and sustainable management. This section investigates the multidimensional strategies driving the circular transformation of urban sludge management, highlighting the recycling and recovery of valuable resources, innovative treatment technologies, and the integration of circular economy practices.

5.4.1 Recycling and Recovery of Resources from Sludge

Circular transformation hinges on the identification and extraction of valuable resources embedded within urban sludge. A key focus is the recovery of organic matter and nutrients present in sludge, which can be utilized to create a closed-loop system. Anaerobic digestion, a widely adopted technique, enables the conversion of organic components in sludge into biogas and digestate (Koskue et al., 2021; Zhang et al., 2016). The biogas produced is a renewable energy source, which can be utilized to power treatment plants or provide electricity to surrounding communities. Moreover, the digestate is a nutrient-rich material that can be employed as a soil conditioner or fertilizer, effectively closing the nutrient loop (Horttanainen et al., 2017; Zhang et al., 2019a).

In addition, the recovery of heavy metals from sludge offers opportunities for resource conservation. Advanced separation techniques, such as electrochemical methods and bioleaching, are being explored to selectively extract metals for potential reuse in various industrial applications, reducing the reliance on virgin resources. Figure 5.1 shows the pyrolysis process used for urban sludge management, while Figure 5.2 shows a potential sludge-to-energy recovery route, and Figure 5.3 shows an anaerobic digestion process in urban sludge management.

5.4.2 Innovative Treatment Technologies for Circular Transformation

The circular transformation of urban sludge management is intrinsically linked to the development of innovative treatment technologies that maximize resource recovery while minimizing environmental impact. One such technology is hydrothermal carbonization (HTC), which involves subjecting sludge to elevated temperatures and pressures to produce biochar and hydrochar. These carbonaceous materials have diverse applications, ranging from soil amendment to energy storage.

Another promising avenue is the integration of membrane technologies, such as forward osmosis and membrane distillation, to concentrate sludge and extract valuable components, as indicated in Table 5.1. The concentrated sludge generated can be further treated or processed for resource recovery, facilitating the reduction of disposal volumes.

5.4.3 Integration of Circular Economy Practices in Sludge Management

The circular transformation of urban sludge management extends beyond technological advances to encompass the overarching principles of the circular economy. Strategies such as product design for circularity, waste-to-energy conversion, and cascading utilization pathways align with circular economy practices (Munasinghe-Arachchige et al., 2020).

For example, reimagining the lifecycle of sludge involves designing products that incorporate sludge-derived materials, thus reducing the demand for virgin resources. Similarly, the adoption of cascading utilization pathways encourages the sequential use of sludge for different purposes, minimizing waste generation, and maximizing resource efficiency (Mandal et al., 2023; Zhu et al., 2022).

This section underscores the shift from conventional linear waste management to circular transformation in urban sludge management. The exploration of resource recycling, innovative treatment technologies, and the integration of circular economy principles sets the stage for understanding how circular transformation can revolutionize water remediation and contribute to sustainable resource utilization.

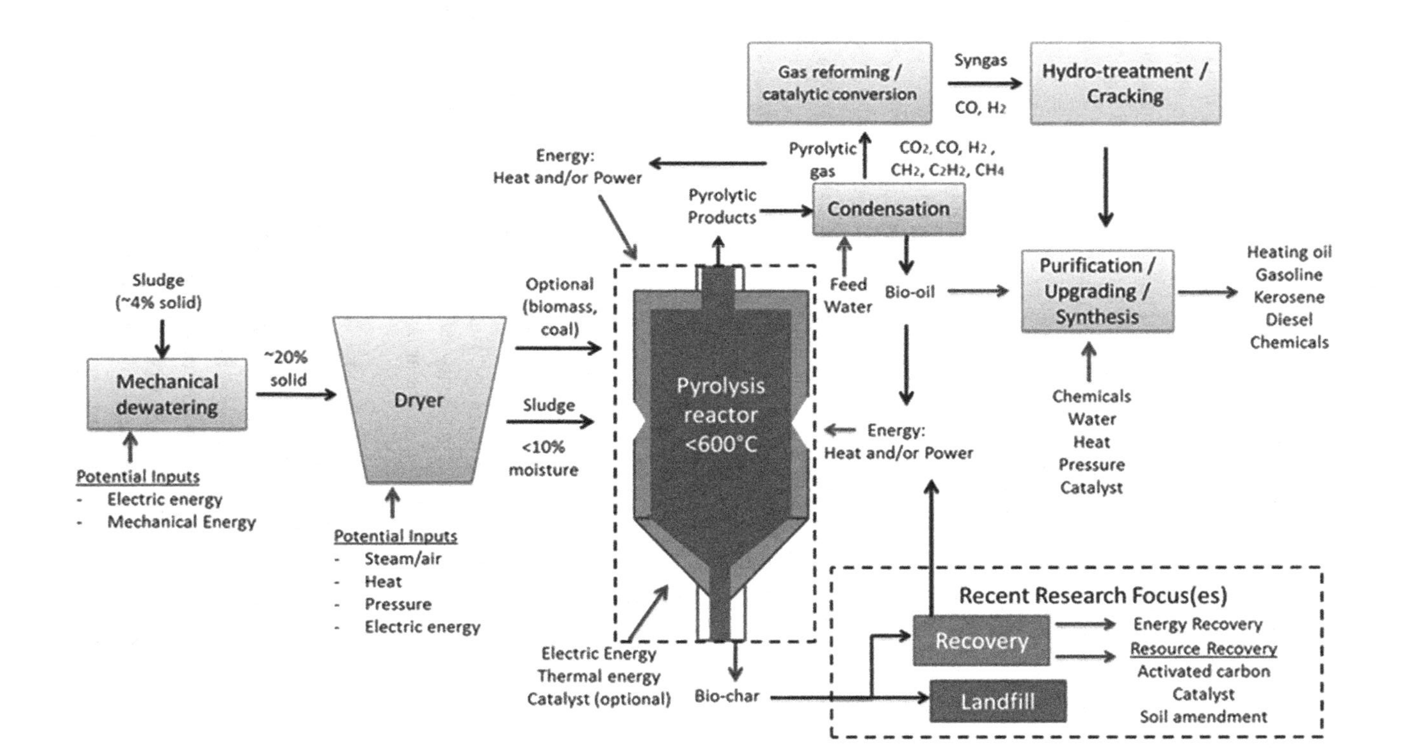

FIGURE 5.1 Schematic of pyrolysis process. Source: https://www.sludgeprocessing.com/features/sludge-to-energy-recovery-methods-an-overview/.

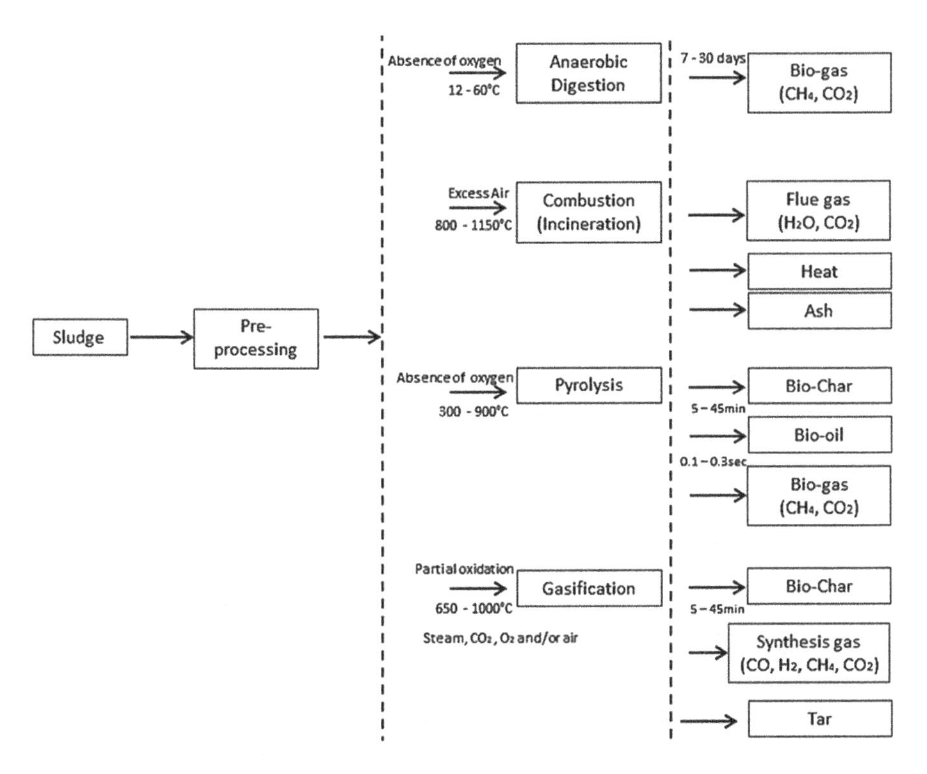

FIGURE 5.2 Potential sludge-to-energy recovery routes. Source: https://www.sludgeprocessing.com/features/sludge-to-energy-recovery-methods-an-overview/.

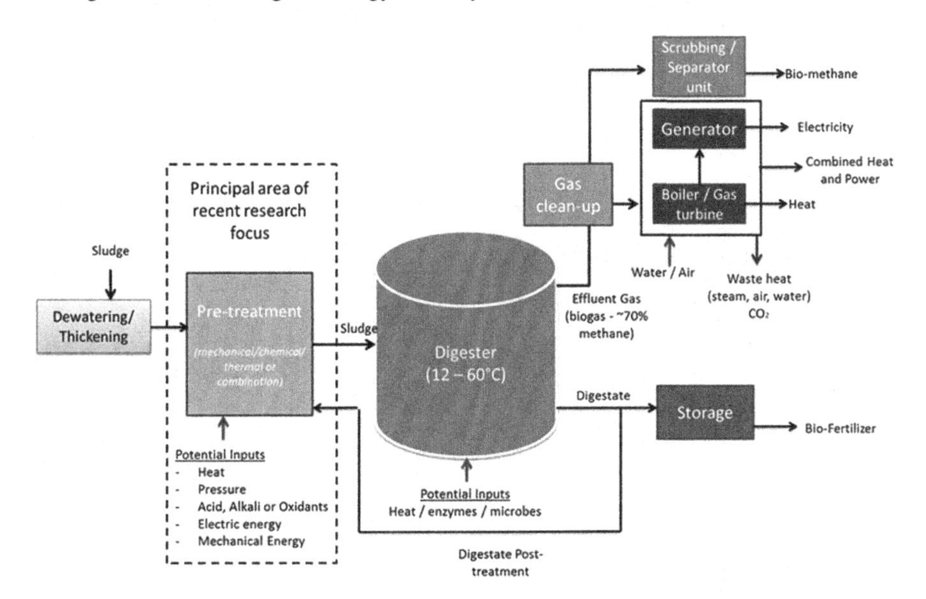

FIGURE 5.3 Schematic of anaerobic digestion process. Source: https://www.sludgeprocessing.com/features/sludge-to-energy-recovery-methods-an-overview/.

TABLE 5.1
Innovative Treatment Technologies for the Circular Transformation of Urban Sludge

Technology	Advantages	Drawbacks	Future Outlooks
Hydrothermal Carbonization (HTC)	• Efficient biochar production. • Pathogen reduction and odor elimination. • Nutrient recovery from solid residue. • Potential energy recovery through biochar combustion. • Versatile treatment for various sludge types.	• High energy input for heating. • Carbon footprint from energy consumption. • Limited scalability for large-scale applications. • Variability in biochar quality. • Dependence on sludge composition for optimal performance.	• Integration with renewable energy sources. • Valorization of biochar in agriculture and construction. • Development of hybrid systems for improved efficiency. • Optimization for specific sludge characteristics. • Assessment of long-term environmental impacts.
Forward Osmosis	• Low energy consumption for separation. • Concentrated brine for resource recovery. • Effective for high-salinity wastewater. • Potential nutrient and metal recovery. • Minimal fouling due to low operating pressure.	• Limited application to specific solutes. • Membrane fouling and degradation over time. • High initial capital and operational costs. • Performance affected by temperature and concentration. • Complex system design and optimization.	• Advances in membrane materials for improved selectivity. • Integration with other separation processes. • Development of hybrid systems for enhanced efficiency. • Application to diverse wastewater streams. • Exploration of potential for direct water reuse.
Membrane Distillation	• High water recovery and solute concentration. • Effective for high-salinity and brackish wastewater. • Energy-efficient with low-temperature operation. • Minimal fouling and scaling compared to traditional membranes. • Suitable for concentrating valuable compounds.	• High energy consumption for heating and cooling. • Membrane wetting and degradation. • Limited application to volatile organic compounds. • Complex system design and optimization. • Costly membrane materials and fabrication.	• Development of anti-fouling and self-cleaning membranes. • Coupling with renewable energy sources. • Integration with low-grade heat recovery systems. • Exploration of hybrid systems for enhanced performance. • Scaling up for industrial applications.

(Continued)

TABLE 5.1 (CONTINUED)
Innovative Treatment Technologies for the Circular Transformation of Urban Sludge

Technology	Advantages	Drawbacks	Future Outlooks
Pyrolysis and Gasification	• High energy recovery and conversion efficiency. • Reduction of sludge volume and disposal. • Biochar production with potential soil amendment. • Syngas production for heat and power generation. • Potential recovery of metals from ash residues.	• Emission of greenhouse gases and pollutants. • Requires careful temperature control and process optimization. • Complex process with variable outputs. • Dependence on feedstock characteristics. • Need for post-treatment of biochar/ash residues.	• Integration with carbon capture and utilization technologies. • Exploration of co-gasification with other waste streams. • Improvement of pyrolysis/gasification reactor designs. • Utilization of syngas for various industrial applications. • Integration with waste-to-energy systems.
Microbial Fuel Cells (MFCs)	• Energy recovery through microbial metabolism. • Simultaneous treatment and energy generation. • Potential for nutrient recovery and removal.	• Low power output and scalability. • Performance affected by environmental factors. • Limited to certain types of organic-rich sludge.	• Enhancement of MFC power output and efficiency. • Integration of MFCs in wastewater treatment plants.
Algae Cultivation	• Nutrient recovery through algal uptake. • Carbon sequestration potential. • Oxygen production and water-quality improvement. • Bioenergy production through algal biomass. • Potential for value-added products (e.g., pigments, bioplastics).	• High land and water requirements. • Algal species selection for specific wastewater. • Risk of algal blooms and operational challenges. • Need for harvesting and dewatering processes. • Limited application to certain wastewater streams.	• Exploration of mixed algal–bacterial systems for enhanced nutrient removal. • Integration of wastewater treatment with biofuel production. • Genetic modification of algae for improved biomass and nutrient uptake. • Optimization of cultivation conditions for higher yields. • Development of cost-effective algal harvesting techniques.

(Continued)

TABLE 5.1 (CONTINUED)
Innovative Treatment Technologies for the Circular Transformation of Urban Sludge

Technology	Advantages	Drawbacks	Future Outlooks
Hydrothermal Liquefaction (HTL)	• High conversion efficiency of sludge into bio-oil. • Production of biofuel and bioproducts from sludge. • Potential for nutrient recovery from liquid products. • Reduction of sludge volume and disposal. • Energy recovery from sludge and liquid products.	• High energy consumption for heating. • Complex system design and optimization. • Management of HTL reaction by-products. • Variability in product yields based on feedstock. • Limited large-scale demonstration projects.	• Integration with algae cultivation for combined biomass processing. • Development of catalysts and process conditions for improved bio-oil quality. • Valorization of bio-oil and biochar in various applications. • Investigation of co-HTL with other waste streams (e.g., food waste). • Assessment of lifecycle environmental impacts and techno-economic feasibility.
Advanced Oxidation Processes (AOPs)	• Efficient removal of recalcitrant organic compounds. • Degradation of emerging contaminants and pathogens. • Reduction of sludge volume and pollutant load. • Potential for resource recovery from treated sludge. • Versatile treatment for various wastewater matrices.	• High energy consumption and operational costs. • Formation of potentially harmful by-products. • Complex reactor design and optimization. • Need for effective catalysts and reaction conditions. • Lack of standardized guidelines for AOP implementation.	• Development of novel AOPs with improved efficiency and selectivity. • Combination of AOPs with other treatment processes for synergistic effects. • Integration of renewable energy sources to reduce energy footprint. • Exploration of AOPs for specific contaminants (e.g., pharmaceuticals). • Adoption of AOPs in decentralized and centralized wastewater treatment systems.

5.5 CASE STUDIES: CIRCULAR SLUDGE MANAGEMENT FOR WATER REMEDIATION

Real-world implementation of circular sludge management strategies serves as a testament to the practicality and effectiveness of circular economy principles in transforming urban sludge from a waste burden into a valuable resource. This section looks into two case studies, each highlighting a distinct city/region where circular sludge management practices have been successfully adopted, shedding light on their unique approaches and outcomes. Figure 5.4 shows a circular sludge management scheme in China.

5.5.1 Case Study 1: China – Successful Circular Sludge Management

In Northeast, East, Central, and North China, a paradigm shift in urban sludge management has been achieved through an integrated circular approach. Leveraging the principles of the circular economy, provinces in China have established a comprehensive system for sludge treatment, resource recovery, and reuse (Wei et al., 2020; MOHURD, 2019). This holistic approach encompasses advanced treatment technologies, stakeholder engagement, and innovative utilization pathways.

A central feature of this case study is the utilization of anaerobic digestion for organic matter degradation and biogas production. The produced biogas not only powers the digestor plant itself but also contributes surplus energy to the local grid. The resulting nutrient-rich digestate is skilfully converted into high-quality organic fertilizer, which not only minimizes waste but also enhances soil health and agricultural productivity in the surrounding region.

5.5.2 Case Study 2: Poland – Implementing Circular Practices

In Cracow, Lodz, Gdansk, Gdynia, Szczecin, and Kielce, in Poland, circular economy practices have been successfully integrated into urban sludge management, focusing on maximizing resource recovery and reducing environmental impact. This case study highlights the adoption of innovative treatment technologies, such as hydrothermal carbonization (HTC) and thermal hydrolysis, to valorize sludge into bioenergy, biochar, and other value-added products.

Of particular note is the city's strategic partnership with local industries to create a symbiotic relationship between sludge generation and resource demand. By providing industries with recovered materials from sludge, such as metals and biochar, the city has facilitated closed-loop material cycles and reduced the need for raw material extraction (Lipinska, 2018).

5.5.3 Lessons Learned and Best Practices

These case studies offer valuable insights into the key factors that contribute to the success of circular sludge management practices. Stakeholder collaboration, regulatory support, and technological innovation emerge as common threads. Furthermore,

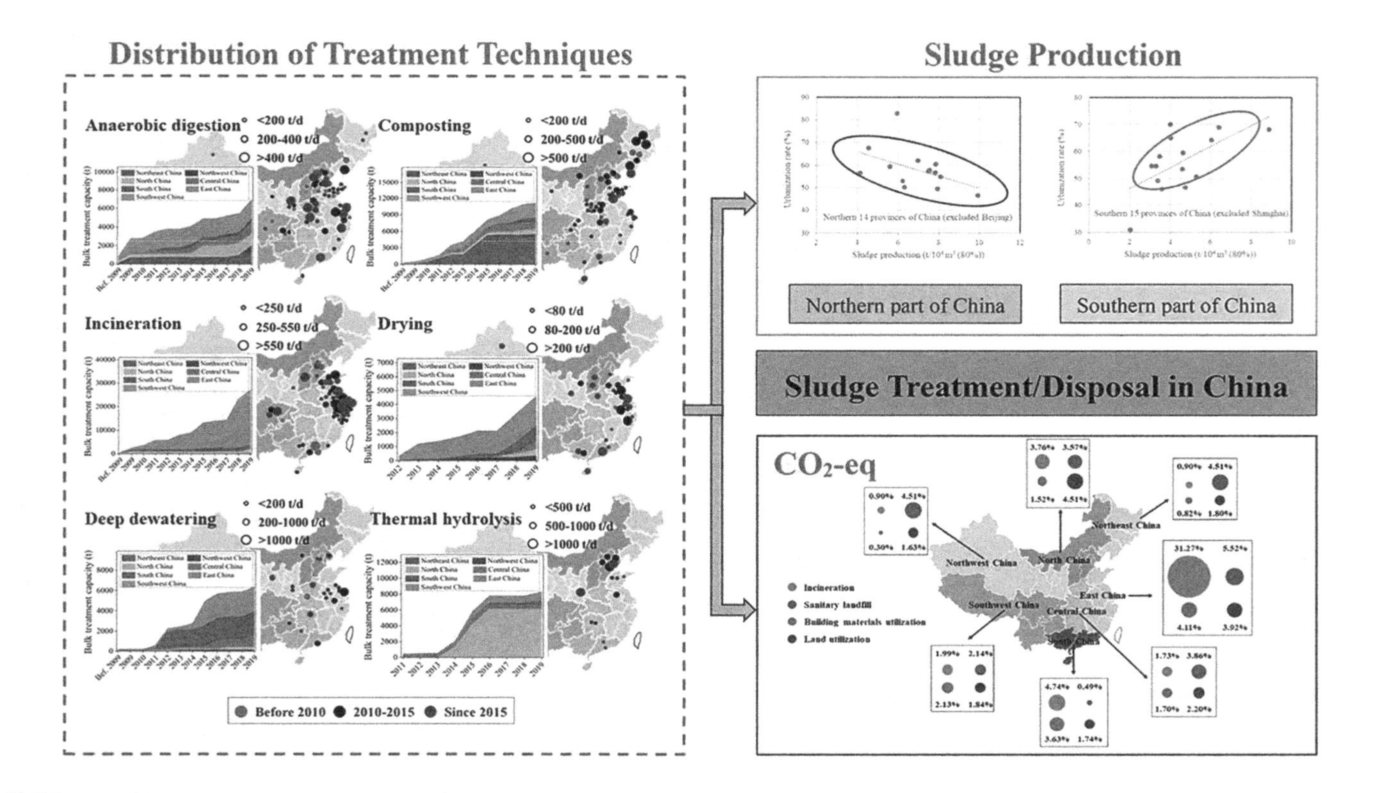

FIGURE 5.4 Circular sludge management in China (Wei et al., 2020).]

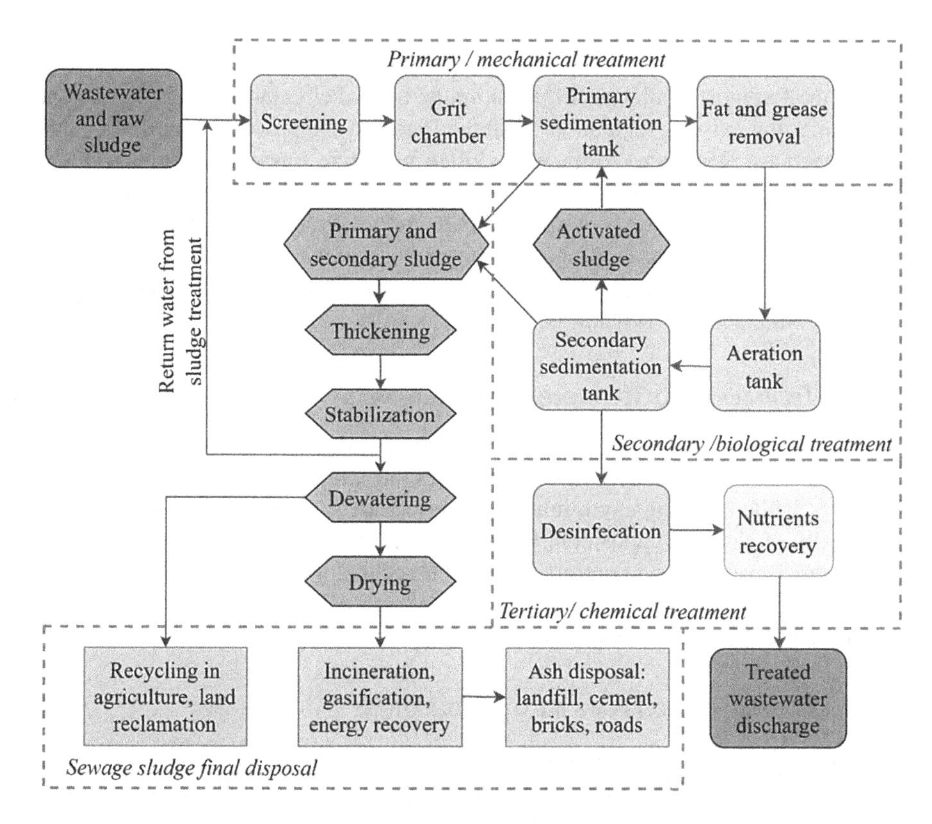

FIGURE 5.5 Wastewater and sewage sludge treatment methods in the EU (Đurdevi et al., 2022).

the experiences of Campos dos Goytacazes, Brazil, and Istanbul, Turkey, underscore the importance of tailoring circular economy solutions to the unique socio-economic and environmental context of each city (Areias et al., 2020; Guven et al., 2020) These case studies serve as inspiring models for cities and regions worldwide, showcasing the transformative potential of circular sludge management in water remediation. By embracing circular practices, these locales have not only mitigated the environmental impacts of urban sludge but also harnessed its latent value to contribute to a more sustainable and resource-efficient future as indicated in Figure 5.5.

5.6 CHALLENGES AND OPPORTUNITIES

As cities strive to incorporate circular economy principles into their urban sludge management strategies, they encounter a dynamic landscape of challenges and opportunities that necessitate a comprehensive understanding of the interplay between technical, regulatory, and economic dimensions. This section navigates through the multifaceted nature of challenges while shedding light on the promising opportunities that arise from embracing circular transformation in urban sludge management.

5.6.1 Regulatory and Policy Challenges

One of the foremost challenges in transitioning toward circular sludge management lies in aligning existing regulations with the innovative approaches required by circular practices. Regulatory frameworks often prioritize linear waste management models, which can hinder the adoption of circular strategies (Mayer et al., 2019; Mininni et al., 2015). Overcoming this challenge necessitates collaborative efforts between local authorities, environmental agencies, and stakeholders to advocate for regulatory flexibility and create an enabling environment for circular sludge management (Nikolaou and Tsagarakis, 2021).

5.6.2 Technical and Technological Challenges

The implementation of circular practices in urban sludge management demands the deployment of advanced treatment technologies that can effectively recover valuable resources while ensuring environmental safety (Barros et al., 2020). However, the complexity of sludge composition, variability, and contaminants presents technical challenges. Developing and optimizing reliable technologies that can handle diverse sludge characteristics and efficiently recover resources pose ongoing research and engineering challenges (Đurdevic et al., 2019; Kumar et al., 2021).

5.6.3 Economic and Financial Considerations

Circular sludge management represents a paradigm shift that requires initial investments in technology, infrastructure, and workforce training. While the potential for resource recovery and reduced waste disposal costs offers long-term economic benefits, the transition phase can strain municipal budgets. Addressing this challenge entails developing comprehensive cost–benefit analyses, exploring funding mechanisms, and incentivizing private sector participation to facilitate the economic viability of circular transformation (Zikakis et al., 2019).

5.6.4 Opportunities for Collaborative Initiatives

The challenges posed by circular sludge management are counterbalanced by compelling opportunities for collaboration across sectors and stakeholders. Circular economy principles inherently encourage collaboration between municipalities, industries, research institutions, and communities (Lu, 2019). By fostering partnerships, cities can leverage combined expertise, resources, and knowledge-sharing to collectively address challenges and develop innovative solutions for sustainable sludge management (Zheng et al., 2020).

These challenges also pave the way for innovative business models, public–private partnerships, and joint ventures that promote resource efficiency and environmental stewardship. Collaborative initiatives can yield synergistic benefits that extend beyond sludge management, contributing to broader urban sustainability goals, including reduced ecological footprints and enhanced resilience against environmental challenges (Abuşoğlu et al., 2017).

In conclusion, the challenges and opportunities presented by the circular transformation of urban sludge management underscore the complex and evolving nature of this endeavor. Addressing these challenges requires a concerted effort from policymakers, researchers, industries, and communities to collectively drive the adoption of circular practices, overcome obstacles, and capitalize on the promising prospects that lie ahead. As cities navigate this transformative journey, they stand to reap not only environmental and economic rewards but also contribute to the global transition toward a more sustainable and circular future.

5.7 FUTURE DIRECTIONS AND RESEARCH NEEDS

The realm of urban sludge management is continuously evolving, fueled by the dynamic nature of circular economy principles, advances in technology, and the ever-growing need for sustainable water remediation practices. As we look ahead, this section delves into the future directions that circular sludge management might take, the research gaps that warrant exploration, and the potential innovations that could shape the landscape of urban water remediation.

5.7.1 Emerging Trends in Circular Sludge Management

In the coming years, circular sludge management is poised to witness several pivotal trends that will redefine its practices. One of the most prominent trends is the integration of digital technologies and data-driven approaches. Smart monitoring systems, real-time analytics, and predictive modeling hold the promise of enhancing operational efficiency, optimizing resource recovery processes, and facilitating informed decision-making in sludge management. Furthermore, the convergence of circular practices with concepts like the Internet of Things (IoT) and Artificial Intelligence (AI) could unlock new avenues for resource optimization and waste reduction (Chang et al., 2020; Chen et al., 2019).

Another emerging trend is the emphasis on decentralized sludge treatment and resource recovery systems. By bringing treatment closer to the source of sludge generation, cities can minimize transportation costs, reduce environmental impacts, and enable efficient resource extraction at a local scale. Decentralization aligns well with circular economy principles, fostering a more resilient and adaptive approach to urban sludge management.

5.7.2 Research Gaps and Areas for Further Study

Despite significant progress, several research gaps persist in the domain of circular sludge management. One crucial area is the comprehensive assessment of environmental and health risks associated with circular practices. While resource recovery is a key objective, it is imperative to ensure that the resources recovered meet stringent quality standards and pose no threat to the environment or human health (Lu et al., 2019; Mailler et al., 2017). Research should also focus on lifecycle assessments, enabling a holistic understanding of the environmental footprint of circular sludge management systems compared with traditional linear approaches (Chen and Kuo, 2016).

Additionally, there is a need for a deeper exploration of socio-economic aspects related to circular sludge management. Stakeholder engagement, community perceptions, and socio-economic impacts require careful consideration to ensure the equitable distribution of benefits and the acceptance of circular practices among diverse urban populations (Elmi et al., 2020).

5.7.3 Potential Innovations and Advancements

The future of circular sludge management holds the promise of groundbreaking innovations that can revolutionize the field. Advanced treatment technologies, such as microbial fuel cells, anaerobic digestion, and thermal hydrolysis, are ripe for further research and development to enhance resource recovery efficiency and reduce energy consumption (Amesho et al., 2024; Fijalkowski et al., 2017).

Innovative approaches, like nanotechnology-enabled sludge treatment, biorefinery concepts, and the incorporation of circular principles into urban planning and design could pave the way for transformative advances in urban sludge management (Collivignarelli et al., 2019). Moreover, the integration of blockchain technology for transparent tracking of resource flows and quality could enhance the traceability and accountability of circular practices.

In conclusion, the path toward circular transformation in urban sludge management is one that brims with potential and possibilities. As emerging trends, research needs, and innovations converge, cities have the opportunity to drive the evolution of circular sludge management practices, realizing both environmental sustainability and resource efficiency goals. The journey ahead requires steadfast commitment to interdisciplinary research, collaboration, and knowledge-sharing to shape a future where circular economy principles seamlessly intertwine with water remediation practices, paving the way for resilient and harmonious urban environments.

5.8 CONCLUSION

In the face of burgeoning urbanization and escalating water pollution concerns, the realm of urban sludge management stands at a critical juncture, striving for innovative and sustainable solutions. The journey we have undertaken through this exploration of circular transformation in urban sludge management underscores the pivotal role that circular economy principles play in reshaping the landscape of water remediation. As we conclude this chapter, let us recapitulate the key points that have emerged, emphasize the profound importance of circular transformation, and extend a resolute call to action for the adoption of sustainable and circular sludge practices.

Throughout this discourse, we have looked into the multifaceted dimensions of urban sludge management, unraveled the diverse sources and characteristics of urban sludge, and navigated the intricate tapestry of circular economy principles within the context of water remediation. We have witnessed how circular transformation manifests in the recycling and recovery of valuable resources from sludge and explored innovative treatment technologies that hold the promise of efficient and sustainable resource extraction. Case studies have illuminated successful instances of circular sludge management, underscoring the tangible benefits of this approach. Moreover, we have examined the challenges that beset the path toward circularity, ranging from regulatory intricacies to economic considerations.

5.8.1 Importance of Circular Transformation in Urban Sludge Management

The significance of circular transformation in urban sludge management reverberates across environmental, economic, and societal spheres. By embracing circular economy principles, cities have the potential to transcend traditional linear approaches, unraveling a tapestry of benefits. Circular sludge management stands as a linchpin for achieving resource efficiency, minimizing waste generation, and alleviating the burden on overtaxed ecosystems. Notably, the recovery of valuable resources from sludge not only mitigates environmental impact but also ushers in economic gains and engenders a symbiotic relationship between water remediation and resource recovery.

5.8.2 Call to Action for Sustainable and Circular Sludge Practices

In the dawn of this transformative era, we extend a resounding call to action for industries, governments, researchers, and stakeholders alike to champion the cause of sustainable and circular sludge practices. The realization of a circular future hinges on collective dedication to innovation, collaboration, and knowledge dissemination. It is imperative that regulatory frameworks be calibrated to incentivize circular sludge management, streamlining pathways for implementation. Technological advancement must be underpinned by rigorous research, with an unwavering focus on environmental integrity, resource recovery efficiency, and public health safeguards.

Institutions and organizations must embrace a proactive ethos, fostering interdisciplinary collaborations and knowledge exchange to surmount the challenges that lie ahead. By forging partnerships that transcend traditional boundaries, we can expedite the deployment of cutting-edge treatment technologies, elevate stakeholder engagement, and weave circular sludge management into the fabric of urban planning and policy formulation.

As we stand on the cusp of a watershed moment in urban sludge management, let our collective resolve fuel the transition toward circularity. By recognizing the symbiotic relationship between water remediation and resource recovery, and by harnessing the potential of circular economy principles, we can chart a course toward a future where urban sludge is transformed from a burden into a valuable asset, and where sustainable water management thrives in harmony with the circular economy.

REFERENCES

Abuşoğlu, A., Özahi, E., Kutlar, A.I., Al-jaf, H., 2017. Life cycle assessment (LCA) of digested sewage sludge incineration for heat and power production. *J. Clean. Prod.* 142, 1684–1692. https://doi.org/10.1016/j.jclepro.2016.11.121.

Amesho, K.T.T., Edoun, E.I., Kadhila, T., Shangdiar, S., Iikela, S., Pandey, S., Chinglenthoiba, C., Lani, M.N., 2024. *Technologies to Convert Waste to Bio-oil, Biochar, and Biogas*. Waste Valorisation for Bioenergy and Bioproducts: Biofuel, Biogas, and Value-Added Products, Elsevier Press (2024), 63–90, https://doi.org/10.1016/B978-0-443-19171-8.00011-0

Areias, I.O.R., Vieira, C.M.F., Colorado, H.A., Delaqua, G.C.G., Monteiro, S.N., Azevedo, A.R.G., 2020. Could city sewage sludge be directly used into clay bricks for building construction? A comprehensive case study from Brazil. *J. Build. Eng.* 31, 101374.

Barros, M.V., Salvador, R., de Francisco, A.C., Piekarski, C.M., 2020. Mapping of research lines on circular economy practices in agriculture: From waste to energy. *Renew. Sustain. Energy Rev.* 131, 109958.

Chang, Z.Y., Long, G.C., Zhoua, J.L., Ma, C., 2020. Valorization of sewage sludge in the fabrication of construction and building materials: A review. *Resour. Conserv. Recy.* 154(1), 130–134.

Chen, H., Yan, S.H., Ye, Z.L., Meng, H.J., Zhu, Y.G., 2012. Utilization of urban sewage sludge: Chinese perspectives. *Environ. Sci. Pollut. R.* 19(5), 1454–1463.

Chen, M.X., Zheng, Y., Zhou, X.M., Li, L., Wang, S.D., Zhao, P.Q., Lu, L.C., Cheng, X., 2019. Recycling of paper sludge powder for achieving sustainable and energy-saving building materials. *Constr. Build. Mater.* 229, 116874.

Chen, Y.C., Kuo, J., 2016. Potential of greenhouse gas emissions from sewage sludge management: A case study of Taiwan. *J. Clean. Prod.* 129, 196–201.

Chojnacka, K., Skrzypczak, D., Szopa, D., Izydorczyk, G., Moustakas, K., Witek-Krowiak, A., 2023. Management of biological sewage sludge: Fertilizer nitrogen recovery as the solution to fertilizer crisis. *J. Environ. Manag.* 326, 116602.

Collivignarelli, M.C., Abba, A., Miino, M.C., Torretta, V., 2019. What advanced treatments can be used to minimize the production of sewage sludge in WWTPs. *Appl. Sci.* 9(13), 2650.

Domini, M., Bertanza, G., Vahidzadeh, R., Pedrazzani, R., 2022. Sewage sludge quality and management for circular economy opportunities in Lombardy. *Appl. Sci.* 12, 10391. https://doi.org/10.3390/app122010391.

Đurdevic, D., Blecich, P., Juric, Z., 2019. Energy recovery from sewage sludge: The case study of Croatia. *Energies* 12, 1927.

Đurdevic, D., Žikovic, S., Blecich, P., 2022. Sustainable sewage sludge management technologies selection based on techno-economic environmental criteria: Case study of Croatia. *Energies* 15, 3941. https://doi.org/10.3390/en15113941.

Elmi, A., Alkhaldy, A., Alolayan, M., 2020. Sewage sludge land application: Balancing act between agronomic benefits and environmental concerns. *J. Clean. Prod.* 150, 119512.

Fijalkowski, K., Rorat, A., Grobelak, A., Kacprzak, M.J., 2017. The presence of contaminations in sewage sludge – The current situation. *J. Environ. Manage.* 203, 1126–1136.

Gherghel, A., Teodosiu, C., De Gisi, S., 2019. A review on wastewater sludge valorisation and its challenges in the context of circular economy. *J. Clean. Prod.* 228, 244–263.

Guven, H., Tanik, A., 2020. Water-energy nexus: Sustainable water management and energy recovery from wastewater in eco-cities. *Smart Sustain. Built Environ.* 9, 54–57.

Hao, X., Qi Chen, Q., van Loosdrecht, M.C.M., Li, J., Jiang, H., 2020. Sustainable disposal of excess sludge: Incineration without anaerobic digestion. *Water Research* 170, 115298. https://doi.org/10.1016/j.watres.2019.115298.

Hao, X.D., Li, J., van Loosdrecht, M.C.M., Jang, H., Liu, R.B., 2019a. Energy recovery from wastewater: Heat over organics. *Water Res.* 161, 74e77. https://doi.org/10.1016/j.watres.2019.05.106.

Hao, X.D., Wang, X.Y., Liu, R.B., Li, S., van Loosdrecht, M.C.M., 2019b. Environmental impacts of resource recovery from wastewater treatment plants. *Water Res.* 160, 268e277. https://doi.org/10.1016/j.watres.2019.05.068.

Havukainen, J., Nguyen, M.T., Hermann, L., Horttanainen, M., Mikkilä, M., Deviatkin, I., Linnanen, L., 2016. Potential of phosphorus recovery from sewage sludge and manure ash by thermochemical treatment. *Waste Manag.* 49, 221–229.

Horttanainen, M., Deviatkin, I., Havukainen, J., 2017. Nitrogen release from mechanically dewatered sewage sludge during thermal drying and potential for recovery. *J. Clean. Prod.* 142, 1819–1826.

Khalili, A., Jamshidi, S., Khalesidoust, M., Vesali, M., 2017. Evaluation of sewage sludge for incineration (case study: Arak wastewater treatment plant). *Environ. Energy Econ. Res.* 1, 249e258. https://doi.org/10.22097/eeer.2017.47251.

Koskue, V., Freguia, S., Ledezma, P., Kokko, M., 2021. Efficient nitrogen removal and recovery from real digested sewage sludge reject water through electroconcentration. *J. Environ. Chem. Eng.* 9, 106286.

Kumar, V., Srivastava, S., Thakur, I.S., 2021. Enhanced recovery of polyhydroxyalkanoates from secondary wastewater sludge of sewage treatment plant: Analysis and process parameters optimization. *Bioresour. Technol. Rep.* 15, 100783.

Lamastra, L., Suciu, N.A., Trevisan, M., 2018. Sewage sludge for sustainable agriculture: Contaminants' contents and potential use as fertilizer. *Chem. Biol. Technol. Agric.* 5, 10.

Lipinska, D., 2018. The water-wastewater-sludge sector and the circular economy. *Comp. Econ. Res.* 21, 121–137.

Lu, J., 2019. Carbon footprint and reduction potential of Chinese wastewater treatment sector, Master thesis of University of Science and Technology of China (in Chinese).

Lu, J., Wang, X., Liu, H., Yu, H., Li, W., 2019. Optimizing operation of municipal wastewater treatment plants in China: The remaining barriers and future implications. *Environ. Int.* 129, 273–278.

Mailler, R., Gasperi, J., Patureau, D., Vulliet, E., Delgenes, N., Danel, A., Deshayes, S., Eudes, V., Guerin, S., Moilleron, R., 2017. Fate of emerging and priority micropollutants during the sewage sludge treatment: Case study of Paris conurbation. Part 1: Contamination of the different types of sewage sludge. *Waste. Manage.* 59(59), 379–393.

Mandal, S., Sundaramurthy, S., Arisutha, S. et al., 2023. Generation of bio-energy after optimization and controlling fluctuations using various sludge activated microbial fuel cell. *Environ Sci Pollut Res* 30, 125077–125087 . https://doi.org/10.1007/s11356-023-26344-3.

Mayer, F., Bhandari, R., Gath, S., 2019. Critical review on life cycle assessment of conventional and innovative waste-to-energy technologies. *Sci. Total. Environ.* 672, 708–721.

Mininni, G., Blanch, A.R., Lucena, F., Berselli, S., 2015. EU policy on sewage sludge utilization and perspectives on new approaches of sludge management. *Environ. Sci. Pollut. R.* 22(10), 7361–7374.

MOHURD, 2019. Chinese statistical yearbook of urban and rural construction (In Chinese). http://www.mohurd.gov.cn/xytj/tjzljsxytjgb/jstjnj.

Munasinghe-Arachchige, S.P., Cooke, P., Nirmalakhandan, N., 2020. Recovery of nitrogen-fertilizer from centrate of anaerobically digested sewage sludge via gas-permeable membranes. *J. Water Process Eng.* 38, 101630.

Munasinghe-Arachchige, S.P., Nirmalakhandan, N., 2020. Nitrogen-fertilizer recovery from the centrate of anaerobically digested sludge. *Environ. Sci. Technol. Lett.* 7, 450–459.

Nguyen, M.D., Thomas, M., Surapaneni, A., Moon, E.M., Milne, N.A., 2022. Beneficial reuse of water treatment sludge in the context of circular economy. *Environ. Technol. Innov.* 28, 102651. https://doi.org/10.1016/j.eti.2022.102651.

Nikolaou, I.E., Tsagarakis, K.P., 2021. An introduction to circular economy and sustainability: Some existing lessons and future directions. *Sustain. Prod. Consum.* 28, 600–609.

Pikaar, I., Guest, J., Ganigué, R., Jensen, P., Rabaey, K., Seviour, T., Trimmer, J., et al., 2022. *Resource Recovery from Water: Principles and Application.* Pikaar, I., Guest, J., Ganigué, R. and Paul, J. (eds). London: IWA Publishing.

Preisner, M., Smol, M., Horttanainen, M., Deviatkin, I., Havukainen, J., et al., 2022. Indicators for resource recovery monitoring within the circular economy model implementation in the wastewater sector. *J. Environ. Manag.* 304, 114261.

Raheem, A., Sikarwar, V.S., He, J., Dastyar, W., Dionysiou, D.D., Wang, W., Zhao, M., 2018. Opportunities and challenges in sustainable treatment and resource reuse of sewage sludge: A review. *Chem. Eng. J.* 337, 616–641.

Ramm, K., 2021. Considerations related to the application of the EU water reuse regulation to the production of snow from reclaimed water. *Circ. Econ. Sustain.* 2, 569–587.

Rorat, A., Courtois, P., Vandenbulcke, F., Lemiere, S., 2019. Sanitary and environmental aspects of sewage sludge management. In: Prasad, M.N.V., de Campos Favas, P.J., Vithanage, M. and Mohan, S.V. (eds) *Industrial and Municipal Sludge*, 1st ed. Oxford: ButterworthHeinemann, pp. 155–180.

Singh, V., Phuleria, H.C., Chandel, M.K., 2020. Estimation of energy recovery potential of sewage sludge in India: Waste to watt approach. *J. Clean. Prod.* 276, 122538. https://doi.org/10.1016/j.jclepro.2020.122538.

Smol, M., 2023. Circular economy in wastewater treatment plant—Water, energy and raw materials recovery. *Energies* 16, 3911. https://doi.org/10.3390/en16093911.

Wan, C., Ding, S., Zhang, C., Tan, X., Zou, W., Liu, X., Yang, X., 2017. Simultaneous recovery of nitrogen and phosphorus from sludge fermentation liquid by zeolite adsorption: Mechanism and application. *Sep. Purif. Technol.* 180, 1–12.

Wang, H., Xiao, K., Yang, J., Yu, Z., Yu, W., Xu, Q., Wu, Q., Liang, S., et al., 2020. Phosphorus recovery from the liquid phase of anaerobic digestate using biochar derived from iron–rich Sludge: A potential phosphorus fertilizer. *Water Res.* 174, 115629.

Wei, L., Qin, K., Ding, J., Xue, M., Yang, C., Jiang, J., Zhao, Q., 2019. Optimization of the co-digestion of sewage sludge, maize straw and cow manure: Microbial responses and effect of fractional organic characteristics. *Sci. Rep.* 9(1), 1e10.

Wei, L., Zhu, F., Li, Q., Xue, C., Xia, X., Yu, H., Zhao, Q., Jiang, J., Bai, S., 2020. Development, current state and future trends of sludge management in China: Based on exploratory data and CO_2-equivaient emissions analysis. *Environ. Int.* 144, 106093. https://doi.org/10.1016/j.envint.2020.106093.

Werle, S., Sobek, S., 2019. Gasification of sewage sludge within a circular economy perspective: A polish case study. *Environ. Sci. Pollut. Res.* 26, 35422–35432.

Yoshida, H., ten Hoeve, M., Christensen, T.H., Bruun, S., Jensen, L.S., Scheutz, C., 2018. Life cycle assessment of sewage sludge management options including longterm impacts after land application. *J. Clean. Prod.* 174, 538e547. https://doi.org/10.1016/j.jclepro.2017.10.17.

Zhang, F., Huang, X.Y., Shi, Y., Zhong, J.M., Ding, Z.H., 2019a. Study on deep dewatering and resource utilization of sludge in municipal sewage treatment plant. In: Wu, F. and Zhou, P. (eds) *Advances in Energy Science and Environment Engineering* Iii. Springer.

Zhang, Q., Yang, W.N., Ngo, H.H., Guo, W., Jin, P., Dzakpasu, M., Yang, S., Wang, Q., Wang, X.C., Ao, D., 2016. Current status of urban wastewater treatment plants in China. *Environ. Int.* 92, 11–22.

Zhao, X.D., Yang, J.Z., Tu, C.Q., Zhou, Z., Wu, W., Chen, G., Yao, J., Ruan, D.N., Qiu, Z., 2019. A full-scale survey of sludge landfill: Sludge properties, leachate characteristics and microbial community structure. *Water Sci. Technol.* 80(6), 1185–1195.

Zheng, X., Liu, T., Guo, M.H., Li, D., Gou, N., Cao, X., Qiu, X.P., Li, X.L., Zhang, Y.Z., Sheng, G.P., Pan, B.Z., Gu, A.Z., Li, Z.B., 2020. Impact of heavy metals on the formation and properties of solvable microbiological products released from activated sludge in biological wastewater treatment. *Water Res.* 179, 115895.

Zhu, Y., Zhao, Q., Li, D., Li, J., Guo, W., 2022. Performance comparison of phosphorus recovery from different sludges in sewage treatment plants through Pyrolysis. *J. Clean. Prod.* 372, 133728.

Zikakis, D., Chauzy, J., Droubogianni, I., Georgakopoulos, A., 2019. Why applying THP on waste activated sludge makes sense: Psyttalia – Athens case study. *Water Pract. Technol.* 14(4), 921–930.

6 Unveiling the Ecological Footprint of Pharmaceuticals and Personal Care Products (PPCPs) as Emerging Contaminants and Management Strategies

Aditi Mishra
Department of Botany, University of Allahabad, Prayagraj, Uttar Pradesh, India

Sneha Rai
Department of Botany, University of Allahabad, Prayagraj, Uttar Pradesh, India

Girjesh Kumar
Department of Botany, University of Allahabad, Prayagraj, Uttar Pradesh, India

Suvendu Manna
Sustainability Cluster, UPES, Dehradun, Uttarakhand, India

Vaibhav Srivastava
Department of Botany, University of Allahabad, Prayagraj, Uttar Pradesh, India

DOI: 10.1201/9781003432869-6

6.1 INTRODUCTION

The rapid advances in resource development, pharmaceutical & healthcare systems, and technology have led to the generation of a great variety of chemicals and compounds. Unfortunately, this has also resulted in the identification of numerous substances that can potentially harm the environment and living organisms. The growing presence of these hidden compounds poses a significant threat to our ecosystem. Emerging contaminants (ECs) refer to a diverse and expanding range of human-made substances that have been recently detected as significant water pollutants despite not having been commonly recognized in water sources until more recently (Gomes et al., 2018, 2020; Khan et al., 2019; Ouda et al., 2021). Examples of such substances include pharmaceuticals and personal care products (PPCPs), flame retardants, surfactants, pesticides, plasticizers, nanomaterials, additives, and chemicals suspected to disrupt the endocrine system ("endocrine disrupters"). These compounds often remain unmetabolized and are released into sewers and water treatment plants, further exacerbating their environmental risks. PPCPs include a wide range of substances, such as prescription drugs and over-the-counter medications, dietary supplements, soaps, shampoos, diagnostic agents, and fragrances. These substances, in various forms including intact, partially metabolized, and fully metabolized residues, enter the environment primarily through human excretion. Personal care products are also released into the environment through activities like cleaning, grooming, and bathing (Richardson et al., 2005; Ellis, 2006). Pharmaceutical products, extensively utilized in human healthcare and animal husbandry, possess targeted biological effects. However, they also pose a significant environmental concern. Whether from patients taking medications or animals receiving veterinary drugs, pharmaceutical contamination stems primarily from these sources. So, it is crucial to explore ways to mitigate the impact and safeguard our ecosystems from these unintended prescriptions for pollution (Zuccato et al., 2006). From 2008 to the present, pharmaceuticals have received increasing attention as potential bioactive chemicals in the environment. Pharmaceuticals, a long-standing presence in water, are now being recognized as potential hazards to ecosystems (Fent et al., 2006). While their environmental levels have only recently been quantified and acknowledged, it is crucial to shed light on these hidden dangers and take proactive measures to safeguard our precious ecosystems. So, it is crucial to delve into the depths and address the emerging challenges of pharmaceutical contamination. Understanding ECs is crucial for several reasons, as their widespread occurrence and persistence raise concerns about their potential to disrupt ecological balances and harm sensitive organisms. They can accumulate in the food chain, leading to bioaccumulation and biomagnification, posing risks to both wildlife and human consumers. Moreover, they have been detected in water bodies, soils, sediments, and even the atmosphere, highlighting their extensive distribution and the need for effective monitoring and management strategies. Their effects can be far-reaching and multifaceted. Ecologically, they can disrupt the reproductive, physiological, and behavioral processes of aquatic and terrestrial organisms, potentially leading to population declines and biodiversity loss. Human health can also be affected, as exposure to specific contaminants has been associated with endocrine disruption, carcinogenicity, developmental abnormalities, and other adverse health outcomes. Furthermore, their presence in water sources presents challenges for providing safe drinking water to communities worldwide. Against these backdrops, the present chapter critically assesses different aspects of PPCPs, including origin sources, fates, risk assessments, and various management strategies.

6.2 EMERGING CONTAMINANTS (ECS): OCCURRENCE AND PHYSIOCHEMICAL BEHAVIOR

Certain compounds that are not fully regulated or are even unregulated but have substantial drawbacks for human health and the environment are referred to as emerging contaminants (ECs). These ECs are primarily found in wastewater originating from various industrial sectors. Extensive literature monitoring has revealed the presence of a variety of ECs in the environment and water sources. These include UV filters, stimulants, antibiotics, artificial sweeteners, repellents, plasticizers, and many more. Pharmaceutical-based ECs significantly impact the environment, particularly in aquatic ecosystems. Some ECs, such as anti-inflammatory drugs, ß-blockers, and antidepressants, are present in higher concentrations, posing challenges for complete removal due to their molecular properties and resistance to biological degradation (Petrie et al., 2013). Therapeutic groups most frequently detected in water are (i) anti-inflammatories and analgesics like paracetamol, ibuprofen, and diclofenac; (ii) antidepressants particularly benzodiazepines; (iii) antiepileptics specifically carbamazepine; (iv) lipid-lowering drugs such as fibrates; (v) ß-blockers like atenolol and metoprolol; (vi) antiulcer drugs and antihistamines like ranitidine and famotidine; (vii) antibiotics covering tetracyclines, b-lactams, penicillins, sulfonamides, chloramphenicol, and imidazole derivatives; (viii) other substances including cocaine, barbiturates, amphetamines, opiates, and other narcotics. Notably, the majority of pharmaceuticals have a molecular mass below 500 Da.

Pharmaceuticals possess distinct characteristics that set them apart from conventional industrial chemical contaminants .Pharmaceuticals are composed of complex biomolecules, exhibiting significant variability in structure, functionality, and shape. Unlike typical industrial chemicals, pharmaceuticals encompass a wide range of molecular properties due to their diverse therapeutic purposes. Pharmaceuticals are also polar molecules, often containing multiple ionizable groups. The pH of the surrounding medium influences the degree of ionization and its properties. Additionally, pharmaceuticals exhibit lipophilic properties, meaning they have an affinity for lipid-based substances, while some also possess moderate solubility in water. This combination of polar and lipophilic characteristics contributes to their unique behavior in various biological and environmental systems. Certain pharmaceuticals exhibit a notable persistence in the environment, exceeding a year or more. Pharmaceuticals such as erythromycin, cyclophosphamide, and sulfamethoxazole can persist for an extended period. Furthermore, some pharmaceuticals, including clofibrinic acid, can last several years, accumulating and becoming biologically active. This prolonged persistence and the potential for biological activity contribute to the concern surrounding their environmental impact. Once administered, pharmaceutical molecules undergo absorption, distribution, and metabolism within the body. These metabolic reactions can lead to modifications in their chemical structure, resulting in metabolites with varying properties. This metabolic transformation plays a vital role in determining the efficacy, toxicity, and elimination of pharmaceuticals from the body.

In summary, pharmaceuticals possess several unique characteristics that differentiate them from conventional industrial chemical contaminants. These include their diverse and complex molecular nature, polar and lipophilic properties, potential environmental persistence, and susceptibility to metabolic transformations within the body. Understanding these distinctive features is essential for assessing their environmental fate and potential risks, and designing effective strategies for their management. On the other hand, due to their extensive applications and usage, PPCPs, preservatives, antimicrobial agents, and sunscreen agents are found at high concentrations of >1000ng/L. Physicochemical properties play an important role in describing the physical form and behavior of various ECs. The mobility of these contaminants is heavily influenced by the phase in which they are present in the environment. For instance, certain organic contaminants, like toluene, benzene, and polychlorinated biphenyls, are insoluble in aqueous media and persist for longer periods than contaminants miscible with water. The fate and transport of pharmaceutical contaminants are significantly influenced by various physicochemical properties, such as solubility, melting temperature, chemical structure, evaporation rate, boiling point, sorption potential, volatility, and the reactivity of inorganic contaminants to form complexes.

6.3 POTENT SOURCES AND SIGNIFICANT ROUTES

The specific characteristics of different geographic zones, including topography, climate, and land-use patterns, contribute to unique environmental risks and the presence of specific types of contaminants in those regions. The production and environmental impact of contaminants are closely tied to their usage and disposal practices. Contaminants can be generated through various sources, including wastewater treatment plants, soil, medical waste, manure, improper handling of raw materials, animal husbandry, and industrial activities. Municipal wastewater treatment plants, for instance, serve as a significant source of pharmaceutical contaminants since they are primarily made to remove biodegradable nitrogenous, phosphorous, and carbon compounds. Consequently, numerous pharmaceutical contaminants reach the environment through these treatment plants, resulting in the presence of pharmaceuticals and personal care products (PPCPs) in the water systems of various countries, including the USA, Japan, Spain, Finland, and the UK. These PPCPs are typically found in concentrations varying from nanograms per liter (ng/L) to micrograms per liter (μg/L), as reported in studies conducted in those countries (Lindqvist et al., 2005). Among the pharmaceutical contaminants, antibiotics hold particular significance. They enter the environment through human and animal excreta, specifically urine and feces, ultimately finding their way into water bodies. This pathway of contamination highlights the role of both human and animal populations in contributing to the presence of antibiotics in the environment (Tran et al., 2015).

Antimicrobial, as well as antifungal agents like triclocarban and miconazole, are commonly found in household items such as soaps, shampoos, and toothpaste. These compounds enter the domestic sewage system when these products are used and discarded, eventually contaminating aquatic environments. This can pose risks to aquatic ecosystems and contribute to the development of antimicrobial resistance. Proper disposal, effective wastewater treatment, and the use of environmentally friendly alternatives are important for reducing their environmental impact. Likewise, organic UV filters, such as oxybenzone, 4-methylbenzylidene camphor, and octocrylene, commonly found in various personal care products including sunscreen, have become significant pollutants in the aquatic environment. These chemicals can enter the aquatic biome through direct usage and indirect pathways. Their widespread presence in products like cosmetics, lotions, sprays, and sunscreens highlights the need for increased awareness and measures to mitigate their impact on aquatic ecosystems. When we swim or bathe in lakes and rivers and engage in activities like washing off, laundering, and showering, we unknowingly introduce organic UV filters like 4-methylbenzylidene camphor, oxybenzone, and octocrylene into the aquatic realm. These indirect routes of contamination contribute to the presence of these contaminants in our water sources, emphasizing the importance of mindful actions to protect our aquatic ecosystems (Tsui et al., 2014). Modern research has revealed that engineered nanomaterials, although difficult to detect, can exhibit contaminant-like behavior (Bour et al., 2015). These contaminants and other pollutants enter the environment and aquatic ecosystems via various pathways, including sewage treatment, industrial effluents, leaching from landfills, and precipitation.

6.3.1 Pharmaceutical Industries

Pharmaceuticals pose a growing concern as contaminants, with traces of these substances found in water sources worldwide (Patel et al., 2019). Pharmaceuticals are a diverse group of chemicals extensively used for diagnostics and treatments (Patel et al., 2019; Tran et al., 2015)). The pharmaceutical industry is recognized as a significant source of emerging contaminants, including counterfeit medicines, hormones, and antibiotics. These emerging contaminants represent a critical environmental challenge. The significance of these contaminants stems from their occurrence in aquatic ecosystems, spoilage of freshwater sources, and potential implications for the environment and human well-being (Sharma et al., 2019). Pharmaceuticals enter the environment through various pathways, including human waste, improper disposal of medications, and their use in agriculture. In intensive animal farming, medications can indirectly impact the environment by applying manure and slurry as organic fertilizers, ultimately transferring these contaminants to living organisms via the food chain. This highlights the interconnectedness of pharmaceutical contamination and its potential consequences for ecosystems and human exposure. Pharmaceuticals commonly detected in drinking water and wastewater encompass a range of substances such as fluoxetine, lipid-lowering drugs, antacids, diclofenac, steroids, analgesics, antibiotics, anti-inflammatory drugs, stimulants, oral contraceptives, nitroglycerin, and propranolol (Naushad et al., 2019; Snyder, 2008). The occurrence of these pharmaceuticals in water sources underscores the need for monitoring and addressing their potential impacts on both environmental and human health. Illegal drugs have been detected in aquatic environments since 2004, appearing

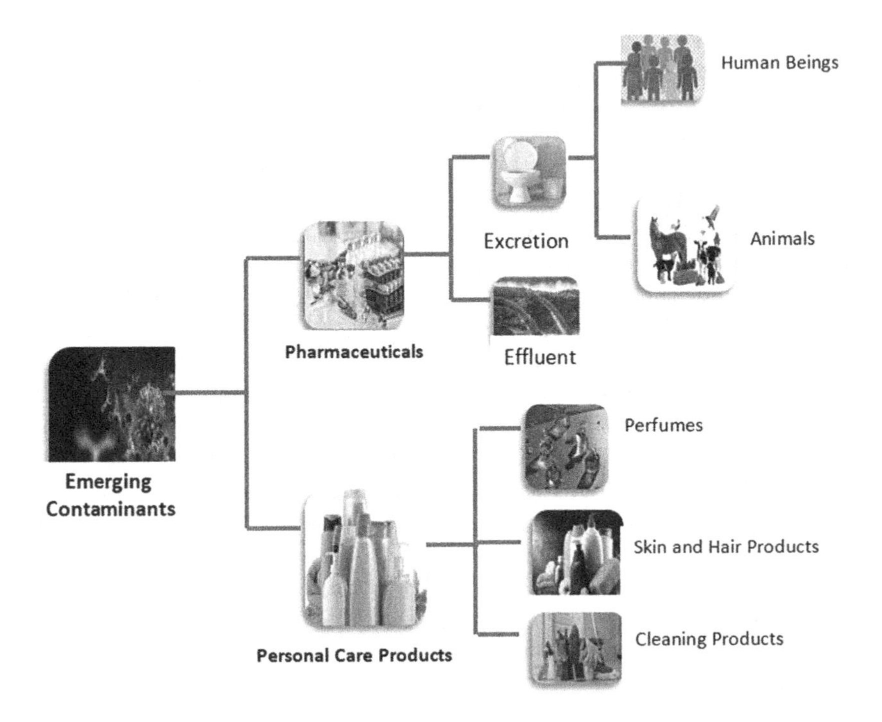

FIGURE 6.1 Routes of emerging contaminants (Khan et al., 2022).

in groundwater, wastewater, and surface water in the following years. Triclosan, a common antibiotic in toothpaste and soaps, has also been observed in the air. The environmental impact of hormones is noteworthy due to their ability to exert androgenic and estrogenic effects on biodiversity, making them significant contaminants (Behera et al., 2011; Naushad and Alothman, 2015). The presence of these substances in aquatic ecosystems highlights their potential implications for the environment and the need for monitoring and mitigation measures.

6.3.2 Personal Care Products

Personal care products (PCPs) encompass various chemicals used for various purposes, including nutrition, cleaning, and beauty. This category includes cosmetics, skincare, haircare, cleaning products, and fragrances. These products are extensively utilized worldwide, resulting in their continuous release into the atmosphere in substantial quantities (Kim et al., 2016). Individual consumer behavior contributes to the growing number of emerging contaminants released in the environment each day, reflecting an exponential increase in the usage of diverse consumer goods, particularly in the beauty and personal care sector (Kasprzyk-Hordern et al., 2009). The widespread use and disposal of PCPs emphasize the importance of understanding their environmental impact and exploring sustainable alternatives.

Most of these substances are bioactive and are categorized according to their capacity to persist and bioaccumulate, posing risks to both human health and the environment (Juliano and Magrini, 2017). Emerging contaminants, commonly found in skincare products, have the potential to reach hazardous levels in the environment, posing a public safety concern (Sharma et al., 2019; Wang and Wang, 2016). Cosmetics have been identified as a significant source of emerging contaminants, with ZnO and TiO_2 nanoparticles being prominent potential pollutants. The following emerging pollutants are commonly found in personal care products: microplastics, TiO_2 nanoparticles, polydimethylsiloxane, UV filters, insect repellents, disinfectant contaminants like triclosan, fragrance contaminants like tonalide, phantolide, preservatives such as diethyl phthalate, ZnO nanoparticles, octinoxate parabens, and butylparaben (Dhanirama et al., 2012). Polycyclic musks are commonly utilized as cost-effective fragrances in a variety of personal care products, including shampoos, hair products, laundry detergents, and cosmetics (Abedi et al., 2018). Their external application on the skin surface increases the likelihood of their presence in larger quantities in the environment in unmetabolized forms (Gani and Kazmi, 2017). Since cosmetics are used regularly, even trace amounts of these contaminants can pose risks to humans and the entire ecosystem, underscoring the importance of addressing their presence and potential impacts.

6.3.3 Transformation Products and Metabolites

When exposed to natural and engineered environments, emerging contaminants can undergo various transformative processes like biodegradation, oxidation, hydrolysis, reduction, and photolysis. These processes reduce contaminant levels but can also give rise to transformation products (TPs) that can be even more toxic than the original compounds. This raises concerns about the ecological risks posed by both the parent compounds and their transformation products (Escher and Fenner, 2011). For example, certain pharmaceuticals, like sulfamethazole, desvenlafaxine, and estrogens, can become biologically active again after being excreted as conjugated metabolites. Likewise, the substances formed during both the intermediate and final stages of bisphenol A (BPA) oxidation during persulfate oxidation treatment have been seen to be more toxic than BPA itself (Olmez et al., 2013).

The generation and characteristics of TPs from emerging contaminants depend on the environment and the pollutant's properties. For example, the production of perfluoroodanoic acid (PFOA) and perfluorooctanesulfonic acid (PFOS) in wastewater treatment plants (WWTPs) from their precursor substance, perfluoroalkyl acid (PFAA) is influenced by treatment methods and temperature (Guerra et al., 2014). Research on metabolites and degradation products, especially in pharmaceuticals, has significantly increased in recent years. Disinfection by-products (DBPs) are a special type of TP formed when disinfectants react with natural organic matter in water, leading to health concerns. Currently, only a small fraction of present DBPs is monitored by environmental agencies.

6.4 RAMIFICATIONS AND FATE OF EMERGING CONTAMINANTS

Emerging contaminants (ECs), naturally occurring or synthetically, have been linked to potential health and environmental drawbacks. The lack of regulatory

TABLE 6.1
Characteristics of Pharmaceuticals, Personal Care Products (PPCPs), and Pesticides

Pharmaceutical Types	Emerging contaminants(ECs)	Molecular weight(Mw)	Octanol coefficient (Log K_{ow})	Acid dissociation constant, pKa	Influent concentration in ng L^{-1}	Effluent concentration in ng L^{-1}
Analgesics and Anti-inflammatories	Ibuprofen	206.3	3.97	4.91	2265	40
	Diclofenac	296.1	0.7	4.2	131	24
	Paracetamol	151.16	046–0.49	9.86	1746–43223	25–4319
Antibiotics	Azithromycin	749	4.02	7.34	2020	60.1–980
	Metronidazole	171.6	–0.02	2.38	42	28
Endocrine Disrupters	Propylparaben	180.2	3.04	7.9	2180	19.7
	Methylparaben	152.15	1.96	8.4	334	11
Anti-inflammatories	Ketorolac	255.27	2.15	4	800	504
Hormones	Estradiol	272.2	4.01	10.46	4	0
	Estrone	270.4	3.13	10.34	47	6
	Estriol	288.4	2.8	10.38	415	0
Psychiatric drugs	Caffeine	194.19	–0.07	10.4	2349	18
	Diazepam	284.7	3.08	3.4	76	1
β-blockers	Propranolol	259.3	3.48	9.45	9.56	8.3
Pesticides	DDT	215.7	2.61	1.7	0.025	41.3
	Metolachlor	283.8	3.13	–1.34	530	15
Biocides	Hydroxybiphenyl I	170.21	3.27	9.8	700	100
Bactericides	Triclosan	289.6	4.76	7.8	10	734
Sunscreen ingredients	Oxybenzone	228.2	3.79	-	2616.8	772
	Octocrylene	361.5	6.9	-	-	153
Fragrances	Galaxolide	258.4	5.9	-	4839.3	4721
	Musk xylene	297.3	4.4	-	83.3–120	25–36

Source: Khan et al.(2022)

limits for these contaminants in wastewater exacerbates the environmental and health risks they pose. These contaminants can enter the food chain through wastewater in agriculture, directly impacting humans and animals (Vithanage et al., 2014). Many emerging contaminants and their derivatives exhibit poor biodegradability, negatively impacting environmental ecosystems (Bilal et al., 2019). Therefore, to evaluate the environmental toxicity of emerging contaminants (ECs), monitoring of their effects on the water bodies and air, using explicit indicators, is crucial. The crustacean *Daphnia magna* is an organismwidely usedfor evaluating environmental toxicity due to its mobility (Bilal et al., 2018). In this bioassay, ECs with concentrations between 10 and 100mg/L are classified as harmful, 1 to 10mg/L as toxic, and <1mg/L as highly toxic to aquatic life (Cleuvers, 2003). However, this method has limitations because of the range of species used for testing and various toxicological endpoints. Despite all these limitations, this bioassay provides a straightforward means of evaluating the toxicity potential of tested ECs. Certain anti-inflammatory pharmaceuticals (like acetaminophen, ibuprofen, and carbamazepine) have been found to be highly toxic to marine life. The occurrence of these antibiotics in water significantly impacts the activity of microbes and the effectiveness of sewage treatment (Gonzalez-Martinez et al., 2014).

Surface water has been the primary focus of research on emerging contaminants (ECs) due to the high EC concentration, and the diverse nature and susceptibility of surface waters to environmental degradation. ECs pose significant environmental challenges, including bacterial toxicity, impacts on crops, wild animals, invertebrates, and fish, and the suppression of fungal, algal, and bacterial growth (Maurer-Jones et al., 2013). These challenges highlight the need for comprehensive understanding and effective mitigation strategies to address the environmental impacts of ECs on various ecosystems and organisms. These ECs have been linked to various adverse consequences, as outlined below:

a) Immune and reproductive system disruption in Baltic seals exposed to organochlorine, leading to significant changes.
b) Gonadal development alterations and eggshell degradation in birds exposed to dichloro-diphenyl-trichloroethane (DDT), affecting their reproductive health.
c) Disruption of reproductive endocrine function in aquatic organisms exposed to ECs present in industrial sewage.
d) Decline in marine gastropod populations due to masculinization, impacting their reproductive capabilities.

These examples highlight the far-reaching effects of ECs on different species and emphasize the importance of addressing and mitigating their effect on the environment and organisms.

6.5 IMPACT ON HEALTH ISSUES

A wide range of compounds in the environment act as contaminants, posing risks to both human health and the ecosystem. These contaminants include everyday substances such as pharmaceuticals, perfluorinated compounds, nanomaterials, and gasoline additives (Richardson and Ternes, 2014). The most concerning health effects associated with these contaminants, known as PPCPs, include disruptions to growth activities, the hypothalamus-pituitary-thyroid axis development, and reproductive mechanisms in both humans and animals (Corsini et al., 2014). These concerns highlight the urgent need to address and mitigate the impact of these contaminants on human and environmental well-being. Perfluorinated compounds (PFCs), including perfluoro-octanesulfonate (PFO), pose significant dangers as emerging contaminants. Their association with cancer and liver adenomas highlights their toxic potential (Klaunig et al., 2012). Moreover, PFCs have harmful effects on human health, such as infertility, reduced sperm production, and thyroid complications (Shrestha et al., 2015). The presence of PFCs in the environment necessitates immediate attention because of their detrimental impact on both human well-being and ecosystem integrity.

MTBE, a widely used gasoline additive, poses significant risks as an emerging environmental pollutant. Its association with kidney cancer, uterine issues, and elevated blood urea nitrogen levels raises concerns for human health (de Peyster et al., 2003). Additionally, MTBE has detrimental effects on lung health, including the potential to cause cancer. Despite ongoing research efforts, our understanding of the interactions and impacts of certain emerging contaminants on the environment remains incomplete. Vigilance is necessary to address the potential threats posed by these substances and further enhance our knowledge in this field.

6.6 RISK ASSESSMENT

Emerging contaminants are representing a global issue presently. Therefore, it is important to have accurate knowledge about them in order to eliminate their associated risks. There is a shortage of knowledge pertaining to the toxicological effects of ECs and their environmental fate and transport. Ongoing studies are a useful source of knowledge and have shed light on the harmful effects of various ECs on biota, but they also reveal a number of shortcomings:

a) Lack of knowledge regarding the behavior, toxicity, and fate of these contaminants in the environment.
b) Insufficient analytical techniques for EC identification.
c) Insufficient knowledge on their effect on the health of animals and humans.

6.6.1 Monitoring

Advanced analytical methods must be used to develop techniques to quantify these emerging contaminants, including research that would help us to know the possible pathways for human and animal exposure. Monitoring their fate in the environment would be a better option to know the consequences and, for this, visualizing the combined effect of several contaminants is an important aspect. The use of biosensors and targeted ultra-performance liquid chromatography (UPLC) can be one of those methods (Naidu et al., 2016).

6.6.2 Regulation

Policy compliance in terms of production, application, and disposal is mandatory. These policies can target long-term as well as short-term management methods, providing appropriate methods for remediation. Source minimization and appropriate disposal methods should be adopted to minimize the potential hazard of ECs. In order to suppress synergistic interactions, heterogeneous organic contaminants should not be mixed. Also, switching to green alternatives and better treatment methods should be adopted (Stuart et al., 2012)

6.7 RISK MITIGATION BY TREATMENT

6.7.1 Phase-changing Technologies

To make the contaminants easier to handle and treat, they can be moved from one phase to another, or, to state it another way, from liquid to solid (Mirasole et al., 2016). These techniques are used to remove a range of contaminants. Many methods falling under this heading are used individually or in combination to separate these contaminants.

6.7.2 Activated Carbon

The amorphous form of carbon known as activated carbon exhibits significant internal porosity and, as a result, strong absorptivity. The surface area and porosity of activated carbon are key determinants of its adsorption capability. The amount of chemical adsorbed per unit mass (or volume) of the adsorbent is known as its adsorption capacity. The use of activated carbon, a versatile adsorbent to remove contaminants, is one of the most popular techniques. Using activated carbon provides approximately 90% efficiency in removing contaminants (Baccar et al., 2012). Adsorption-based systems can be used along with additional treatment methods in a logical order to get appropriate results according to our requirements. As an illustration, the removal of contaminants has been achieved, using a combination of three different ways (activated carbon, ultrafiltration, and coagulation) (Acero et al., 2012).

6.7.3 Biochar

Biochar, a charcoal-based substance created by pyrolyzing biomass at high temperatures under low-oxygen conditions, is utilized for the purpose of the adsorption of harmful substances. The conditions under which biochar is produced determine how well pollutants are adsorbed. Despite sharing certain similarities with activated carbon in terms of properties, biochar has not surpassed it in linked systems. Although biochar is very similar in properties to activated carbon, it cannot easily be coupled with other methods.

6.7.4 Clay

Adsorbents made up of clay minerals and their modified derivatives can be utilized to remove the majority of chemical pollutants from aqueous solutions. Clay materials have unique qualities due to their small particle sizes and distinctive crystal structures. These characteristics include limited permeability, swelling behavior, catalytic abilities, cation exchange capabilities, and plastic behavior when wet. For example, bentonite clay, an abundant and inexpensive adsorbent, cannot extract harmful heavy metal ions from water (Wahab et al., 2019). Depending on the particular organic makeup, a single type of clay may generate varied removal efficiencies. It is evident that the properties of the adsorbent material determine how effective the removal process will be. The presence of iron, nitrogen, or other minerals may also affect the rate. In the case of metal oxide-pillared clays, another fascinating aspect is the development of coupled processes that make use of the adsorptive properties of clay in conjunction with ions capable of producing reactions within the porous matrix. This porous matrix, which can range in size between 1.49 and 3.2 nm, gives the clay mineral enough room to perform catalysis (Tireli et al., 2014).

6.7.5 Carbon Nanotubes

Carbon nanotubes (CNTs) are extended cylindrical molecules made up of a hexagonal arrangement of hybridized carbon atoms. They can be created by rolling up either a single sheet of graphene (single-walled carbon nanotubes) or several graphene sheets (multi-walled carbon nanotubes).

The surface area significantly impacts how well carbon nanotubes remove contaminants. As the surface area of nanotubes changes depending on whether single- or multi-walled structures are present, this can result in different clearance rates for the same pollutant. Combining this method with adsorptive properties with other reactive nanoparticles is crucial, but this can provide us with a better and more advanced methodology (Rodriguez-Narvaez et al., 2017).

6.7.6 Biological Processes

Organic compounds can be totally changed into less harmful materials using an attractive process known as biological treatment; it is regarded to be economical and ecologically advantageous when compared to physical or chemical procedures for eliminating impurities. The fundamental sewage treatment system, which includes activated sludge and biological trickling filters, may quickly transform aqueous organic molecules into biomass, which is subsequently differentiated from the aqueous phase by the process of settlement (Johnson and Sumpter, 2001). However, these treatments are not capable of converting all the contaminant biomass. For example, removing estrogen using this methodology is highly efficient, but the rates may vary. It has been reported that natural and synthetic estrogen removal and the removal of nonylphenol derivatives can be accomplished under mesophilic and thermophilic environments employing anaerobic methods (Rodriguez-Narvaez et al., 2017). The biological fixation rate relies on various elements, including pH, oxygen content, temperature, and others, to ensure the growth of nitrifying bacteria. The microbial population grows on inert packings, such as rock or shaped polystyrene, in attached growth processes, which are defined by trickling filters. A thin liquid film of wastewater runs down and through the reactor after being sprayed over the packing at the top. Air circulates between the packing's layers, is transmitted to the fluid film, and finally to the biofilm, where the degradative reactions occur.

6.8 CONCLUSION AND FUTURE PROSPECTS

It is obvious that the current lifestyle followed by human beings uses chemicals in almost every field, which has led to the transformation of chemicals into contaminants. This is a matter of concern, and it is clearly stated that the incorporation of contaminants into the environment is going to cause drastic effects in the future. That's why it is essential to consider all the aspects and focus on reducing the use and having a suitable disposal method and policies for treatment in a sustainable manner at appropriate places to mitigate the risks caused by them. There is a need to regulate these contaminants and protect the groundwater. Deciding a threshold value for all the known contaminants would help manage the production and treatment. We must focus on the toxicological effects, bioaccumulation, and persistence of such compounds, designing a risk-based approach to determine the fate of these emerging contaminants. It would be convenient to have both primary- and secondary-level treatment methodologies and short- and long-term objectives for managing these contaminants.

It is equally important to consider that environmental exposure is not due to a single contaminant but rather to mixtures containing a number of contaminants plus other chemicals. A more convenient option would be to reduce the overall input of ECs into the environment by using methods such as better chemical handling and use, "green" alternatives, less EC use during manufacturing, less waste, proper waste disposal, less chemical discharge, and appropriate water treatment systems. Even while phase-changing methods are effective for treating wastewater, they are not entirely effective when there is a low concentration of contaminants in the water. Since they can degrade these contaminants in the aqueous phase, these methods can be utilized as concentration pretreatments, operating in succession with other treatments. The main motive should be to reduce the effects of chemicals that are now categorized as emerging contaminants.

REFERENCES

Abedi, G., Talebpour, Z., Jamechenarboo, F. (2018) "The survey of analytical methods for sample preparation and analysis of fragrances in cosmetics and personal care products," *TrAC Trends in Analytical Chemistry*, 102, 41–59.

Acero, J.L. et al. (2012) "Coupling of adsorption, coagulation, and ultrafiltration processes for the removal of emerging contaminants in a secondary effluent," *Chemical Engineering Journal*, 210, pp. 1–8. Available at: https://doi.org/10.1016/j.cej.2012.08.043.

Baccar, R., Sarr`a, M., Bouzid, J., Feki, M., Bl´anquez, P. (2012) "Removal of pharmaceutical compounds by activated carbon prepared from agricultural by-product," *Chemical Engineering Journal*, 211, 310–317.

Behera, S. K., Kim, H. W., Oh, J. E., & Park, H. S. (2011) "Occurrence and removal of antibiotics, hormones and several other pharmaceuticals in wastewater treatment plants of the largest industrial city of Korea," *Science of the total environment*, *409*(20), 4351–4360.

Bilal, M., Adeel, M., Rasheed, T., Zhao, Y. and Iqbal, H.M. (2019) "Emerging contaminants of high concern and their enzyme-assisted biodegradation–a review," *Environment International*, 124, pp. 336–353.

Bilal, M., Rasheed, T., Iqbal, H. M., Hu, H., Wang, W., & Zhang, X. (2018) "Horseradish peroxidase immobilization by copolymerization into cross-linked polyacrylamide gel and its dye degradation and detoxification potential," *International Journal of Biological Macromolecules*, 113, 983–990.

Bour, A., Mouchet, F., Silvestre, J., Gauthier, L. and Pinelli, E. (2015) "Environmentally relevant approaches to assess nanoparticles ecotoxicity: A review," *Journal of hazardous materials*, 283, pp.764–777

Cleuvers, M. (2003) "Aquatic ecotoxicity of pharmaceuticals including the assessment of combination effects," *Toxicology Letters*, 142, 185–194.

Corsini, E., Luebke, R.W., Germolec, D.R., DeWitt, J.C. (2014) "Perfluorinated compounds: Emerging POPs with potential immunotoxicity," *Toxicology Letters*, 230, 263–270.

de Peyster, A., MacLean, K.J., Stephens, B.A., Ahern, L.D., Westover, C.M., Rozenshteyn, D. (2003) "Subchronic studies in sprague-dawley rats to investigate mechanisms of MTBE-induced leydig cell cancer," *Toxicological Sciences*, 72, 31–42.

Dhanirama, D., Gronow, J., Voulvoulis, N. (2012) "Cosmetics as a potential source of environmental contamination in the UK," *Environmental Technology*, 33, 1597–1608.

Ellis, J. B. (2006) "Pharmaceutical and personal care products (PPCPs) in urban receiving waters," *Environmental pollution*, *144*(1), 184–189.

Escher, B. I., & Fenner, K. (2011) "Recent advances in environmental risk assessment of transformation products," *Environmental Science & Technology*, 45(9), 3835–3847.

Fent, K., Weston, A.A., Caminada, D. (2006) "Ecotoxicology of human pharmaceuticals," *Aquatic Toxicology*, 76, 122–159.

Gani, K.M., Kazmi, A.A. (2017) "Contamination of emerging contaminants in Indian aquatic sources: First overview of the situation," *Journal of Hazardous, Toxic, and Radioactive Waste*, 21, 04016026.

Gomes, I.B., Maillard, J.Y., Simões, L.C. and Simões, M. (2020) "Emerging contaminants affect the microbiome of water systems—Strategies for their mitigation," *NPJ Clean Water*, 3(1), p.39.

Gomes, I. B., Simões, L. C., & Simões, M. (2018) "The effects of emerging environmental contaminants on Stenotrophomonas maltophilia isolated from drinking water in planktonic and sessile states," *Science of the Total Environment*, 643, 1348–1356.

Gonzalez-Martinez, A., Rodriguez-Sanchez, A., Martinez-Toledo, M., Garcia-Ruiz, M.-J., Hontoria, E., Osorio-Robles, F., Gonzalez–Lopez, J. (2014) "Effect of ciprofloxacin antibiotic on the partial-nitritation process and bacterial community structure of a submerged biofilter," *Science of the Total Environment*, 476, 276–287.

Guerra, P., Kim, M., Shah, A., Alaee, M., Smyth, S.A. (2014) "Occurrence and fate of antibiotic, analgesic/antiinflammatory,and antifungal compounds in five wastewater treatment processes," *Science of The Total Environment*, 473–474, 235e243.http://dx.doi.org/10.1016/j.scitotenv.2013.12.008.

Johnson, A.A.C. and Sumpter, J.P. (2001) "Removal of endocrine-disrupting chemicals in activated sludge treatment works," *Environmental Science & Technology*, 35(24), pp. 4697–4703. Available at: https://doi.org/10.1021/es010171j.

Juliano, C., Magrini, G.A. (2017) "Cosmetic ingredients as emerging pollutants of environmental and health concern. A Mini-Review," Cosmetics, 4.

Kasprzyk-Hordern, B., Dinsdale, R.M., Guwy, A.J. (2009) "The removal of pharmaceuticals, personal care products, endocrine disruptors and illicit drugs during wastewater treatment and its impact on the quality of receiving waters," *Water Research*, 43, 363–380.

Khan, S., Dan, Z., Haiyan, H. (2019) "Adsorption mechanism of Pb(II) and Ni(II) from aqueous solution by TiO2 nanoparticles: Kinetics, isotherms and thermodynamic studies," *Desalination and Water Treatment*, 155, 237–249.

Khan, S., Naushad, M., Govarthanan, M., Iqbal, J. and Alfadul, S.M. (2022) "Emerging contaminants of high concern for the environment: Current trends and future research," *Environmental Research*, 207, p. 112609.

Kim, E., Jung, C., Han, J., Her, N., Park, C. M., Jang, M., ... & Yoon, Y. (2016) "Sorptive removal of selected emerging contaminants using biochar in aqueous solution," *Journal of Industrial and Engineering Chemistry, 36*, 364–371.

Klaunig, J. E., Hocevar, B. A., & Kamendulis, L. M. (2012) "Mode of action analysis of perfluorooctanoic acid (PFOA) tumorigenicity and human relevance," *Reproductive toxicology*, 33(4), 410–418.

Lindqvist, N., Tuhkanen, T., and Kronberg, L. (2005) "Occurrence of acidic pharmaceuticals in raw and treated sewages and in receiving waters," *Water Research*, 39, 2219–2228.

Maurer-Jones, M.A., Gunsolus, I.L., Murphy, C.J., Haynes, C.L. (2013) "Toxicity of engineered nanoparticles in the environment," *Analytical Chemistry*, 85, 3036–3049

Mirasole, C. et al. (2016) "Liquid chromatography-tandem mass spectrometry and passive sampling: Powerful tools for the determination of emerging pollutants in water for human consumption," *Journal of Mass Spectrometry*, 51(9), pp. 814–20. Available at: https://doi.org/10.1002/jms.3813.

Naidu, R. et al. (2016) "Emerging contaminants in the environment: Risk-based analysis for better management," *Chemosphere*, 154, pp. 350–357. Available at: https://doi.org/10.1016/j.chemosphere.2016.03.068.

Naushad, M., Alothman, Z.A. (2015) "Separation of toxic Pb2+ metal from aqueous solution using strongly acidic cation-exchange resin: Analytical applications for the removal of metal ions from pharmaceutical formulation," *Desalination and Water Treatment*, 53, 2158–2166.

Naushad, M., Sharma, G., Alothman, Z.A. (2019) "Photodegradation of toxic dye using Gum Arabic-crosslinked-poly(acrylamide)/Ni(OH)2/FeOOH nanocomposites hydrogel," *Journal of Cleaner Production*, 241, 118263.

Olmez-Hanci, T., Arslan-Alaton, I., & Genc, B. (2013) "Bisphenol A treatment by the hot persulfate process: Oxidation products and acute toxicity," *Journal of Hazardous Materials*, 263, 283–290.

Ouda, M., Kadadou, D., Swaidan, B., Al-Othman, A., Al-Asheh, S., Banat, F., & Hasan, S. W. (2021) "Emerging contaminants in the water bodies of the Middle East and North Africa (MENA): A critical review,". *Science of the Total Environment, 754*, 142177.

Patel, M., Kumar, R., Kishor, K., Mlsna, T., Pittman Jr., C. U., & Mohan, D. (2019) "Pharmaceuticals of emerging concern in aquatic systems: Chemistry, occurrence, effects, and removal methods, *Chemical Reviews*," *119*(6), 3510–3673.

Petrie, B., McAdam, E.J., Scrimshaw, M.D., Lester, J.N., Cartmell, E. (2013) "Fate of drugs during wastewater treatment," *TrAC Trends in Analytical Chemistry*, 49, 145e159. http://dx.doi.org/10.1016/j.trac.2013.05.007.

Richardson, B. J., Lam, P. K., & Martin, M. (2005) "Emerging chemicals of concern: Pharmaceuticals and personal care products (PPCPs) in Asia, with particular reference to Southern China," *Marine pollution bulletin*, *50*(9), 913–920.

Richardson, S. D., & Ternes, T. A. (2014) "Water analysis: Emerging contaminants and current issues," *Analytical chemistry*, 86(6), 2813–2848.

Rodriguez-Narvaez, O.M. et al. (2017) "Treatment technologies for emerging contaminants in water: A review," *Chemical Engineering Journal*, 323, pp. 361–380. Available at: https://doi.org/10.1016/j.cej.2017.04.106.

Sharma, B. M., Bečanová, J., Scheringer, M., Sharma, A., Bharat, G. K., Whitehead, P. G., ... & Nizzetto, L. (2019) "Health and ecological risk assessment of emerging contaminants (pharmaceuticals, personal care products, and artificial sweeteners) in surface and groundwater (drinking water) in the Ganges River Basin, India," *Science of the Total Environment*, *646*, 1459–1467.

Shrestha, S., Bloom, M.S., Yucel, R., Seegal, R.F., Wu, Q., Kannan, K., Rej, R., Fitzgerald, E.F. (2015) "Perfluoroalkyl substances and thyroid function in older adults," *Environment International*, 75, 206–214.

Snyder, S.A. (2008) "Occurrence, treatment, and toxicological relevance of EDCs and pharmaceuticals in water," *Ozone: Science & Engineering*, 30, 65–69.

Stuart, M.E. et al. (2012) "Review of risk from potential emerging contaminants in UK groundwater," *Science of the Total Environment*, 416, pp. 1–21. Available at: https://doi.org/10.1016/j.scitotenv.2011.11.072.

Tireli, A.A. et al. (2014) "Fenton-like processes and adsorption using iron oxide-pillared clay with magnetic properties for organic compound mitigation," *Environmental Science and Pollution Research*, 22(2), pp. 870–881. Available at: https://doi.org/10.1007/s11356-014-2973-x.

Tran, N., Drogui, P., Brar, S.K. (2015) "Sonochemical techniques to degrade pharmaceutical organic pollutants," *Environmental Chemistry Letters*, 13, 251–268.

Tsui, M. M., Leung, H. W., Lam, P. K., & Murphy, M. B. (2014) "Seasonal occurrence, removal efficiencies and preliminary risk assessment of multiple classes of organic UV filters in wastewater treatment plants," *Water Research*, *53*, 58–67.

Vithanage, M., Rajapaksha, A.U., Tang, X., Thiele-Bruhn, S., Kim, K.H., Lee, S.-E., Ok, Y. S. (2014) " Sorption and transport of sulfamethazine in agricultural soils amended with invasive-plant-derived biochar," *Journal of Environmental Management*, 141, 95–103.

Wahab, N. et al. (2019) "Synthesis, characterization, and applications of silk/bentonite clay composite for heavy metal removal from aqueous solution," *Frontiers in Chemistry*, 7. Available at: https://doi.org/10.3389/fchem.2019.00654.

Wang, J., & Wang, S. (2016) "Removal of pharmaceuticals and personal care products (PPCPs) from wastewater: A review," *Journal of Environmental Management*, *182*, 620–640.

Zuccato, E., et al. _(2006) "Pharmaceuticals in the environment in Italy: Causes, occurrence, effects, and control," *Environmental Science and Pollution Research*, 13, 15–21.

7 Recent Advances in Potential Application of Biomass Waste and Biomass-derived Materials for Bioremediation of Pollutants from Wastewater

Lopamudra Das
Department of Chemical Engineering, MJP Rohilkhand University, Uttar Pradesh, India
School of Advanced Studies on Industrial Pollution Control Engineering, Jadavpur University, West Bengal, India

Papita Das
School of Advanced Studies on Industrial Pollution Control Engineering, Jadavpur University, West Bengal, India
Department of Chemical Engineering, Jadavpur University, West Bengal, India

Avijit Bhowal
School of Advanced Studies on Industrial Pollution Control Engineering, Jadavpur University, West Bengal, India
Department of Chemical Engineering, Jadavpur University, West Bengal, India

DOI: 10.1201/9781003432869-7

7.1 INTRODUCTION

Harmful toxic materials present in wastewater have turned out to be an important threat to human health, contributing to the scarcity of clean water (Manna et al., 2015). Surface and groundwater are usually polluted through industrial discharges containing hazardous materials or by the inaccessibility of adequate appropriate remediation operations. Water is also contaminated by reckless anthropogenic activities, such as discarding garbage into open waterbodies (Thines et al., 2017). Although many efficient materials have been researched for their use in wastewater purification systems, not all of those materials are economic and ecologically protective and hence have not been adopted in the wide-ranging decontamination processes of polluted water. Therefore, there is a crucial requirement for production of the eco-friendly, easily operated, low-cost, and efficient materials for wastewater decontamination.

Biomass can be suitable materials for real-world use in the pollutant removal process from contaminated water, as they are recyclable, biocompatible, green, and abundantly available. Biomass as a natural resource is gaining attention in the research field as a source of materials for use in the field of the water decontamination process. Biomass is largely the direct or indirect derivatives of plants/vegetation, such as harvested agro-wastes, forestry waste, marine plants, etc. (Tsyntsarski et al., 2015; Abou Oualid et al., 2020). First-generation biomass is mainly edible food sources including corn, sugarcane, vegetable oils, etc. Second-generation biomass material consists of agro-waste/by-products, industrial sludge, domestic waste, waste from forestry, and metropolitan landfills. Third-generation biomass is mostly represented seaweed, cyanobacteria, and macro/microalgae (Jin et al., 2014Taniya et al., 2017). Applications of enhanced biomass in the remediation of hazardous pollutants from wastewater are very extensive due to the variety of materials and the ease of modification. Several scientists have established notable techniques for biomass modification to develop novel efficient processes in the water purification field to enhance the effectiveness of pollutant removal (Yadav et al., 2013; Abou Oualid et al., 2020). High specific surface area and porosity of the biomass materials are significant indices for effective removal of impurities from contaminated water (Das et al., 2021a; Lian and Xing, 2017). There are different types of natural biomass materials, some of which can be utilized after simple cleaning whereas others require modification before being used and hence must be treated before use. To increase the specific surface area, the non-carbon constituents are removed from the biomass, increasing the carbon content (Ahmad et al., 2006). The products developed following the carbonization process are referred to as having a low specific surface area due to the nonexistence of an activation agent during activation. Hence, either activation of biomass is needed or biomass containing an activator element can be selected as raw materials to achieve high efficiency. Plant-based biomass has high contents of cellulosic and lignocellulosic material (95–98%), which can be utilized to derive biopolymer and bioplastic materials (Kumar et al., 2017). Eco-friendly bioplastic material are derived from renewable biomaterial or biomass, such as plant/crop waste, micro-/macroalgae, etc., rather than fossil fuel. In the past few years, extensive studies have been conducted for the determination of the desirable characteristics and features of biomass material for use in the wastewater decontamination process.

Here, different type of biomass will be reviewed, with discussion and overview on carbonaceous materials, biopolymers, and bioplastics derived from biomass material for the application to water decontamination, in concert with the latest approaches.

7.2 DIFFERENT TYPES OF BIOMASS AND BIOMASS WASTE-DERIVED MATERIALS

Biomass is defined as the total weight or total extent of living species of one animal or plant or of all the communal species (generally, related to the unit area or volume of habitation). Generally, biomass material consists of mostly organic substance in combination with a smaller fraction of inorganic elements. Biomass materials are fundamentally grouped into four main categories: agro-waste, municipal waste, industrial waste, and forestry waste (Abou Oualid et al., 2020). These materials are primarily classified into various groups including algae, peels of crops or fruits, fungal cultures, cellulose/lignocellulosic materials, sewage sludge, and waste from aquatic organisms, domestic waste, sludge from industrial discharge material, plant waste, etc. (Cai et al., 2015; Taniya et al., 2017). The composition of the plant-based biomass is mostly cellulose, hemicellulose, and lignin. Agro-waste obtained from plant-based biomass precursors can be modified by changing the size distribution, structure, specific surface area, and moisture content through a number of effective physical techniques, such as grinding, cutting, thermal drying, scorching, carbonization, pyrolysis, and so on. In chemical modification techniques, the structure and surface properties of the biomass precursor is directly modified by eliminating or incorporating new functional elements which improve the active adsorptive sites on the precursor surface.

7.2.1 Carbon-based Materials

The carbonization or pyrolysis process of inert biomass results in the production of biochar and these carbonaceous materials are porous solid forms. Biochar is generally derived from natural resources that are naturally widely available, for example agro waste/by-product or crop waste, animal waste, organic domestic solid waste, algae, sewage sludge, composts, and forestry waste. Due to its high porosity, biochar materials are considered to be very effective in the adsorption process and in the field of sewage purification. Traditionally, activated carbon is produced using coal (Kopac and Toprak, 2007), coconut shell (Liang et al., 2020), peat (Tsyntsarski et al., 2015), etc. Hence, plant-based raw materials or lignocellulosic materials, with a high carbon content, are suitable for potential activation. Carbonaceous material or activated biochar production using biomass material has notable benefits of cost-effectiveness, biocompatibility, and high efficiency in comparison with conventional activated carbon. Agro-waste-based biomass provides a widespread source of precursor substances intended for the production of activated biochar because of its abundance and ease of accessibility (Tran et al., 2021). Biomass-derived carbonaceous materials have high adsorptive capacity, low production costs, easy preparation methods, and eco-friendly features (Tang et al., 2017). Among the commercially available carbonaceous materials, activated carbon, graphene, and carbon nanotubes are widespread carbon-based adsorbents studied widely in wastewater treatments because of their extremely flexible structures, along with topographies. Considerable research has been done on the synthesis of biomass-derived carbonaceous material, such as Liu et al. (2010), who synthesised activated carbon using bamboo as the raw material by chemical activation with phosphoric acid, and their study revealed the microporous structure of the external surface of prepared material (Liu et al., 2010). Yuliusman and coworkers reported the preparation of activated carbon using palm shells by activation with KOH and $ZnCl_2$ (Yuliusman et al., 2017). Furthermore, Sun et al. and colleagues investigated the production of activated carbon from waste cellulose fibres (Sun et al., 2019). In another study, nanocarbons including graphene sheet and carbon nanotubes were fabricated using rice husk as the raw material by applying a microwave plasma irradiation technique (Wang et al., 2015). Taniya et al. synthesized graphene quantum dots (PS-FLG) using biomass wastes such as peanut shell by chemical activation (using KOH) and mechanical exfoliation techniques (Taniya et al., 2017).

7.2.2 Biopolymers and Bioplastic Materials

In general, biomass-based biopolymers and bioplastics are mainly derived from reusable or recyclable biomass material for assembling them as a part of the circular economy. Moreover, biomass provides equivalent replacements for the synthesis of fossil-derived polymers which reduces the emission of greenhouse gases and thus promotes a more sustainable civilization. The advantageous approaches for synthesis of vinyl monomers, carboxylic acids, alcohols, and rubbers from biomass material have been widely studied by many researchers and scientists. Numerous studies of biomass have revealed that natural polysaccharides can be used as a polymer matrix in an impregnation process by electrostatic interaction, owing to the existence of hydrogen bonds in structure, with good stability (Das et al., 2021b). Zhang and coworkers synthesised cellulose nanofibers/chitosan composite aerogels through a chemical crosslinking process (Zhang et al., 2021). The literature has reported that biopolymers and their derivatives have high efficiency in the removal of hazardous impurities from contaminated aqueous solution. Naturally obtained biopolymers based on polysaccharides (for example, starch, cellulose, chitosan/chitin, alginate) have a distinct role in the treatment of polluted water due to their desirable uptake efficiency, fast kinetics, and favorable recyclability (Vakili et al., 2014). Composite material blending of various biopolymers with other natural additives, such as starch, cellulose, alginate, chitin, lignin, or silica, can also play vital roles in the wastewater treatment process.

Bioplastic is generally defined as a biodegradable plastic derivative of recyclable/renewable natural resources. Traditional plastics are mainly derived from petrochemical resources, though the demand of their synthesis from renewable and biodegradable biomass is rising prominently worldwide (Song et al., 2009). For increasing the mechanical strength and storage capacity of bioplastic, lignocellulosic fibres and lignin can be used to reinforce bioplastic materials instead of conventional synthetic non-biodegradable fibres (such as fiberglass and carbon fibers) as they are biodegradable, more eco-friendly, abundant, and inexpensive (Yang et al., 2019). Some well-known widely used biodegradable biopolymers for fabrication of bioplastics are polyhydroxyalkanoates, polylactide, polybutylene succinate, etc. (Meereboer et al., 2020; Philp et al., 2013). Furthermore, polyhydroxyalkanoates are mainly derived from agricultural/

crop waste and food wastes, such as wheat fiber/straw, rice husk, sugarcane bagasse, and peel of fruits/vegetable (Gowda and Shivakumar, 2014). Arikan and Ozsoy (2011) synthesized starch-based bioplastics using waste sludge from the industrial wastewater from a potato processing unit (Arikan and Ozsoy, 2011). The foremost end-of-life option of biomass-based plastics are reutilizing/recycling, burning, composting, landfill, etc. (Vert et al., 2012; Algieri et al., 2017). Also, there are some commonly used non-biodegradable bioplastics such as polyethylene terephthalate, polytrimethylene terephthalate, and thermoplastic polyester elastomers which are derived from nonbiodegradable biomass resources (Rahman and Bhoi, 2021).

7.3 APPLICATION OF BIOMASS/BIOMASS-DERIVED MATERIALS IN WASTEWATER TREATMENT PROCESS

With developing technologies, various industries discharge worrying amounts of toxic pollutants in their effluent stream, which then affect the ecosystem. There is a crucial necessity to treat the contaminated industrial sewages before clearing the unprotected waterbodies to prevent toxic effects on living organisms. Among all the wastewater treatment processes, adsorption is considered to be a reasonable technique because of its low operating cost, high uptake capacity, recycling and regeneration capability of the adsorbent, and ability to show a high level of water purification without generation of any secondary sludge (Das et al., 2021).

Biomass and its derivatives have been considered to be biosorbents in wastewater purification process because of their abundantly availability, low price, and the existence of several functional elements/groups. Exclusively, agro-waste-derived biomass can be used as biosorbents for wastewater treatment due to its recyclability, effectiveness, easy accessibility, cost effectivity, chemical and thermal stability, and eco-friendliness (Mo et al., 2018). The high porous structure and higher specific surface area of these biosorbents, along with several functional groups, enhance their capabilities in remediation of water from harmful pollutants, i.e., dyes, heavy metals, fluoride, and oils. In order to build a sustainable circular economy in the wastewater purification process, of prime importance is the recyclability of a natural resource or waste material available in large amounts. Nowadays, researchers are paying more attention to the modification of biomass to develop novel efficient biosorbents (Yang et al., 2019; Wang and Li, 2013).

7.3.1 Dye Removal

A dye is a chemical substance that imparts color when applied to a substance (such as cloths, papers, leathers, etc.). Dyes and dyeing products are used in numerous sectors including the textile, plastic, rubber, paper, print, and leather industries. Dyes can be natural or synthetic. Dye concentration in effluents of the textile industries generally vary from 10 to 200 mg/L (Islam, 2020). Synthetic dyes have complex chemical structure, photolytic stability, and are non-biodegradable substances (Sudha et al., 2018). Consumption of dye-contaminated water may cause tumor malignancy, chromosomal abnormalities, allergic reactions in eyes and skin, respirational pain, etc. (Puvaneswari et al., 2006; Sudha et al., 2018).

An extensive use of biomass and biomass waste is the decontamination of hazardous dye-contaminated polluted water (Das et al., 2020a, 2020b). Several research studies have described the potential effectiveness of biomass material in the dye-removal process, such as Gong et al. (2005), who studied the feasibility of using biomass products in the decolorization process for the elimination of methylene blue (MB), brilliant cresyl blue, and neutral red dyes from polluted water. In addition, Tran and coworkers evaluated the performance of hydrothermally carbonized KOH-activated carbon derived from coffee husks in a MB dye remediation study and estimated the highest uptake capacity obtained as 367.65 mg/g (Tran et al., 2021). Tuli and colleagues established that the phosphoric acid-treated activated carbon, prepared using tea waste, had a noteworthy specific surface area (850.58 m^2/g) and a high biosorption efficiency (98%) toward MB dye adsorption (Tuli et al., 2020). Peng's group conveyed the excellent adsorption properties of a cellulose-clay composite hydrogel in the treatment of MB-contaminated water, where the removal efficiency achieved was greater than 95% (Peng et al., 2016b). In another research study, highly porous nitrogen-doped carbon biosorbents were produced using waste cellulose fibres by employing a spray-drying process, and the biosorption potential in the remediation process of methyl orange (MO) dye from contaminated water was determined (Sun et al., 2019). Abou Oualid and coworkers reported the characterization process and application of the *Codium decorticatum* gtreen macroalga in wastewater treatment for bioremediation of Congo red and Crystal violet dyes. In their study, the highest adsorption capacities obtained were 278.46 mg/g and191.01 mg/g for Congo red and Crystal violet dye, respectively (Abou Oualid et al., 2020). Zhu et al. (2011) developed a Fe3O4-impregnated composite using cellulose and activated carbon (cellulose/Fe3O4/activated carbon) to decontaminate Congo red-containing wastewater (Zhu et al., 2011).

7.3.2 Heavy Metal Removal

With industries including metal plating services, print and paper manufacturing, mining, fertilizer production, tanneries, and the manufacture of batteries and insecticides, etc., different heavy metals are being discharged into waterbodies through industrial effluents (Ahmad et al., 2006). Heavy metals are hazardous, non-biodegradable, and show higher tendency for accumulation in living organisms. Therefore, treatment of heavy metal-containing wastewater has become one of the most important challenges to scientists and researchers worldwide. Several research studies have reported on the development and application of biomass waste and biomass-derived materials in a heavy-metal-removal process from wastewater, such as Zhou et al. (2017), who investigated the production of biochar from banana peels, and successfully applied the biochar in the wastewater treatment process for the remediation of lead (Pb^{2+}) (highest uptake capacity obtained as 359 mg/g). Dos Santos et al. (2011) reported the effectivess of three types of biosorbent derived from sugarcane bagasse for remediation of Cu^{2+} ions, where chemical modification was employed using sodium hydroxide, citric acid, or sodium hydroxide. In their study, the highest uptake capacity value was obtained for citric acid-modified adsorbent and it was 31.53 mg/g (Long et al. 2021) prepared a sodium hydroxide-treated sawdust from *Michelia figo* wood for the treatment of lead-contaminated water and found a lead removal efficiency of 96.39% using a biosorbent dosage of 10 g/L. Luo et al. (2018a) examined the synthesis and performance of pyrolyzed corncob biochar on the heavy metal elimination process from cadmium (Cd^{2+})-contaminated polluted water. Carbonization took place at 350C°C, 450°C, and 550°C temperatures under a N_2 atmosphere, and the highest uptake capacity for Cd^{2+} removal was 85.65 mg/g using prepared biochar at 350°C (Luo et al., 2018a). A study by Zhu and coworkers prepared a magnetic microporous chitosan-based hydrogel (CS–g–P(AA)) and evaluated its potential in the bioremediation of heavy metals from cadmium- (Cd^{2+}) and lead- (Pb^{2+}) contaminated aqueous solutions. Experimental results from this study reported that the highest biosorption capacities of the biosorbent developed were 308.84 mg/g and 695.22 mg/g for Cd^{2+} and Pb^{2+}, respectively (Zhu et al., 2016). In another study, Lu et al. (2015) prepared alginate-based hydrogel material, employing the liquid–liquid phase separation method for the remediation of multiple ions of heavy metals such as cupper (Cu^{2+}), silver (Ag^{+}), and iron (Fe^{3+}), with maximum achievable uptake capacity estimated as 54.9 mg/g for Cu^{2+}, 82.8 mg/g for Ag^{+}, and 135.5 mg/g for Fe^{3+} ions (Lu et al., 2015). Tavker et al. (2021) extracted nanofibrillated cellulose of 44–50 nm crystal size from waste orange peel, using alkali treatment and acid hydrolysis techniques. They also investigated the feasibility of using the prepared material in the remediation process for the removal of chromium and cadmium from wastewater. Results showed that the maximum uptake efficiency obtained was 47% for chromium and 83.49% for cadmium (Tavker et al., 2021)

7.3.3 Fluoride Removal

Fluorine is a highly electronegative reactive chemical element with atomic number 9 and it is the lightest element in the halogen group (Sivasankar, 2016). Fluoride exists in different amounts in the Earth's crusts, rock-forming minerals, and volcanic ash. The fluoride concentration in groundwater is subject to the chemical, physical, and geographical characteristics of the underground layer of water-bearing permeable sedimentary rock. Fluoride can be emitted from effluents of different manufacturing industries such as power stations, semiconductor industries, steel and aluminum industries, and glass, plastic, and cement production industries. Intake of fluoride over permissible limits (permissible limit 1–1.5 mg/L) causes fluorosis disorder, generative decay, and skeletal tissue damage. Metabolically inactive/dead microbial biomass and plant-based biomass are very cost effective, eco-friendly, and efficient materials to treat fluoride-contaminated water as they are biodegradable and need less processing to enhance the fluoride removal rate or efficiency (Robledo-Peralta et al., 2022). As a result of the existence of various functional elements, such as amine, carboxyl, hydroxyl, and phosphoric acid, on the biomass surface, adsorption of fluoride ions becomes easy by combining with those functional groups. Over the past few years, several research studies have described treatment methods aimed at the reduction of excessive fluoride concentrations in fluoride-contaminated water, using modified lignocellulosic biomass materials. For instance, Yadav et al. (2013) studied the efficiency of wheat straw, sawdust, and activated carbon derived from sugarcane in the wastewater treatment process intended to eliminate fluoride ions present in polluted water. Results of their study exhibited that the highest achievable uptake efficiency were 56.4 % for prepared activated carbon, 49.8% for sawdust, and 40.2% for wheat straw, using a biomass dosage of 4 g/L at pH 6 (Yadav et al., 2013). Paudyal and coworkers examined the synthesis and application of Zr-loaded orange waste gel in the reduction of fluoride concentration in wastewater. Outcomes of this research showed that the maximum estimated adsorption capacity achieved was 22.8 mg/g at pH 2–4 using dosages of 1.6 g/L (Paudyal et al., 2012). Cai and colleagues reported the defluoridation process by applying Al/Fe oxide-impregnated tea waste while approximate fluoride biosorption capacities were estimated as 3.83 mg/g for the untreated tea waste, 10.47 mg/g for Tea–Fe, 13.79 for Tea–Al, and 18.52 mg/g for Tea–Al–Fe with removal efficiency of more than 90% at pH 4 (Cai et al., 2015). Mwakabona et al. (2019) fabricated Fe(III)-loaded sisal fibre to treat a fluoride-contaminated aqueous solution, and the highest uptake efficiency obtained was 53.4% at pH 2 while the biosorption capacity was estimated as 0.2 mg/g, using a 15 g/L adsorbent dosage (Mwakabona et al., 2019). In another study, Srinivasulu and Pindi (2021) evaluated the performance behavior of activated tamarind seeds in the defluoridation process and the maximum removal efficiency was 94% while highest biosorption capacity was determined as 1.79 mg/g under neutral pH conditions (Srinivasulu and Pindi, 2021).

7.3.4 Oily Wastewater Treatment

Nowadays, the treatment technology of oily wastewater or oil-containing sewage has turned out to be a significant research topic, and development is aimed at sustainable development because of the extensive production of oil-containing wastewater and the frequent occurrence of oil spills in industrial effluents. Harmful and carcinogenic elements/compounds are introduced into aquatic environment through oily wastewater, which can destroy or hamper marine life as well as the ecosystem. In several studies, the material development and treatment processes were described on the treatment of oily wastewater, such as Zhang and coworkers, who synthesized cross-linked cellulose nanofibers/chitosan composite aerogel by employing a freeze-drying technique and also investigated utilization techniques producing aerogel in oil-water separation. Results confirmed that the oil adsorption capacity of the prepared composite was three times higher than the composite without added chitosan (Zhang et al., 2021). A study by Vlaev et al. (2011) demonstrated that pyrolyzed rice husk ash had considerable potential for treatment of oily wastewater due to having high specific surface area and an amazingly porous structure. The value for maximum uptake capacity by the prepared adsorbent was 6.22 g/g for crude oil and 5.02 g/g for diesel oil (Vlaev et al., 2011). Bazargan et al. (2014) investigated the production of alkali-treated rice husk and its use in the oily wastewater treatment process. By alkali treatment of rice husk at high temperature, silica was removed from the rice husk surface and the maximum uptake capacity by the prepared material was 19 g/g. Yue and coworkers synthesized a recyclable cellulose-based membrane, cellulose/LDH (layered double hydroxide), by employing a hydrothermal reaction and hydrophobic modification technology in order to treat oily polluted water. The study revealed that the prepared membrane showed excellent adsorption efficiency, high recyclability, and remarkable hydrophobicity under acidic conditions which makes it a suitable candidate to adsorb or remove oils from wastewater (Yue et al., 2017).

7.4 CONCLUSION AND THE FUTURE DIRECTION OF BIOMASS UTILIZATION

The removal of different pollutants from wastewater is a challenging task. Biomass and biomass-derived materials can provide effective and eco-friendly treatment processes for wastewater decontamination and can be part of a sustainable circular economy. The present chapter shows that biomass wastes have a notable effectiveness in adsorbing dyes, heavy metals, fluoride ions, and oils from polluted wastewater. However, the use of crop-based and microorganism-based biomass wastes in India is not very high, while the discarding of enormous quantities of such waste leads to negative impacts on the environment. To further develop the application of biomass waste, waste discarding management needs to be executed carefully. Recycling of biomass wastes is considered to be the best solution from that point for sustainable development and to decrease the consumption of renewable resources.

In this chapter, the latest research, openings and opportunities, sources and advances in using material production from biomass wastes for efficient wastewater treatment have been discussed. Unfortunately, most of the biomass studies reported by researchers are mainly based on small- and lab-scale studies, preparing adsorbent materials which are generally invalid for remediation of mixed pollutants or complex elements associated with real polluted water/industrial effluents. Research indicates that the results obtained from lab-based experiments can significantly deviate from those achieved during large-scale applications or using real mixed wastewater.

7.5 ACKNOWLEDGMENT

The authors are very grateful to the Department of Chemical Engineering, Jadavpur University, West Bengal, India, and the School of Advanced Studies on Industrial Pollution Control Engineering, Jadavpur University, West Bengal, India.

REFERENCES

Abou Oualid, H., Abdellaoui, Y., Laabd, M., El Ouardi, M., Brahmi, Y., Iazza, M., Abou Oualid, J., 2020. Eco-efficient green seaweed codium decorticatum biosorbent for textile dyes: Characterization, mechanism, recyclability, and RSM optimization. *ACS Omega* 5. https://doi.org/10.1021/acsomega.0c02311

Ahmad, A., Senapati, S., Khan, M.I., Kumar, R., Sastry, M., 2006. Extra-/intracellular biosynthesis of gold nanoparticles by an alkalotolerant fungus, *Trichothecium* sp. *J. Biomed. Nanotechnol.* 1. https://doi.org/10.1166/jbn.2005.012

Algieri, C., Donato, L., Giorno, L., 2017. Tyrosinase immobilized on a hydrophobic membrane. *Biotechnol. Appl. Biochem.* 64. https://doi.org/10.1002/bab.1462

Bazargan, A., Tan, J., Hui, C.W., McKay, G., 2014. Utilization of rice husks for the production of oil sorbent materials. *Cellulose.* https://doi.org/10.1007/s10570-014-0203-9

Bello, O.S., Adegoke, K.A., Olaniyan, A.A., Abdulazeez, H., 2015. Dye adsorption using biomass wastes and natural adsorbents: Overview and future prospects. *Desalin. Water Treat.* https://doi.org/10.1080/19443994.2013.862028

Cai, H.M., Chen, G.J., Peng, C.Y., Zhang, Z.Z., Dong, Y.Y., Shang, G.Z., Zhu, X.H., Gao, H.J., Wan, X.C., 2015. Removal of fluoride from drinking water using tea waste loaded with Al/Fe oxides: A novel, safe and efficient biosorbent. *Appl. Surf. Sci.* https://doi.org/10.1016/j.apsusc.2014.11.164

Das, L., Das, P., Bhowal, A., Bhattachariee, C., 2020a. Treatment of malachite green dye containing solution using bio-degradable Sodium alginate/NaOH treated activated sugarcane baggsse charcoal beads: Batch, optimization using response surface methodology and continuous fixed bed column study. *J. Environ. Manage.* https://doi.org/10.1016/j.jenvman.2020.111272

Das, L., Das, P., Bhowal, A., Bhattacharjee, C., 2021a. Enhanced biosorption of fluoride by extracted nanocellulose/polyvinyl alcohol composite in batch and fixed-bed system: ANN analysis and numerical modeling. *Environ. Sci. Pollut. Res.* https://doi.org/10.1007/s11356-021-14026-x

Das, L., Saha, N., Ganguli, A., Das, P., Bhowal, A., Bhattacharjee, C., 2021b. Calcium alginate–bentonite/activated biochar composite beads for removal of dye and Biodegradation of dye-loaded composite after use: Synthesis, removal, mathematical modeling and biodegradation kinetics. *Environ. Technol. Innov.* https://doi.org/10.1016/j.eti.2021.101955

Das, L., Saha, N., Das, P., Bhowal, A., Bhattacharya, C., 2020b. Application of synthesized nanocellulose material for removal of malachite green from wastewater, in: *Recent Trends in Waste Water Treatment and Water Resource Management.* https://doi.org/10.1007/978-981-15-0706-9_2

Das, R., Lindström, T., Sharma, P.R., Chi, K., Hsiao, B.S., 2021. Nanocellulose for sustainable water purification. *Chem. Rev.* https://doi.org/10.1021/acs.chemrev.1c00683

Dos Santos, V.C.G., De Souza, J.V.T.M., Tarley, C.R.T., Caetano, J., Dragunski, D.C., 2011. Copper ions adsorption from aqueous medium using the biosorbent sugarcane Bagasse in Natura and chemically modified. *Water. Air. Soil Pollut.* https://doi.org/10.1007/s11270-010-0537-3

Gong, R., Li, M., Yang, C., Sun, Y., Chen, J., 2005. Removal of cationic dyes from aqueous solution by adsorption on peanut hull. *J. Hazard. Mater.* 121, 247–250. https://doi.org/10.1016/j.jhazmat.2005.01.029

Gowda, V., Shivakumar, S., 2014. Agrowaste-based Polyhydroxyalkanoate (PHA) production using hydrolytic potential of Bacillus thuringiensis IAM 12077. *Brazilian Arch. Biol. Technol.* https://doi.org/10.1590/S1516-89132014000100009

Islam, S., 2020. A study on the solutions of environment pollutions and worker's health problems caused by textile manufacturing operations. *Biomed. J. Sci. Tech. Res.* https://doi.org/10.26717/bjstr.2020.28.004692

Jin, H., Capareda, S., Chang, Z., Gao, J., Xu, Y., Zhang, J., 2014. Biochar pyrolytically produced from municipal solid wastes for aqueous As(V) removal: Adsorption property and its improvement with KOH activation. *Bioresour. Technol.* https://doi.org/10.1016/j.biortech.2014.06.103

Kopac, T., Toprak, A., 2007. Preparation of activated carbons from Zonguldak region coals by physical and chemical activations for hydrogen sorption. *Int. J. Hydrogen Energy.* https://doi.org/10.1016/j.ijhydene.2007.08.002

Kumar, R., Sharma, R.K., Singh, A.P., 2017. Cellulose based grafted biosorbents – Journey from lignocellulose biomass to toxic metal ions sorption applications – A review. *J. Mol. Liq.* https://doi.org/10.1016/j.molliq.2017.02.050

Lian, F., Xing, B., 2017. Black carbon (Biochar) in water/soil environments: Molecular structure, sorption, stability, and potential risk. *Environ. Sci. Technol.* https://doi.org/10.1021/acs.est.7b02528

Liang, Q., Liu, Y., Chen, M., Ma, L., Yang, B., Li, L., Liu, Q., 2020. Optimized preparation of activated carbon from coconut shell and municipal sludge. *Mater. Chem. Phys.* https://doi.org/10.1016/j.matchemphys.2019.122327

Liu, Q.S., Zheng, T., Wang, P., Guo, L., 2010. Preparation and characterization of activated carbon from bamboo by microwave-induced phosphoric acid activation. *Ind. Crops Prod.* https://doi.org/10.1016/j.indcrop.2009.10.011

Long, M., Jiang, H., Li, X., 2021. Biosorption of Cu2+, Pb2+, Cd2+ and their mixture from aqueous solutions by Michelia figo sawdust. *Sci. Rep.* https://doi.org/10.1038/s41598-021-91052-2

Lu, T., Xiang, T., Huang, X.-L., Zhao, W.-F., Zhang, Q., Zhao, C.-S., 2015. Post-crosslinking towards stimuli-responsive sodium alginate beads for the removal of dye and heavy metals. *Carbohydr. Polym.*, 133:587–595. https://doi.org/10.1016/j.biortech.2018.03.075

Luo, M., Lin, H., Li, B., Dong, Y., He, Y., Wang, L., 2018. A novel modification of lignin on corncob-based biochar to enhance removal of cadmium from water. *Bioresour. Technol.* https://doi.org/10.1016/j.biortech.2018.03.075

Manna, S., Roy, D., Saha, P., Adhikari, B., 2015. Defluoridation of aqueous solution using alkali-steam treated water hyacinth and elephant grass. *J. Taiwan Inst. Chem. Eng.* https://doi.org/10.1016/j.jtice.2014.12.003

Meereboer, K.W., Misra, M., Mohanty, A.K., 2020. Review of recent advances in the biodegradability of polyhydroxyalkanoate (PHA) bioplastics and their composites. *Green Chem.* https://doi.org/10.1039/d0gc01647k

Mo, J., Yang, Q., Zhang, N., Zhang, W., Zheng, Y., Zhang, Z., 2018. A review on agro-industrial waste (AIW) derived adsorbents for water and wastewater treatment. *J. Environ. Manage.* https://doi.org/10.1016/j.jenvman.2018.08.069

Mwakabona, H.T., Mlay, H.R., Van der Bruggen, B., Njau, K.N., 2019. Water defluoridation by Fe(III)-loaded sisal fibre: Understanding the influence of the preparation pathways on biosorbents' defluoridation properties. *J. Hazard. Mater.* https://doi.org/10.1016/j.jhazmat.2018.08.088

Paudyal, H., Pangeni, B., Inoue, K., Matsueda, M., Suzuki, R., Kawakita, H., Ohto, K., Biswas, B.K., Alam, S., 2012. Adsorption behavior of fluoride ions on Zirconium(IV)-loaded orange waste gel from aqueous solution. *Sep. Sci. Technol.* https://doi.org/10.1080/01496395.2011.607204

Peng, N., Hu, D., Zeng, J., Li, Y., Liang, L., Chang, C., 2016. Superabsorbent cellulose-clay nanocomposite hydrogels for highly efficient removal of dye in water. *ACS Sustain. Chem. Eng.* https://doi.org/10.1021/acssuschemeng.6b02178

Philp, J.C., Bartsev, A., Ritchie, R.J., Baucher, M.A., Guy, K., 2013. Bioplastics science from a policy vantage point. *N. Biotechnol.* https://doi.org/10.1016/j.nbt.2012.11.021

Puvaneswari, N., Muthukrishnan, J., Gunasekaran, P., 2006. Toxicity assessment and microbial degradation of azo dyes. *Indian J. Exp. Biol.*, 44(8), 618–626.

Rahman, M.H., Bhoi, P.R., 2021. An overview of non-biodegradable bioplastics. *J. Clean. Prod.* https://doi.org/10.1016/j.jclepro.2021.126218

Robledo-Peralta, A., Torres-Castañón, L.A., Rodríguez-Beltrán, R.I., Reynoso-Cuevas, L., 2022. Lignocellulosic biomass as sorbent for fluoride removal in drinking water. *Polymers (Basel).* https://doi.org/10.3390/polym14235219

Sivasankar, V., Darchen, A., Omine, K., Sakthivel, R., 2016. Fluoride: A world ubiquitous compound, its chemistry, and ways of contamination, in: *Surface Modified Carbons as Scavengers for Fluoride from Water.* https://doi.org/10.1007/978-3-319-40686-2_2

Song, J.H., Murphy, R.J., Narayan, R., Davies, G.B.H., 2009. Biodegradable and compostable alternatives to conventional plastics. *Philos. Trans. R. Soc. B Biol. Sci.* https://doi.org/10.1098/rstb.2008.0289

Srinivasulu, D., Pindi, P.K., 2021. Activated tamarind seed coat: A green biosorbent to remove fluoride from aqueous solutions. *Water Sci. Technol. Water Supply.* https://doi.org/10.2166/WS.2021.037

Sudha, M., Saranya, A., Selvakumar, G., Sivakumar, N., 2018. Microbial degradation of Azo Dyes: A review. *Int. J. Curr. Microbiol. Appl. Sci.*, 3, 670–690.

Sun, Y., Chen, A., Sun, W., Shah, K.J., Zheng, H., Zhu, C., 2019. Removal of Cu and Cr ions from aqueous solutions by a chitosan based flocculant. *Desalin. Water Treat.* https://doi.org/10.5004/dwt.2019.23953

Tang, W., Zhang, Y., Zhong, Y., Shen, T., Wang, X., Xia, X., Tu, J., 2017. Natural biomass-derived carbons for electrochemical energy storage. *Mater. Res. Bull.* https://doi.org/10.1016/j.materresbull.2016.12.025

Tavker, N., Yadav, V.K., Yadav, K.K., Cabral-Pinto, M.M.S., Alam, J., Shukla, A.K., Ali, F.A.A., Alhoshan, M., 2021. Removal of cadmium and chromium by mixture of silver nanoparticles and nano-fibrillated cellulose isolated from waste peels of citrus sinensis. *Polymers (Basel).* https://doi.org/10.3390/polym13020234

Thines, R.K., Mubarak, N.M., Nizamuddin, S., Sahu, J.N., Abdullah, E.C., Ganesan, P., 2017. Application potential of carbon nanomaterials in water and wastewater treatment: A review. *J. Taiwan Inst. Chem. Eng.* https://doi.org/10.1016/j.jtice.2017.01.018

Tran, T.H., Le, H.H., Pham, T.H., Nguyen, D.T., La, D.D., Chang, S.W., Lee, S.M., Chung, W.J., Nguyen, D.D., 2021. Comparative study on methylene blue adsorption behavior of coffee husk-derived activated carbon materials prepared using hydrothermal and soaking methods. *J. Environ. Chem. Eng.* https://doi.org/10.1016/j.jece.2021.105362

Tsyntsarski, B., Stoycheva, I., Tsoncheva, T., Genova, I., Dimitrov, M., Petrova, B., Paneva, D., Cherkezova-Zheleva, Z., Budinova, T., Kolev, H., Gomis-Berenguer, A., Ania, C.O., Mitov, I., Petrov, N., 2015. Activated carbons from waste biomass and low rank coals as catalyst supports for hydrogen production by methanol decomposition. *Fuel Process. Technol.* https://doi.org/10.1016/j.fuproc.2015.04.016

Tuli, F.J., Hossain, A., Kibria, A.K.M.F., Tareq, A.R.M., Mamun, S.M.M.A., Ullah, A.K.M.A., 2020. Removal of methylene blue from water by low-cost activated carbon prepared from tea waste: A study of adsorption isotherm and kinetics. *Environ. Nanotechnol. Monit. Manag.* https://doi.org/10.1016/j.enmm.2020.100354

Vakili, M., Rafatullah, M., Salamatinia, B., Abdullah, A.Z., Ibrahim, M.H., Tan, K.B., Gholami, Z., Amouzgar, P., 2014. Application of chitosan and its derivatives as adsorbents for dye removal from water and wastewater: A review. *Carbohydr. Polym.* https://doi.org/10.1016/j.carbpol.2014.07.007

Vlaev, L., Petkov, P., Dimitrov, A., Genieva, S., 2011. Cleanup of water polluted with crude oil or diesel fuel using rice husks ash. *J. Taiwan Inst. Chem. Eng.* https://doi.org/10.1016/j.jtice.2011.04.004

Wang, L., Li, J., 2013. Adsorption of C.I. Reactive Red 228 dye from aqueous solution by modified cellulose from flax shive: Kinetics, equilibrium, and thermodynamics. *Ind. Crops Prod.* https://doi.org/10.1016/j.indcrop.2012.05.031

Wang, Z., Ogata, H., Morimoto, S., Ortiz-Medina, J., Fujishige, M., Takeuchi, K., Muramatsu, H., Hayashi, T., Terrones, M., Hashimoto, Y., Endo, M., 2015. Nanocarbons from rice husk by microwave plasma irradiation: From graphene and carbon nanotubes to graphenated carbon nanotube hybrids. *Carbon.* https://doi.org/10.1016/j.carbon.2015.07.037

Yadav, A.K., Abbassi, R., Gupta, A., Dadashzadeh, M., 2013. Removal of fluoride from aqueous solution and groundwater by wheat straw, Sawdust and activated bagasse carbon of sugarcane. *Ecol. Eng.* https://doi.org/10.1016/j.ecoleng.2012.12.069

Yang, H., Ye, S., Zhou, J., Liang, T., 2019. Biomass-derived porous carbon materials for supercapacitor. *Front. Chem.* https://doi.org/10.3389/fchem.2019.00274

Yue, X., Li, J., Zhang, T., Qiu, F., Yang, D., Xue, M., 2017. In situ one-step fabrication of durable superhydrophobic-superoleophilic cellulose/LDH membrane with hierarchical structure for efficiency oil/water separation. *Chem. Eng. J.* https://doi.org/10.1016/j.cej.2017.07.026

Yuliusman, N., Afdhol, M.K., Amiliana, R.A., Hanafi, A., 2017. Preparation of activated carbon from palm shells using KOH and ZnCl2 as the activating agent, in: *IOP Conference Series: Earth and Environmental Science.* https://doi.org/10.1088/1755-1315/75/1/012009

Zhang, M., Jiang, S., Han, F., Li, M., Wang, N., Liu, L., 2021. Anisotropic cellulose nanofiber/chitosan aerogel with thermal management and oil absorption properties. *Carbohydr. Polym.* https://doi.org/10.1016/j.carbpol.2021.118033

Zhou, N., Chen, H., Xi, J., Yao, D., Zhou, Z., Tian, Y., Lu, X., 2017. Biochars with excellent Pb(II) adsorption property produced from fresh and dehydrated banana peels via hydrothermal carbonization. *Bioresour. Technol.* https://doi.org/10.1016/j.biortech.2017.01.074

Zhu, Y., Zheng, Y., Wang, F., Wang, A., 2016. Fabrication of magnetic macroporous chitosan-g-poly (acrylic acid) hydrogel for removal of Cd2+ and Pb2+. *Int. J. Biol. Macromol.* https://doi.org/10.1016/j.ijbiomac.2016.09.005

8 Rainwater Harvesting and Its Application to Water Resources

Savita R. Seetimani
Resource Person, Karnataka State Forest Academy, Gungaragatti Dharwad, Karnataka, India

Mahantesh M. Kurjogi
Multi-Disciplinary Research Unit, Karnataka Institute of Medical Sciences, Hubli, Karnataka, India

8.1 INTRODUCTION

The global population is estimated to reach 8.5 billion by 2030 (https://www.un.org/en/global-issues/population), with ever-increasing urbanization and industrialization. On the other hand, climate change has had a noticeable impact on water resources which are being depleted every year. Globally, 45% of agricultural lands are under periodic water shortages (Ashraf and Foolad, 2007). Around 70% of the world's poor reside in rural areas and are frequently dependent on sources of income based on rainfall. The improvement of rainwater management is essential for raising labor returns and reducing poverty (Hatibu et al., 2006; Sharma et al., 2008). Hunger, poverty, and access to water are all closely related; the majority of the hungry and impoverished reside in areas where access to water is a significant barrier to the production of food. Therefore, harvesting water helps to break the link between hunger, poverty, and water (Rockstrom et al., 2007). In addition to the ever-increasing population, excessive demands for water for domestic, agricultural, and industrial usage are also increasing. Several methods have been reported to minimize water usage in the agricultural sector and modern lifestyle but, to date, no technology is available to generate artificial water to feed the global need. In this context, adopting a rainwater harvesting system for the management of freshwater is crucial in areas of low rainfall, as utilization of every raindrop is precious. On the other hand, the implementation of rainwater harvesting systems in high-rainfall areas is also important from an economic point of view. Hence, rainwater harvesting is one of the efficient water-conserving methods suitable for all climatic conditions. Therefore, rainwater harvesting is the collective term for diversified interventions to use rainfall water through collection and storage, either in soil or in man-made dams, tanks, or containers, to bridge dry spells and

 DOI: 10.1201/9781003432869-8

droughts. The effect is increased retention of water in the landscape, enabling the management and use of water for multiple purposes (Jennie, 2009).

Rainwater Harvesting (RWH) is one of the oldest and simplest techniques used to save water. It is evident that rainwater harvesting has been part of ancient human civilization and identity. An archeological site examination in China indicated that rainwater harvesting technology may date back almost 6,000 years. Ruins of cisterns built as early as 2000 BC for storing runoff from hillsides for agricultural and domestic purposes are still standing in Israel (Gould and Nissen-Petersen, 1999). Similarly prominent water harvesting systems were discovered in the cities of the Indus Valley civilization and were found to have flourished about 3000–1500 BC (Upinder, 2008).

In recent times, India witnessed several extreme rain events like floods, hurricanes, and storms while, on the other hand, it was also experienced severe drought due to failure of monsoon rainfall. So, analyzing rainfall variation and the implementation of rainwater harvesting systems has become an integral part of better management of rainwater. Furthermore, it is observed that, in India, even though the annual average rainfall (1170 mm) is higher than the global average rainfall (800 mm), insufficient water is available to meet the demands of the growing population. Due to a lack of appropriate rainwater harvesting systems, most of the rainwater falling during monsoon rainfall is diverted to the fluvial system which finally ends up in seawater. In addition, very little rainwater is left above the ground's surface for groundwater recharge; as a result, tube wells have failed to discharge the water (Suresh, 2012).

However, the practice of RWH is not greatly emphasized in many parts of modern India after the construction of dam and irrigation projects. Therefore, recently, the Government of India has taken several important initiatives and programs for the conservation and management of groundwater including effective implementation of rainwater harvesting in the country (url:http://jalshaktidowr.gov.in/sites/default/files/Steps_to_control_water_depletion_Feb2021).

In the broadest sense, the term "rainwater harvesting" generally refers to a technology used to collect and store rainwater from rooftops, rock catchments, and land surfaces in order to provide water resources for humans, animals, or agricultural usage. In this process, the collected rainwater could be directly diverted for utilization or allowed to infiltrate the soil to recharge the groundwater.

8.2 WATER HARVESTING TECHNIQUES BROADLY CLASSIFIED INTO TWO TYPES

8.2.1 Roof Harvesting or Roof-based Rainwater Harvesting

The system of catching the rainfall where it falls is known as rainwater harvesting and, in rooftop harvesting system, the roof becomes the catchment from where the rainwater is collected. Rainwater harvesting systems could be installed in both new and existing buildings. The general requirements of harvested roof water are the convenience, quantity, and quality of rainwater collected which mainly play a key role in this equation by making rainwater available for usage. The quantity refers to the amount of obtainable water from the roof surface; it depends on the area of catchment, the roof texture, and the average annual rainfall of the region, which is normally very small as compared to the volume of rainfall on the open ground surface. The quality of harvested rainwater depends mainly on the season of water harvest, the material used for roof covering, and the complexity of the RWH system (Suresh, 2012). Harvested water can be stored either in a tank or diverted to an artificial groundwater recharge system. This method is simple and cost-effective and helps in enhancing the groundwater level of the area. Furthermore, the rainwater collected is classified into two categories based on the usage of roof-based rainwater harvesting.

8.2.1.1 Direct Use Storage

Rainwater collected from any building roof is conveyed to a storage tank. It is necessary to check the quality of the roof water before connecting the roof to the storage tank. The harvested water may be directly used for gardening and other domestic usages. The roof surface is the catchment for yielding the runoff in the rainwater harvesting process. In this process, runoff water is conveyed to the storage tank by a guttering system and a downpipe and filtering system can also be provided ahead of the storage tank to keep the roof water clean. Rainwater harvesting systems, especially those based on rooftop rainwater harvesting, are the most cost-effective and reliable method of supplying water for drinking and sanitation; even in remote places, its safe use can be guaranteed with a minimal additional expense like chlorination, ceramic filters, and solar disinfection (van Koppen et al., 2005). Overall, a roof water harvesting system requires three essential components such as a roof or catchment area, a water storage tank or collection system, and a utilization system.

8.2.1.1.1 Roof or Catchment

This is the prepared surface area on which the precipitation is incident, the runoff from which is collected. Catchment might be the roof top area of households, buildings, or designated ground area. A roof or catchment should have the characteristic of very low level of rainwater absorption, the surface of the roof must be solid, rather vegetative, so that it could yield a very high proportion of the incident rainwater. Therefore, generally, roofs prepared of metal sheets, tiles, or asbestos sheets are the most ideal catchment for this type of rainwater harvesting system, whereas roof surfaces with grass or any other plant materials, like palm/coconut leaf, are usually not recommended for this system. When there is enough rainfall, the anticipated volume of harvested rainwater should ideally be at least equal to the annual rate of intended consumption per person (Suresh, 2012).

8.2.1.1.2 Water Storage Tank or Collection System

A collection system is the arrangement made to collect and store rainfall with least quantitative loss. In general, it consists of a collection pipe covering the catchment area and a storage tank if the water is not to be used immediately. Collection systems are established in such a way that runoff is collected and stored by gravity. The storage structure might be a small plastic tank for household use or a concrete tank for larger-scale use. The size of the water storage tank affects how well the RWH system functions because larger storage tanks function better than smaller ones. A 500-liter water storage tank is likely to overflow during heavy rain, wasting a significant amount of runoff and maybe running out of water quickly. The storage tank may be a tank, drum, or cistern as water storage or collecting tanks, which can be installed above ground, below ground, or partially below ground. However, the basic specifications for any water storage tanks for better performance of RWH system are as follows:

1. Water losses from seepage and evaporation must be less than 5% of the daily requirement, and provision must be made to handle any overflow.
2. The storage tank should not be hazardous or dangerous for users.
3. If the water is to be used for drinking purposes, the tank should be fully covered to ensure the supply of high-quality water.
4. Ventilation should be provided in the tank to avoid the proliferation of anaerobic microorganisms.
5. Preferably, the tank should also be cost effective, eco-friendly, and long-lasting with hassle-free entry and exit of water. Furthermore, cleaning and monitoring should be easy and low-cost.

The amount of rainwater that is actually harvested from the roof surface depends on the size of the roof and the intensity of the rainfall. The approximate volume of roof water harvested can be calculated using the following equation (Suresh, 2012):

$$Q = C \times R \times A$$

where R = total annual rainfall (mm), A = guttered roof area (m^2), and C = coefficient of runoff.

The coefficient of runoff (from 0 to 1) accounts for the evaporation loss from the roof area and the losses from the roof and storage point.

The values for the runoff coefficient for different roof surfaces are given in Table 8.1.

It is necessary to gather information on annual rainfall as well as approximate roof size in order to determine the amount of water that may be anticipated from the RWH system. The following equation describes the difference between the harvested water and the utilizable volume of roof water.

$$U = E \times Q$$

where U = usable volume of harvested roof water, Q = quantity of roof water, and E = storage efficiency (<1). The value of E depends on tank size, climate, and method by which the water is drawn.

TABLE 8.1
Runoff Co-efficient for different types of roof surface.

S.No	Roof surface	Runoff coefficient
1	Galvanozed iron sheet	>0.9
2	Tile (glazed)	0.6–0.9
3	Asbestos sheet	0.8–0.9
4	Organic (thatch, plam)	0.2

8.2.1.1.3 Utilization System

This is the arrangement needed to make use of the stored rainwater for productive purposes. It usually consists of a distribution system that directs water to the point-of-use through a channel or pipes. If there is no possibility for gravity flow, an electric pump may also be added as part of a utilization system.

8.2.1.2 Groundwater Recharge

Many different types of structures could be used to replenish groundwater aquifers so that rainfall will percolate into the earth rather than draining away off the surface. Geologically advantageous planning and construction are required, with the recharge structures being constructed using permeable soil, murrum (gravelly lateritic material), etc. The structures must not be built in locations with rock, slopes, or impenetrable clayey soils. Enough measures must be implemented to prevent the intrusion of sewage water and polluted urban surface runoff into recharging structures. Recharge structures are intended to be planned and built in places with significant groundwater improvement for a variety of functions. The following are some common recharging techniques.

8.2.1.2.1 Filling-up Bore Wells

Rainwater collected on the building's roof is channeled into a filter tank by way of drain pipes. After settlement, purified water is delivered to defunct bore wells to potentially recharge them by filling up depleted deep aquifers (https://housing.com/news/different-rain-water-harvesting-methods/).

8.2.1.2.2 Recharge Pits

A recharge pit also helps to replenish groundwater by recharging the deep aquifers. Recharge pits are small pits with holes spaced out at regular intervals that can be built using stone or brick masonry walls to aid in the infiltration of water or to refuel a borewell. The pit's capacity is determined by the catchment area, rainfall intensity, and the rate of soil recharge. Usually, the pits have dimensions of 1 to 2 m width and 2 to 3 m depth.

8.2.1.2.3 Trenches for Recharge

Recharge trenches are built by excavating the ground and then replacing it with porous materials like rocks and pebbles. The length of the trench is selected based on the anticipated amount of runoff, where the width varies from 0.5 to 1.0 m and depth of the recharge trench would ideally be 1.5 m. This method of recharge may be useful for roadside drains, modest residences, playgrounds, and parks.

8.2.2 Runoff Harvesting

Rainwater can flow away as surface runoff, and, in this technique, runoff water is directed into a small pond or reservoir that has been constructed to collect runoff water using earthen bunds or embankments. Ground catchment techniques offer greater possibility for collecting water from a larger surface area than rooftop catchment techniques. The potential sources from which we could capture runoff are as follows.

FIGURE 8.1 Dugout farm pond.

8.2.2.1 Dugout Ponds

Dugout ponds are constructed for the purpose of receiving runoff from the catchment during the monsoon season; these ponds are built by excavating the soil from the ground surface at the lower or intermediate reaches of the hill slope. The pond's size is determined by the catchment area, water requirement, and annual rainfall totals. To account for evaporation and seepage losses, the storage capacity should be at least twice the total volume of water needed.with 10% additional storage space made available for sediment buildup. A typical representation of dugout farm pond is shown in Figure 8.1.

8.2.2.2 Embankment-type Farm Ponds

Embankment-type farm ponds are generally constructed across the stream or water courses to store the runoff water. Such ponds are made up of an earthen dam, whose size is determined by the volume of water to be stored. For embankment and dugout-cum-embankment, an earthwork ratio of 5 to 20 could be attained depending on the site characteristics.

8.2.2.3 Percolation Tanks

Percolation tanks are man-made ponds that submerge a piece of permeable land so that groundwater can recharge through percolation (Figure 8.2). Large campuses with accessible acreage and the right topography may build these ponds. This tank could receive both surface runoff and direct roof runoff.

8.2.2.4 Check Dams

Check dams are small barriers built across the direction of water flow on shallow rivers, nala (small drainage channel), and streams for water harvesting (Figure 8.3). The main goal of check dam construction is to minimize the slope steepness of the gully

FIGURE 8.2 Percolation tank.

FIGURE 8.3 Check dam.

bed by building a series of checks at regular intervals. In order to absorb runoff and force it to percolate through the soil profile, impediments can be placed in the flow channel to slow the speed of the running water. Check dams could provide protective irrigation in times of moisture stress to increase the productivity of the crops. The water stored in the dam could be made available for livestock and domestic needs and also to retain water flow in the nala (small drainage channel) for a greater number of days.

8.3 *IN-SITU* MOISTURE CONSERVATION TECHNIQUES

Harvesting rainwater where it moves into the soil profile is known as *in-situ* water harvesting or *in-situ* moisture conservation. The small quantity of the rainwater could be harvested through *in-situ* moisture conservation techniques. In arid and semi-arid regions, this technique combines the advantages of water harvesting and mulching and thus reduces evaporation and finally runoff. The following soil moisture conservation techniques are generally used under rainfed conditions.

8.3.1 Contour Bunding

Contour bunding refers to the construction of small embankment-like structures ("bunds") at predetermined intervals across a land slope. Bunds store water between them that percolates into the soil and flattensthe slope of the land, thus reducing the velocity of surface runoff. Contour bunding is applicable for land with a slope of 2–8% and, in low-rainfall areas, for moisture conservation.

8.3.2 Bench Terracing

Bench terracing is the construction of step-like fields on hill slopes along the contours by half cutting and half filling. Long uninterrupted slopes could be converted by bench terracing into a number of small platforms which could be used for cultivation. Bench terraces act to decrease the rate of runoff, thus decreasing soil erosion and increasing soil moisture.

8.3.3 Contour Trenching

Contour trenching is a method of constructing staggered or continuous trenches along the contours on a sloped field to capture and infiltrate runoff water. Contour trenching is done predominantly on land with a steep gradient and shallow soil depth. Trenches should be continuous in medium- to high-rainfall areas with 0.2–0.5% slopes, as in the case of graded bunds. In low-rainfall areas, trenches might be discontinuous and arranged in a staggered manner (Figure 8.4).

8.3.4 Contour Farming

Contour farming is a kind of farming that conserves the soil and water by creating ridges to facilitate water break (Figure 8.5). Furrows that don't follow the natural contours of the soil when it rains produce speedy runoff because they provide an easy path for water to flow downward. Furthermore, the contour lines create a water break that lessens the enlargement of rills and gullies during heavy rainfall, allowing water to percolate into the soil.

8.3.5 Conservation Furrows

The opening of furrows parallel to rainfed crop rows across the slope of land with a country plough results in conservation furrows. During runoff, the rainwater gets collected and infiltrates into the soil within these furrows and is accessible to the crop for a longer duration compared to no-furrow (Figure 8.6).

FIGURE 8.4 Staggered trench.

FIGURE 8.5 Contour farming.

8.3.6 Mulching

The protective covering of the soil surface around the plants by living or non-living mulch to achieve reduced water evaporation makes for more efficient crop production and is called mulching (Chakraborty et al., 2008; Kader et al., 2017). Mulching

FIGURE 8.6 Conservation furrows.

optimizes water use and helps to enhance crop yield under rainfed conditions (Yu et al., 2018). The principal objectives of mulching are to control soil erosion, save water in low-rainfall regions, and to reduce water vapor loss, weed problems, and nutrient loss (Van Derwerken and Wilcox, 1988).

There are two types of mulching (Figure 8.7). The first type is the inorganic mulches, such as gravel or plastic films the latter being completely impermeable to water, while the second type is the natural or organic mulches, such as leaves, straw, grass clippings, and compost. Mulching reduced soil erosion and loss of water throughout the soil surface by avoiding quick evaporation. In this manner, mulching plays a fruitful part in conserving water (Sharma and Bhardwaj, 2017). Similarly, it

FIGURE 8.7 Inorganic (plastic) and organic (natural) mulching.

helps to control temperature variations and improves physical, chemical, and biological properties of the soil by supplementing nutrients to the soil (Dilip et al., 1990).

8.4 ADVANTAGES OF RAINWATER HARVESTING

Rainwater Harvesting (RWH) has several advantages not only for human being but also for the effective use of renewable resources, which supports the management of ecosystems and the management of adaptation to climate change. The simple and easy implementation of rainwater harvesting systems is an added advantage of this technology. Furthermore, this technology is suitable for all types of climate, is cost-effective and highly decentralized, and can be executed at the individual and community levels. Rainwater is known to be a pure form of water, free from any contagious microorganisms and organic matter, which can be harvested and used as potable water. Rainwater harvesting technology also helps to reduce the salinity level in saline areas and maintains a balance between the fresh and saline water interface, especially in coastal regions.

In agriculture, rainwater harvesting has shown the potential of doubling food production, compared with the 10% increase achievable from irrigation. Furthermore, the implementation of rainwater harvesting systems also helps in the prevention of soil erosion, reduction of flooding, and preserving of the existing groundwater table through recharge. Another important benefit of rainwater harvesting is reduced salt accumulation and promotion of a good soil environment for plant root growth. This happens when gathered rainwater percolates deep into the soil to dilute available salt in that particular zone, resulting in better root growth and water uptake. Similarly, another positive effect of rainwater harvesting is the increased species diversity among flora and fauna. In addition, reports also highlight the reduced energy requirements which results in reduced CO_2 emissions associated with rainwater harvesting as compared with conventional water supply technologies. An important, but not always mentioned affect, is on the aesthetic and cultural ecosystem services, where rainwater harvesting has improved both rural and urban area vegetation for improved human well-being.

8.5 CONCLUSIONS

This chapter highlights the simplest technology used for water conservation in diverse rainfall regions. It will not only provide the most sustainable and efficient means of water management but also unlock the vista of several other economic pursuits that will empower people at the grass-root level. The different rainwater harvesting interventions discussed in this chapter can have several benefits both on ecosystems and human well-being. These synergies between humans and ecosystem are especially evident when rainwater harvesting initiatives boost rainfed agriculture, through watershed management, and deal with household water supplies in urban and rural regions.

Rainwater harvesting is a local intervention, with primarily local benefits on ecosystems and human livelihoods. Rainwater harvesting interventions should always be compared with alternative water management interventions and infrastructure investments. Measures have to be taken for considering rainfall as a significant renewable resource in water management policies, strategies, and plans. Further rainwater harvesting interventions can be included as potential opportunities in land and water resource management activities for human and ecosystem welfare. Overall, it's time to appreciate the value of rainwater harvesting technology and educate people about the need to address the worldwide water crisis.

REFERENCES

Ashraf, M., and Foolad, M. (2007). Roles of glycine betaine and proline in improving plant abiotic stress resistance. *Enviro Exp Bot.* 59: 206–216.

Barron, J. (2009). Rainwater harvesting: A lifeline for human well-being. United Nations Environment Programme and Stockholm Environment Institute. www.unep.org/depi/.

Chakraborty, D., Nagarajan, S., Aggarwal, P., Gupta, V. K., Tomar, R. K., Garg, R. N., Sahoo, R. N., Sarkar, A., Chopra, U. K., Sarma, K. S. S., and Kalra, N. (2008). Effect of mulching on soil and plant water status and the growth and yield of wheat (Triticum Aestivum L.) in a semi-arid environment. *Agric Water Manag.* 95: 1323–1334.

Dilip, K. G., Sachin, S. S., and Rajesh, K. (1990). Importance of mulch in crop production. *Indian J. Soil Cons.* 18: 20–26.

Gould, J., and Nissen, P. E. (1999). *Rainwater Catchment Systems for Domestic Rain: Design Construction and Implementation.* London: Intermediate Technology Publications, p. 335.

Hatibu, N., Mutabzi, K., Senkondo, E. M., and Msangi, A. S. K. (2006). Economics of rainwater harvesting for crop enterprises in semi-arid areas of East Africa. *Agric. Water Manag.* 80(3): 74–86.

Kader, M. A., Senge, M., Mojid, M. A., and Ito, K. (2017). Recent advances in mulching materials and methods for modifying soil environment. *Soil Tillage Res.* 168: 155–166.

Rockstrom, J., Hatibu, N., Oweis, T. Y., Wani, S. P., Barron, J., Bruggeman, A., Farahani, J., Karlsberg, L., and Qiang, Z. (2007). Managing water in rainfed agriculture. In Molden, D. (Eds.), *Water for Food, Water for Life: A Comprehensive Assessment of Water Management in Agriculture.* London: Earthscan, and Colombo: International Water Management Institute, pp. 315–352.

Sharma, R., and Bhardwaj, S. (2017). Effect of mulching on soil and water conservation – A review. *Agri Revi.* 38(4): 311–315.

Sharma, B. R., Rao, K. V., Vittal, K. P. R., and Amarasinghe, U. (2008). Converting rain into grain: Opportunities for realising the potential rainfed agriculture in India. Proceedings National Workshop of National River Linking Project of India, International Water Management Institute, Colombo (pp. 239–252). http://www.iwmi.cgiar.org/Publications/Other/PDF/NRLP%20Proceeding-2%20Paper%2010.pdf.

Singh, U. (2008). A History of Ancient and Early Medieval India: From the Stone Age to the 12th Century. New Delhi: Pearson Education, p. 155. ISBN 978-81-317-1120-0.

Suresh, R. (2012). *Soil and Water Conservation Engineering.* Standard Publishers Distributors, pp. 591.

Van Derwerken, J. E., and Wilcox, L. D. (1988). Influence of plastic mulch and type and frequency of irrigation on growth and yield of bellpepper. *Hortic. Sci.* 23: 985–988.

Van Koppen, B., Namara, R., and Stafilos-Rothschild, C. C. (2005). Reducing poverty through investments in agricultural water Management: Poverty and gender issues and synthesis of Sub-Saharan Africa Case study Reports. Working paper 101. International Water Management Institute, Colombo.

Yu, Y. Y., Turner, N. C., Gong, Y. H., Li, F. M., Fang, C., Ge, L. J., and Ye, J. S. (2018). Benefits and limitations to straw and plastic film mulch on maize yieldand water use efficiency: A meta-analysis across hydrothermal gradients. *Eur. J. Agron.* 99: 138–147.

9 Circular Economy in Urban Wastewater Management

A Concise Review

Anirban Biswas
Department of Environmental Science, Nabadwip Vidyasagar College, Nabadwip, India

Saroni Biswas
Centre for Sustainable Development and Research, Kolkata, India

9.1 INTRODUCTION

Circular economy principles can transform urban wastewater management in a resource-efficient sustainable process. Worldwide, current wastewater management practices have very low nutrient recovery and reuse options, causing massive environmental difficulties such as eutrophication, climate change, and, to some extent, food and health insecurity (Kjerstadius et al., 2017; Skambraks et al., 2017; Wielemaker et al., 2018; Hoffmann et al., 2020; Öberg et al., 2020). On the other hand, nutrient recovery may be key drivers for sustainability in the circular economy of wastewater management (Hoffmann et al., 2020; Larsen et al., 2021). Circular economy principles applied to urban wastewater management offer significant potential for sustainable resource recovery, reduced environmental impact, and enhanced water resilience. By integrating source separation, advanced treatment technologies, water reuse, resource recovery, and stakeholder collaboration, and any large- scale wastewater treatment facilities, cities can optimize their wastewater management systems and contribute to a more sustainable strategy.

Wastewater contains a few important nutrient elements among other substances, including nitrogen and phosphorus, which are vital plant nutrients. In Finland and Sweden, annually, about 9,800 tons of phosphorus (P) and 78,000 tons of nitrogen (N) are recovered in the wastewater treatment plants (SYKE, 2019; SEI, 2020); in the effluent, approximately 4 and 5% of phosphorus and 34 and 40% of nitrogen is left and released to water bodies in Finland and Sweden, respectively (Naturvårdsverket, 2018; Dagerskog and Olsson, 2020; Lehtoranta et al., 2021). Most of the phosphorus

DOI: 10.1201/9781003432869-9

and nitrogen removed from the wastewater in different (physical, chemical, biological) treatment processes accumulates in the sludge while a small fraction is contained in the wastewater in a plant-available form (Warman and Termeer, 2005). Current common practice is to use the wastewater treatment plant-produced sludge in landfill and or landscaping in certain small areas, whereas most of the sludge is unused, with a small portion being utilized by plants as nutrient sources while the rest are pollutants in waterbodies (Valtanen et al., 2015). There are certain contamination risks to soil, crops, and waterbodies by pathogens, heavy metals, and organic pollutants (e.g., pharmaceuticals and personal care products, etc.) from the sludge (Seleiman et al., 2020), as the current wastewater treatment processes are insufficient to remove them, and these are retained in the sludge (Magnusson and Norén, 2014; Vieno, 2014; Talvitie et al., 2017; Lehtoranta et al., 2021a).

Source separation is an efficient way of recovering nutrients from municipal wastewater. Municipal wastewater contains black water (containing feces and urine) and gray water (from kitchen appliances, showers, etc.), creating problems for nutrient recovery from highly concentrated flows, mixed with harmful substances. Gray water contains a high concentration of heavy metals, which can be separated at source. Metal concentrations in black water are generally much lower than in sewage sludge and can be utilized effectively in agriculture. The black water fraction mainly contains the pathogens in wastewater and the pathogen availability depends on the treatment method used in treating the separated black water (Fidjeland et al., 2015), which is a robust option for safe recycling of plant nutrients.

9.2 CIRCULAR ECONOMY IN WASTEWATER MANAGEMENT

The conventional linear model of wastewater management, targeted toward pollutant removal and discharge, is being replaced by a circular approach of resource recovery, reuse, and recycling (Mo and Zhang, 2013; Rashidi et al., 2015). The principle of the circular economy is based on reducing, reusing, recycling, and recovering resources to minimize waste generation and maximize resource efficiency. In the urban wastewater management, circular economy principles can be implemented at different stages of the wastewater treatment process, in the following seven subsections.

9.2.1 Source Separation and Segregation

One key step in adopting a circular economy approach to wastewater management is source separation (Lennartsson et al., 2009; Malila et al., 2019) and segregation of wastewater streams. By source separation, different streams of wastewater can be treated and processed separately, allowing more efficient resource recovery and recycling.

9.2.2 Advanced Treatment Technologies

Conventional wastewater treatment processes aims at pollutant removal; however, any advanced treatment technologies can extract valuable resources from wastewater like the recovery of nitrogen and phosphorus, which can be used as fertilizers (Drewes and Horstmeyer, 2015). In addition, energy recovery from wastewater, through microbial fuel cells or anaerobic digestion, can contribute to renewable energy production.

9.2.3 Water Recycling

Effective water reuse is environmental, economic, and socially beneficial. Circular economy focuses on water recycling and reuse (Giakoumis et al., 2020). Treated wastewater can further be treated to meet the specific standards and can be used for some non-potable applications, like toilet flushing, irrigation, gardening, and industrial processes. Thus, water recycling leads to reduced freshwater demand.

9.2.4 Resource Recovery

Wastewater contains various valuable resources, including organic matter, nutrients, and metals. Circular economy involves the identification, collection, and processing of waste materials to extract and recover value-added resources like some materials and chemicals (Chen, 2022; Ingrao et al., 2018). For example, organic matter can be converted into biogas through anaerobic digestion, providing a renewable energy source. Nutrient-rich wastewater can be used for agro-irrigation, reducing the need for synthetic fertilizers. Metals and other valuable compounds can be extracted from wastewater sludge through various extraction techniques.

9.2.5 Innovative and Circular Design

A circular economy is the way to deliver a service or product, which is functional and made of the optimum materials considering resource recovery and recyclability and minimizing its negative impact throughout the whole lifecycle (Aho, 2016). This involves considering the entire lifecycle of the infrastructure and equipment, optimizing their durability, modularity and their ease of maintenance and repair. Innovative technologies, such as membrane filtration, electrochemical processes, and bioconversion techniques, can also play a role in improving resource recovery and overall system efficiency.

9.2.6 Integration with Urban Systems

The urban wastewater treatment plants (WWTPs) are an important part of the circular economy for energy production and resource recovery during clean water production (Mo et al., 2013; Rashidi et al., 2015).To achieve a circular economy, integration with other urban systems is crucial. This includes considering the broader urban interactions, such as optimizing the use of resources within the city, promoting eco-industrial symbiosis, and creating synergies with other waste streams. Collaboration between stakeholders, including municipalities, industries, and researchers, is essential to creating a holistic and integrated approach to circular wastewater management.

9.2.7 Stakeholder Engagement and Policy

To drive the adoption of circular economy principles in urban wastewater management, supportive policies and regulations are crucial. Governments can incentivize resource recovery, water reuse, and recycling through financial mechanisms, tax incentives, and regulatory frameworks. Stakeholder engagement is essential to foster collaboration between municipalities, water utilities, industries, and the public. Raising awareness about the benefits of circular wastewater management can encourage behavioral changes and facilitate the transition toward a circular economy.

9.3 METHOD

The review method of the literature included the steps of identification, screening, eligible records inclusion, and the final step, meta-analysis (Figure 9.1).

9.3.1 Search Selection

Published work were systematically searched in the widely used academic bibliographic databases like PubMed (http://www.ncbi.nlm.nih.gov/pubmed), Science Direct (www.sciencedirect.com/), Web of Science (https://webofknowledge.com/),

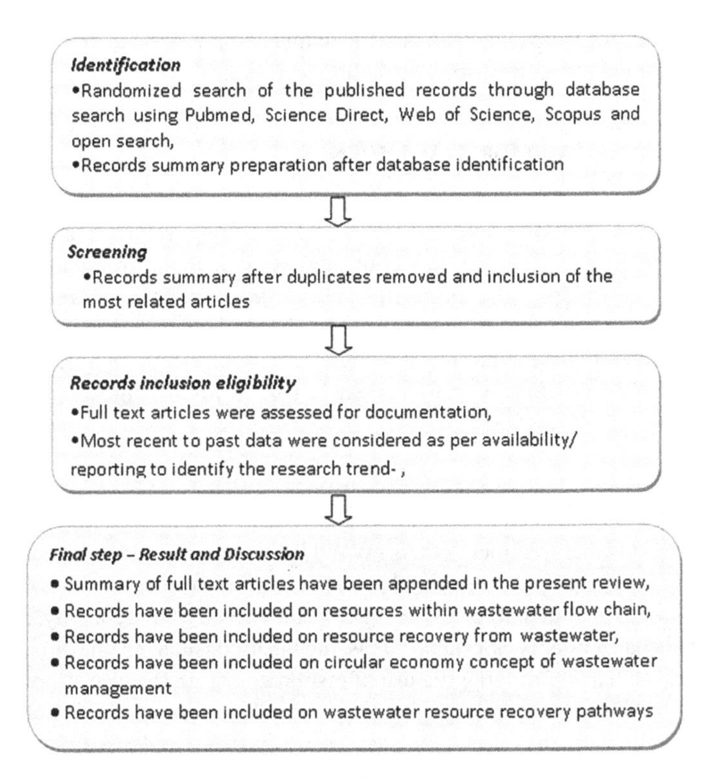

FIGURE 9.1 Review process flowchart and selection criteria.

TABLE 9.1
Database Search Criteria and Revisions Including PubMed, Science Direct, Web of Science, Scopus, and Open-Source

Total records identified through database search including all the databases and open-source	734
Summary of full-text articles appended in the present review excluding duplicates and considering the greatest relatedness	77
Wastewater flow chain	30
Resources within	20
Resource recovery from wastewater	14
Circular economy concept of wastewater management	22
Wastewater resource recovery pathways	38

and Scopus (www.scopus.com/), plus literature searches using Google Scholar and searches of the references of the publications included. To search the database, comprehensive English terms representing "Wastewater" and "Recycling" to "Circular economy" were used, with the Boolean operator "and"; before conducting the full search, a pilot search was made with the required modifications to meet the search criteria. The keywords were searched in each bibliographic database. The searched articles had to meet the following criteria to be included in the review process: (a) the study was published in journals, (b) the study was conducted to identify the resources within the wastewater flow chain, (c) the resource recovery process from municipal wastewater, (d) the circular economy interfacing with resource recovery from wastewater, and (e) the circular economy in urban wastewater management. Articles that met the screening test underwent full-text review by the authors. During these extensive database searches, any duplicate details were excluded.

9.3.2 Documentation, Identification, and Screening

The searched data were categorized on the basis of "selected criteria" to meet the review aim (Table 9.1). Article titles and keywords were checked thoroughly for this selection. Quality assessment of the selected articles was performed by extending the selection criteria, depth of the data, and exact presentation in the results to the discussion section to meet our aim (Biswas, 2019). This selectivity resulted from the enormous numbers of research and review articles available related to wastewater reuse, resource recovery from wastewater, circular economy concept, circular economy in wastewater management, etc., and was required concerning data acceptability and consideration for this particular review.

9.4 RESULT AND DISCUSSION

9.4.1 Resource Recovery at Wastewater Treatment Plants (WWTPs)

Resource recovery is the process of extracting and reusing valuable materials from the wastewater treatment process. This minimizes waste generation, reduces the environmental impact, and optimizes resource utilization. Some common resources that can be recovered from wastewater treatment facilities include:

Nutrients. Depending on wastewater sources, it can contain valuable nutrients like nitrogen and phosphorus which can be recovered and reused as fertilizer.

Water reuse. Depending on the wastewater sources, some part of the wastewater can be used fpr non-potable applications such as some industrial processes, toilet flushing, gardening, car washing, etc., by which freshwater resources can be in less demand.

Metals and chemicals. Sometimes, wastewater treatment processes can recover specific metals and chemicals which are valuable resources.

Bio-solids. The remaining solids, after the wastewater treatment, are called bio-solids which can further be processed into fertilizer and/or soil conditioner.

Biogas. To break down the organic matter in wastewater, anaerobic digestion is a common process that produces biogas, a mixture of methane and carbon dioxide, which can be stored and used as renewable energy source.

Although the scientific community has developed wastewater resource recovery technologies considerably in recent years, their widespread application in municipal wastewater treatment facilities (WWTPs) is still lacking. Both technical and numerous non-technical explanations can be used to explain this. Sustainable urban development heavily relies on wastewater management (UNEP, 2010). Historically, the purpose of wastewater treatment has been to safeguard downstream users' health. In more recent decades, protecting the ecosystem by preventing nutrient poisoning of surface waterways has become an additional goal. As a result, methods for removing phosphorus (P) and nitrogen (N) have been added to WWTPs (Verstraete et al., 2009). The most popular method of treating wastewater is the standard activated sludge (CAS) process, in which aerobic bacteria break down the wastewater's organic content while receiving continual oxygen. Although the CAS method is successful in achieving the legal effluent quality standards, it is not seen to be sustainable due to its limited potential for resource recovery and lack of cost-effectiveness, as well as its high energy requirement and significant environmental impact (Verstraete and Vlaeminck, 2011).

9.4.2 Nutrient Recovery

Nutrient recovery is the process of removing and recycling useful nutrients from wastewater, such as those present in large quantities in wastewater, particularly in domestic sewage and agricultural runoff. Nutrient recovery attempts to recover and use these nutrients in an efficient manner rather than allowing them to be released into water bodies, where they may contribute to water pollution and ecological damage (Hoffmann et al., 2020; Larsen et al., 2021).

There are several technologies and approaches for nutrient recovery in wastewater treatment.

Biological Nutrient Removal (BNR). BNR processes, involving the use of microbes, are commonly used in wastewater treatment plants to remove nutrients. In this process, bacteria are employed to convert nitrogen compounds (ammonia) into nitrogen gas (denitrification) and to convert phosphorus into a form that can be readily precipitated and removed (Wainaina et al., 2021).

Struvite Precipitation. Struvite is a crystalline compound, magnesium ammonium phosphate (MAP), that can be used as a fertilizer in agriculture. It is produced in the WWTPs and facilitated by secondary treatment and anaerobic sludge digestion facilities (Barr and Munch, 2001; Jaffer et al., 2002).

Enhanced Biological Phosphorus Removal. Some processes can selectively encourage the growth of bacteria that can store excess phosphorus as intracellular polyphosphate. This stored phosphorus can then be removed from the wastewater through sludge washing and subsequently recovered.

Membrane Filtration. Membrane filtration processes, viz. ultrafiltration, microfiltration, and reverse osmosis, can be used to separate and concentrate nutrients from wastewater. The concentrated nutrient stream can then be further processed to recover the nutrients (Salman et al., 2022).

Nutrient Recovery through Algal Cultivation. Micro algae have the ability to assimilate nutrients, including nitrogen and phosphorus, from wastewater during growth. Algae can be cultivated in wastewater ponds or bioreactors, allowing for the uptake of nutrients. Afterward, the harvested algae can be used as a nutrient-rich biomass for various applications, such as bioenergy production, animal feed, or soil amendment (Goh et al., 2022).

Worldwide, approximately 20% of manufactured nitrogen and phosphorus is contained in domestic wastewater (Batstone et al., 2015; Matassa et al., 2015), most of which has recovery potential. In the water-scarce city of Monterey (California, USA), a large agricultural area is supplied with almost 80,000 m^3 per day of nutrient-rich reclaimed municipal wastewater to irrigate and fertilise crops. Implementing nutrient recovery technologies in wastewater treatment systems has multiple benefits. It reduces the discharge of nutrients into the environment, thereby mitigating water pollution and eutrophication. Additionally, nutrient recovery can contribute to the production of valuable resources like fertilizers, reducing the reliance on conventional chemical fertilizers. It also promotes sustainability, resource efficiency, and the circular economy by closing the nutrient loop and creating a more sustainable wastewater management approach.

9.4.3 Water Reuse

Water reuse, water reclamation, or recycled water, is the treatment and reuse of wastewater for various purposes, instead of discharging it into the environment. Wastewater treatment plants play a crucial role in the process of water reuse by treating wastewater to a quality suitable for specific non-potable applications.

Some key methods of water reuse from wastewater treatment plants can be summarized thus.

Treatment Process. Wastewater undergoes an advanced treatment process to remove impurities and contaminants before it can be reused. This treatment typically involves primary, secondary, and tertiary treatment processes, which may include physical, chemical, and biological treatment steps. The objective is to achieve a high level of water quality that meets the specific requirements of the intended reuse application.

Non-potable Applications. Recycled water from wastewater treatment plants is commonly used for non-potable purposes, which do not require drinking water quality. Examples of non-potable water reuse applications include:

- Irrigation of agricultural crops, landscaping, parks, and golf courses.
- Industrial processes, such as cooling towers, manufacturing, and construction.
- Toilet flushing in commercial buildings, public facilities, and residential developments.
- Groundwater recharge, where treated wastewater is infiltrated into the ground to replenish groundwater reserves.
- Environmental restoration of wetlands, rivers, and lakes.

Recycled water originated from any WWTPs, in case of any municipality, significantly reduces the freshwater demand as a whole (Verstraete et al., 2009). There is a success story of wastewater reclamation and reuse in the Windhoek city in Namibia, where recycled wastewaters meet 25% of the city's potable water supply (Verstraete et al., 2011). As another example the city of Chennai (India) reuse 40% of the generated wastewater for satisfying 15% of the population demand (IWA, 2018). At Xi'an University in China, water treatment system produces water for non-potable uses like toilet flushing, gardening and waterfront landscaping which count 50% less freshwater usage. Singapore and Israel are continuing with nationwide wastewater reuse schemes. In Israel, around a quarter of the country's water demand is met by treated wastewater, and Singapore has achieved 40% with its wastewater reclamation plant (PUB, 2016).

Wastewater of any municipal WWTP contains only domestic wastewater, small part of industrial water and storm water. Agriculture water usage does not come with this wastewater surge in many Western countries.In Netherlands, WWTPs equate to 20% of the total volume of fresh water (Ranade and Bhandari, 2014). Ultra and micro filtration could reduce these freshwater abstractions by 17%, while reverse osmosis could reduce it by 13%. Osmosis is the only technology which can reclaim a potable quality water supply (Ranade and Bhandari, 2014).

Water recycling from WWTPs is an increasingly important strategy for sustainable water management. By maximizing the use of available water resources, it contributes to water conservation, environmental protection, and long-term water sustainability. Water reuse offers several advantages, including conserving freshwater resources, reducing the demand for potable water, and providing a sustainable solution for water scarcity issues. It can also help alleviate pressure on natural water sources and reduce the discharge of treated wastewater into sensitive ecosystems. However, challenges associated with water reuse include public acceptance, ensuring water quality and safety, addressing potential health risks, and addressing concerns about esthetics and odor.

9.4.4 Energy Recovery

Energy recovery in wastewater treatment refers to the process of capturing and utilizing the energy present in wastewater to generate heat, electricity, or other usable forms of energy. Wastewater treatment plants can produce significant amounts of energy in the form of biogas and thermal energy, which can be harnessed for various purposes, although a detailed understanding of the available energy in any wastewater stream is a critical step to developing resource and energy recovery from any WWTP. Some common methods of energy recovery in wastewater treatment plants are as follows

Biogas Generation. During the treatment process, organic matter in wastewater is broken down through anaerobic digestion, producing biogas (Gude, 2015). Biogas is a mixture of methane and carbon dioxide, which can be captured and used as a renewable energy source. Biogas can be used to generate electricity and heat through combined heat and power (CHP) systems or used directly as a fuel for boilers, engines, or turbines.

Heat Recovery. Wastewater contains thermal energy that can be recovered and utilized. Heat exchangers can capture the heat from wastewater and transfer it to other parts of the treatment process or to heat buildings or nearby facilities. This reduces the need for additional energy sources for heating purposes, thus increasing energy efficiency.

Wastewater contains considerable amounts of chemical energy which can be converted to heat and subsequently to electrical energy. Primary sludge contains almost 66% of the energy entering the WWTPs and the rest enters in the secondary treatment (Shizas and Bagley, 2004). A municipal WWTP holds a marked proportion of the total energy demand of the city being serviced (Schopf et al., 2018). But potential chemical energy acquired in typical municipal wastewater is very high (Wan et al., 2016). In the Netherlands, WWTPs have potential energy in the form of methane, if all the chemical oxygen demand (COD) of the influent were to enter the anaerobic digester and converted into biogas at 80% efficiency. However, only 25% of this maximum potential energy has been used (Frijns et al., 2013). It has been estimated that methane recovered from wastewater would have been a substitute for natural gas consumption in the Netherlands, although attaining a maximum of 1% of demand. If all the methane were converted to electricity, it would have met less than 1% of the total electricity demand of the Netherlands. The sludge contains 80% water, so the dewatering and incineration process consumes a lot of energy, resulting in the low energy-recovery potential of sludge (Frijns et al., 2013). The anaerobic digestion process produces biogas (largely methane) which can be used to compensate for 25–50% of the energy required in the activated sludge process, and some plant modifications may further reduce the considerable energy requirements (Gude, 2015). It has been reported that sometimes a WWTP contains more than ten times the energy required for the treatment (Shizas and Bagley, 2004). Thermal energy contained in WWTP effluent is more than the on-site heat energy demand as these WWTPs have high potential for providing support for industrial purposes (Kretschmer et al., 2016). Overall, the heat energy recovered from Dutch WWTPs has an energy recovery potential approximately ten times higher than the heat derived from recovered methane combustion in a CHP unit (Roest et al., 2010). As the time advanced, minimal loss of the energy as volatiles has also been reported (Heidrich et al., 2011). It is also a matter of fact that not all the available energy in wastewater can be harvested in a usable form (Shizas and Bagley, 2004; Heidrich et al., 2011). An energy balance comparison between the conventional activated sludge (CAS) treatment and anaerobic digestion process ended up commenting that the CAS process requires 25 kW h electrical energy per capita per year and anaerobic digestion produces a net 5 kW h electrical energy. Hence it proves that new and advanced designed WWTPs have greater potential for energy reuse and recovery, otherwise there may be a loss of about a 30 kW h per capita per year equivalent of electrical energy.

If the maximum captured energy from wastewater were available for reuse, then the WWTPs could have been net energy producers rather than consumers. Implementing energy recovery systems in wastewater treatment plants offers several benefits. It reduces the reliance on fossil fuels, lowers operating costs, and decreases the carbon footprint of the treatment process. Energy recovery also provides an opportunity for wastewater treatment plants to become more self-sufficient and potentially even generate revenue through sales of excess energy.

9.4.5 Fertilizer Production Potential

Wastewater treatment plants have the potential to produce fertilizers from the by-products generated during the treatment process. The resulting products can be used to enhance soil fertility and provide essential nutrients for plant growth. Here are some aspects related to the fertilizer potential of wastewater treatment.

Biosolids. Biosolids are the residual solids generated from wastewater treatment processes (Gerba and Pepper, 2009). These organic solids can be further treated and processed to produce fertilizer products. Biosolids are typically rich in nitrogen, phosphorus, and other essential nutrients. Through proper treatment and stabilization processes, biosolids can be transformed into a safe and beneficial fertilizer material. Currently, most US land application utilizes Class B biosolids (Gerba and Pepper, 2009).

Nutrient Recovery. Wastewater treatment processes often include technologies for nutrient recovery. Nitrogen and phosphorus, which are essential plant nutrients, can be extracted and concentrated from the wastewater. These recovered nutrients can be processed into fertilizer products that provide a concentrated source of nutrients for agricultural use.

Struvite Production. Struvite is a crystalline compound composed of magnesium, ammonium, and phosphate. It can be formed as a by-product during wastewater treatment, particularly through the process of struvite precipitation. Struvite can be collected, further processed, and used as a slow-release fertilizer in agriculture.

Compost. Some wastewater treatment plants utilize composting as a means of biosolid treatment and stabilization. Composting involves the controlled decomposition of organic matter and converts nitrogen from unstable ammonia to stable organic forms of nitrogen, reducing the volume of waste (Zhu et al., 2006). The resulting compost can be used as a fertilizer to improve soil quality and enhance plant growth and for overall soil amendment.

In humans, almost all the amount of phosphorus (P) consumed in diets is excreted from the body. It is estimated that, globally, about 17% of all mined mineral P ends up in human excreta. Urban areas are the "hotspots" of P, and urine is the single largest source of the P arising from them (Cordell et al., 2009). In the Flanders region of Belgium, the total P entering the WWTPs is equal to 8% of Flemish industrial P ore imports and 14% of the total fertilizer orthophosphate P used in the region. P could be recovered upon incineration of sludge ash with 90% efficiency (Cornel and Schaum, 2009), such recovery leading to supply potential of 11% of the Flemish fertilizer demand. On the other hand, soluble P recovery as struvite leads to total recovery at between 10 and 50%, depending on the applied treatment processes used (Cornel and Schaum, 2009; Wilfert et al., 2015). As a consequence, the struvite supply potential is lower (at 3%) than the sludge recovery route (Wilfert et al., 2015).

It has been estimated that wastewater N recovery practices can meet 30% of the global N fertilizer demand (Motlagh and Goel, 2014). In the Netherlands, where agriculture is intensive, only 18% of the demand is met by the N that enters WWTPs (Mulder, 2003). In Flanders, 14% of total N fertilizer demand or 4% of industrially fixed N could be met with N recovered from WWTPs. After CAS, only 20% of influent N is retained in the sludge with the currently available technologies (Siegrist et al., 2008; Matassa et al., 2015). The biodrying process of sludge is done in an energetically favorable way and, simultaneously, ammonium sulfate is recovered (Winkler et al., 2013), although it could satisfy only 2% of total N fertilizer demand of the Flemish.

Suitably treated wastewater effluent can be used for irrigation in agriculture. The effluent provides water and nutrients for crops, effectively acting as a form of irrigation and fertilization ("fertigation") simultaneously. This practice reduces the demand for freshwater resources and provides a sustainable source of nutrients for plant growth. It's worth noting that the production and use of fertilizers derived from wastewater treatment by-products require careful consideration of regulatory guidelines and standards to ensure the safety of both the environment and public health. Strict protocols are in place to ensure proper treatment, monitoring, and testing of these materials to meet quality and safety standards.

9.5 FUTURE CHALLENGES

The wastewater circular economy presents several future challenges that need to be addressed for its successful implementation.

Technology and Infrastructure. Establishing the necessary infrastructure and deploying advanced technologies for wastewater treatment and resource recovery can be a significant challenge. Upgrading existing wastewater treatment plants and implementing innovative technologies require substantial investments and technical expertise.

Regulatory Frameworks. Developing comprehensive and adaptable regulatory frameworks that support and encourage the circular economy in wastewater management is crucial. Regulations need to address issues such as water quality standards, resource recovery guidelines, and the safe reuse of treated wastewater.

Community Perception. Public acceptance plays a vital role in the success of the wastewater circular economy. Educating and engaging the public to overcome stigmas associated with wastewater and recycled products is essential. Communicating the safety and benefits of the circular economy approach is crucial for gaining public trust and support.

Innovative Technologies. Continued research and development are necessary to enhance and optimize technologies for resource recovery and water treatment. Advancements in areas such as membrane filtration, nutrient recovery, energy generation, and wastewater monitoring can help overcome existing challenges and improve the efficiency and effectiveness of the circular economy.

Collaboration. Achieving a circular economy in wastewater management requires collaboration among various stakeholders, including governments, water utilities, industries, research institutions, and communities. Establishing partnerships and fostering cooperation among these entities is essential for sharing knowledge, resources, best practices, and overcoming barriers to implementation.

Economic Efficiency. Making the circular economy financially sustainable is a challenge. Identifying viable business models, cost-effective technologies, and potential revenue streams from resource recovery can help ensure the economic feasibility of circular wastewater management.

Water Scarcity and Climate Change. Growing water scarcity due to climate change and population growth is a significant challenge. Implementing the circular economy in wastewater management can help alleviate water stress by maximizing water reuse and minimizing the discharge of treated wastewater. However, adapting to changing climatic conditions and ensuring sufficient water supply for various sectors require integrated planning and management strategies.

Addressing these challenges requires a multidisciplinary approach involving policymakers, engineers, researchers, businesses, and the public. Collaboration, innovation, supportive regulations, and public awareness are key factors in driving the wastewater circular economy forward and realizing its potential for sustainable water management, resource recovery, and environmental protection.

9.6 CONCLUSION

The circular economy of wastewater treatment has historically been mainly the utilization of nutrients either from sludge or directly from wastewater in agriculture. There are currently several projects in Europe that seek to develop the circular economy through holistic ecosystem thinking and cooperation, which is also the starting point for the circular economy in the big picture. Energy recovery is increasingly taking place through biogas or heat recovery, which is important for climate change. In addition, as primary sources become scarcer, resource utilization from wastewater has become an increasingly profitable option. Currently, nutrients, metals, carbon dioxide, alginate, proteins, and cellulose can be recovered from wastewater for industry. Water reuse and recycling is also very important, especially in countries

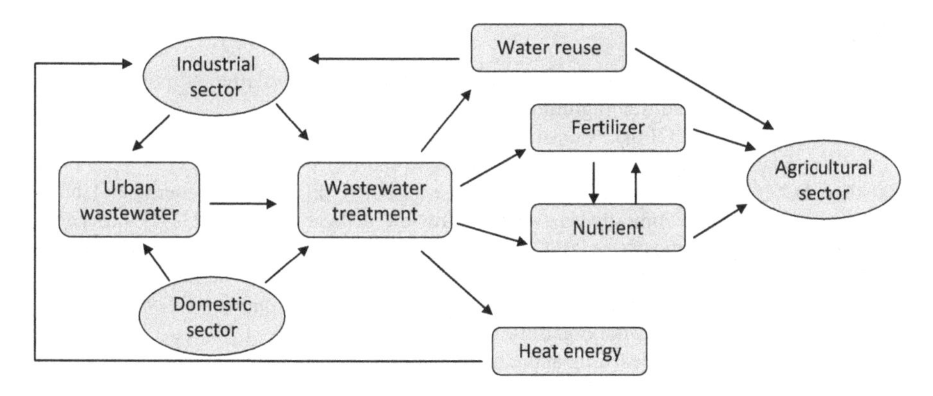

FIGURE 9.2 Wastewater circular economy pathways.

where freshwater sources are scarce. Wastewater treatment plants can also utilize industrial by-products in their process.

The above prospects will be made possible by automatic management. New energy solutions and treatment goals, integrated approaches, advanced wastewater treatment, modeling and process optimization, and scalable technology all need a large amount of data and its mathematical management, different metrics, and extensive use of existing data. In the future, artificial intelligence will use the process and environmental parameters to predict how the processes of a WWTP should be run in order to control the predicted event in the coming days.

Municipal wastewater treatment plants can play an important role in helping cities toward a sustainable future, characterized by circular flow of water, waste, material, and energy (Figure 9.2). WWTPs are beginning to be perceived, not only in the traditional role of wastewater and sewage sludge treatment, but also in the new role associated with resources and energy recovery. WWTPs in the near future are to become "ecologically sustainable" technological systems and a very important nexus in SMART cities.

- Utilities and government authorities should integrate water and wastewater management at strategic, planning, and implementation levels.
- Awareness should be built around wastewater management. The government can help in designing acceptable solutions, certifications, and authorization. Also, financial institutions can help through long-term capital.
- Blended finance models or viability gap funding should be established to boost public–private partnership, sharing equal risks.
- The government should provide financial and land resources and create an enabling business environment for the wastewater management techniques.

CONFLICT OF INTEREST

The authors declare that the review work is solely for academic purposes and there is no commercial or any financial conflict of interest.

REFERENCES

Aho, M. (2016). 'Designing circular economy'. Gaia Telegraph newsletter. http://www.gaia.fi/news-blogs/blogs/designing-circulareconomy.

Barr, K. and Münch, E. (2001). 'Controlled struvite crystallisation for removing phosphorus from anaerobic digester side streams'. *Water Research*, 35(1), 151–159.

Batstone, D.J., Hülsen, T., Mehta, C.M. and Keller, J. (2015). 'Platforms for energy and nutrient recovery from domestic wastewater: A review'. *Chemosphere*, 140, 2–11. https://doi.org/10.1016/j.chemosphere.2014.10.021.

Biswas, A. (2019). 'A systematic review on arsenic bio-availability in human and animals: Special focus on the rice-human system'. *Reviews of Environmental Contamination and Toxicology*. https://doi.org/10.1007/398_2019_28.

Chen, X. (2022). 'Machine learning approach for a circular economy with waste recycling in smart cities'. *Energy Reports*, 8, 3127–3140. https://doi.org/10.1016/j.egyr.2022.01.193.

Cordell, D., Drangert, J.O. and White, S. (2009). 'The story of phosphorus: Global food security and food for thought'. *Global Environmental Change*, 19(2), 292–305.

Cornel, P. and Schaum, C. (2009). 'Phosphorus recovery from wastewater: Needs, technologies and costs'. *Water Science Technology*, 59(6), 1069–1076.

Dagerskog, L. and Olsson, O. (2020). 'Swedish sludge management at the crossroads'. Stockholm: Stockholm Environment Institute.

Drewes, J.E. and Horstmeyer, N. (2015). 'Recent developments in potable water reuse'. In D. Fatta-Kassinos, D. Dionysiou, and K. Kümmerer (eds.), *Advanced Treatment Technologies for Urban Wastewater Reuse*. The Handbook of Environmental Chemistry, vol. 45. Cham: Springer. https://doi.org/10.1007/698_2015_341.

Frijns, J., Hofman, J. and Nederlof, M. (2013). 'The potential of (waste) water as energy carrier'. *Energy Conversion and Management*, 65, 357–363.

Gerba, C.P. and Pepper, I.L. (2009). 'Wastewater treatment and biosolids reuse'. *Environmental Microbiology*, 503–530. https://doi.org/10.1016/B978-0-12-370519-8.00024-9.

Giakoumis, T., Vaghela, C. and Voulvoulis, N. (2020). 'The role of water reuse in the circular economy'. In P. Verlicchi (ed.), *Advances in Chemical Pollution, Environmental Management and Protection*. Elsevier, 5:227–252. https://doi.org/10.1016/bs.apmp.2020.07.013.

Goh, P.S., Ahmad, N.A., Lim, J.W., Liang, Y.Y., Kang, H.S., Ismail, A.F. and Arthanareeswaran, G. (2022). 'Microalgae-enabled wastewater remediation and nutrient recovery through membrane photobioreactors: Recent achievements and future perspective'. *Membranes*, 12, 1094. https://doi.org/10.3390/membranes12111094.

Gude, V.G. (2015). 'Energy and water autarky of wastewater treatment and power generation systems'. *Renewable Sustainable Energy Reviews*, 45, 52–68.

Heidrich, E.S., Curtis, T.P. and Dolfing, J. (2011). 'Determination of the internal chemical energy of wastewater'. *Environmental Science and Technology*, 45(2), 827–832.

Hoffmann, S., Feldmann, U., Bach, P.M., Binz, C., Farrelly, M., Frantzeskaki, N., et al. (2020). 'A research agenda for the future of urban water management: Exploring the potential of nongrid, small-grid, and hybrid solutions'. *Environmental Science & Technology*, 54, 5312–5322. https://doi.org/10.1021/acs.est.9b05222.

Ingrao, C., Faccilongo, N., Di Gioia, L. and Messineo, A. (2018). 'Food waste recovery into energy in a circular economy perspective: A comprehensive review of aspects related to plant operation and environmental assessment'. *Journal of Cleaner Production*, 184, 869–892. https://doi.org/10.1016/j.jclepro.2018.02.267.

IWA, Wastewater Report. (2018). 'The reuse opportunity'. London: The International Water Association.

Jaffer, Y., Clark, T., Pearce, P. and Parsons, S. (2002). 'Potential phosphorus recovery by struvite formation'. *Water Research*, 36(7), 1834–1842.

Kjerstadius, H., Bernstad Saraiva, A., Spångberg, J. and Davidsson, A. (2017). 'Carbon footprint of urban source separation for nutrient recovery'. *Journal of Environment Management*, 197, 250–257. https://doi.org/10.1016/j.jenvman.2017.03.094.

Kretschmer, F., Neugebauer, G., Kollmann, R., Eder, M., Zach, F., Zottl, A., et al. (2016). 'Resource recovery from wastewater in Austria: Wastewater treatment plants as regional energy cells'. *Journal of Water Reuse and Desalination*, 6(3), 421–429.

Larsen, T.A., Gruendl, H. and Binz, C. (2021). 'The potential contribution of urine source separation to the SDG agenda – A review of the progress so far and future development options'. *Environmental Science: Water Research & Technology*. https://doi.org/10.1039/d0ew01064b.

Lehtoranta, S., Malila, R., Fjader, P., Laukka, V., Mustajoki, J. and Aysto, L. (2021a). 'Jätevesien ravinteet kiertoon turvallisesti ja tehokkaasti (Efficient and Safe Nurtient Recovery from Municipal Wastewaters)'. *Suomen ympäristökeskuksen raportteja*, 18, 2021. Helsinki: Finnish Environment Institute [In Finnish].

Magnusson, K. and Norén, F. (2014). 'Screening of microplastic particles in and down-stream a wastewater treatment plant'. Sweden: IVL Swedish Environmental Research Institute. Number C 55 August 2014 Report.

Malila, R., Lehtoranta, S. and Viskari, E.L. (2019). 'The role of source separation in nutrient recovery – Comparison of alternative wastewater treatment systems'. *Journal of Cleaner Production*, 219, 350–358. https://doi.org/10.1016/j.jclepro.2019.02.024.

Matassa, S., Batstone, D.J., Hu Lsen, T., Schnoor, J. and Verstraete, W. (2015). 'Can direct conversion of used nitrogen to new feed and protein help feed the world?' *Environmental Science and Technology*, 49, 5247–5254. https://doi.org/10.1021/es505432w.

Matassa, S., Batstone, D.J., Hülsen, T., Schnoor, J. and Verstraete, W. (2015). 'Can direct conversion of used nitrogen to new feed and protein help feed the world?' *Environmental Science and Technology*, 49(9), 5247–5254.

Mo, W. and Zhang, Q. (2013). 'Energy–nutrients–water nexus: Integrated resource recovery in municipal wastewater treatment plants'. *Journal of Environmental Management*, 127, 255–267.

Motlagh, A.M. and Goel, R.K. (2014). 'Sustainability of activated sludge processes'. In S. Ahuja (ed.), *Water Reclamation and Sustainability*. Elsevier, 391–414. https://doi.org/10.1016/B978-0-12-411645-0.00016-X.

Mulder, A. (2003). 'The quest for sustainable nitrogen removal technologies'. *Water Science Technology*, 48(1), 67–75.

Naturvårdsverket. (2018). 'Advanced wastewater treatment for separation and removal of pharmaceutical residues and other hazardous substances'. Needs, Technologies and Impacts. Bromma: The Swedish Environmental Protection Agency.

Oberg, G., Metson, G.S., Kuwayama, Y. and Conrad, S.A. (2020). 'Conventional sewer systems are too time-consuming, costly and inflexible to meet the challenges of the 21st century'. *Sustainability*, 12, 6518. https://doi.org/10.3390/su12166518.

PUB. (2016). 'Our water, our future'. Singapore's National Water Agency (PUB).

Ranade, V.V. and Bhandari, V.M. (2014). 'Industrial Wastewater Treatment, Recycling, and Reuse: An Overview.' Editor(s): Vivek V. Ranade, Vinay M. Bhandari, Industrial Wastewater Treatment, Recycling and Reuse, Butterworth-Heinemann, 1–80.

Rashidi, H., GhaffarianHoseini, A., GhaffarianHoseini, A., Sulaiman, N.M.N., Tookey, J. and Hashim, N.A. (2015). 'Application of wastewater treatment in sustainable design of green built environments: A review'. *Renewable and Sustainable Energy Reviews*, 49, 845–856.

Roest, K., Hofman, J. and van Loosdrecht, M. (2010). 'De Nederlandse watercyclus kan energie opleveren'. *H2O*, 25/26, 47–50.

Salman, M., Shakir, M. and Yaseen, M. (2022). 'Recent developments in membrane filtration for wastewater treatment'. In T. Karchiyappan, R.R. Karri and M.H. Dehghani (eds.), *Industrial Wastewater Treatment.* Water Science and Technology Library, vol 106. Cham: Springer. https://doi.org/10.1007/978-3-030-98202-7_1.

Schopf, K., Judex, J., Schmid, B. and Kienberger, T. (2018). 'Modelling the bioenergy potential of municipal wastewater treatment plants'. *Water Science Technology*, 77(11), 2613–2623.

SEI. (2020). 'Swedish sludge management at the crossroads'. Sweden: Stockholm Environment Institute'. Policy brief January 2020. https://www.sei.org/wp-content/uploads/2020/01/sei-2020-pb-sludge-crossroads.pdf.

Seleiman, M.F., Santanen, A. and Makela, P.S.A. (2020). 'Recycling sludge on cropland as fertilizer - Advantages and risks'. *Resource Conservation Recycling*, 155, 104647. https://doi.org/10.1016/j.resconrec.2019.104647.

Shizas, I. and Bagley, D.M. (2004). 'Experimental determination of energy content of unknown organics in municipal wastewater streams'. *Journal of Energy Engineering*, 130(2), 45–53.

Siegrist, H., Salzgeber, D., Eugster, J. and Joss, A. (2008). 'Anammox brings WWTP closer to energy autarky due to increased biogas production and reduced aeration energy for N-removal'. *Water Science and Technology*, 57(3), 383–388.

Skambraks, A.K., Kjerstadius, H., Meier, M., Davidsson, A., Wuttke, M. and Giese, T. (2017). 'Source separation sewage systems as a trend in urban wastewater management: Drivers for the implementation of pilot areas in Northern Europe'. *Sustainable Cities and Society*, 28, 287–296. https://doi.org/10.1016/j.scs.2016.09.013.

SYKE. (2019). 'Yhdyskuntien Jätevesien Kuormitus Vesiin'. https://www.ymparisto.fi/fi-FI/Kartat_ja_tilastot/Vesihuoltorapor-tit/Yhdyskuntien_jatevesien_kuormitus_vesiin.

Talvitie, J., Mikola, A., Koistinen, A. and Setala, O. (2017). 'Solutions to microplastic pollution – Removal of microplastics from wastewater effluent with advanced wastewater treatment technologies'. *Water Research*, 123, 401–407. https://doi.org/10.1016/j.watres.2017.07.005.

UNEP. (2010). *Sick Water? The Central Role of Wastewater Management in Sustainable Development: A Rapid Response Assessment*, ed. E. Corcoran. Arendal: UNEP/GRID-Arendal, p. 85.

Valtanen, M., Sillanpaa, N. and Setala, H. (2015). 'Key factors affecting urban runoff pollution under cold climatic conditions'. *Journal of Hydrology*, 529, 1578–1589. https://doi.org/10.1016/j.jhydrol.2015.08.026.

Verstraete, W. and Vlaeminck, S.E. (2011). 'Zero waste water: Shortcycling of wastewater resources for sustainable cities of the future'. *International Journal Sustainable Development World Ecology*, 18(3), 253–264.

Verstraete, W., Van de Caveye, P. and Diamantis, V. (2009). 'Maximum use of resources present in domestic used water'. *Bioresource Technology*, 100(23), 5537–5545.

Vieno, N. (2014). 'Haitalliset Aineet Jätevedenpuhdistamoilla –hankkeen Loppuraportti. Harmful substances at wastewater treatment plant – Final report'. Helsinki: Finnish Water Utilities Association. Vesilaitosyhdistyksen monistesarja nro 34. https://www.vvy.fi/haitta-aineselvitys [In Finnish].

Wainaina, S., Lukitawesa and Taherzadeh, M. (2021). 'Microbial conversion of food waste: Volatile fatty acids platform'. In J. Wong, G. Kaur, M. Taherzadeh, A. Pandey, and K. Lasaridi (eds.), *Current Developments in Biotechnology and Bioengineering.* Elsevier, 205–233. https://doi.org/10.1016/B978-0-12-819148-4.00007-5.

Wan, J., Gu, J., Zhao, Q. and Liu, Y. (2016). 'COD capture: A feasible option towards energy self-sufficient domestic wastewater treatment'. *Scientific Reports*, 6(1), 1–9.

Warman, P.R. and Termeer, W.C. (2005). 'Evaluation of sewage sludge, septic waste and sludge compost applications to corn and forage: Yields and N, P and K content of crops and soils'. *Bioresource Technology*, 96, 955–961. https://doi.org/10.1016/j.biortech.2004.08.003.

Web source 1 'Circular Economy Concept'. www.ellenmacarthurfoundation.org. Retrieved 2023-07-30.

Wielemaker, R.C., Weijma, J. and Zeeman, G. (2018). 'Harvest to harvest: Recovering nutrients with new sanitation systems for reuse in urban agriculture'. *Resources Conservation and Recycling*, 128, 426–437. https://doi.org/10.1016/j.resconrec.2016.09.015.

Wilfert, P., Kumar, P.S., Korving, L., Witkamp, G.J. and van Loosdrecht, M.C.M. (2015). 'The relevance of phosphorus and iron chemistry to the recovery of phosphorus from wastewater: A review'. *Environmental Science and Technology*, 49(16), 9400–9414.

Winkler, M.K.H., Bennenbroek, M.H., Horstink, F.H., van Loosdrecht, M.C.M. and van de Pol, G.J. (2013). 'The biodrying concept: An innovative technology creating energy from sewage sludge'. *Bioresource Technology*, 147, 124–129.

Zhu, N. (2006). 'Composting of high moisture content swine manure with corncob in a pilot-scale aerated static bin system'. *Bioresource Technology*, 97, 1870–1875.

10 Circular Economy to Transfer Urban Water

Arunkumar Yadav
Department of Civil Engineering, Manipal Institute of Technology Bengaluru, Manipal Academy of Higher Education, Manipal, Karnataka, India

Sushruta S. Hakkimane
Department of Biotechnology, Manipal Institute of Technology Bengaluru, Manipal Academy of Higher Education, Manipal, Karnataka, India

Santosh L. Gaonkar
Department of Chemistry, Manipal Institute of Technology, Manipal Academy of Higher Education, Manipal, Karnataka, India

10.1 INTRODUCTION

The whole world is witnessing rapid urbanization. Population growth is also rising (UN-DESA, 2018). The global growth in population is responsible for the rapid consumption of resources.

Water is one of the most important resources required by all living things on the Earth and it is largely employed in the water resources sector, irrigation, power generation, domestic purposes, and industrial production. By 2050 the annual amount of supplies consumed worldwide will total over 90 billion tons (Swilling et al., 2018). Water, a natural and renewable resource, plays an essential part in daily life.

The water cycle is a multi-stage journey. This hydrologic cycle is one of the four major biogeochemical cycles that regularly recycle elements, minerals, and nutrients in the ecosystem, together with water. We know that the hydrologic cycle explains and understands the recycling and distribution of water and each stage is important to the others. However, as a result of human-induced activities and population growth, the hydrologic cycle has been disrupted for the past few decades (Vörösmarty and Sahagian, 2000). Worldwide, an area is classified as "water-stressed" if the per capita available water resources fall between 1000 and 1700 m^3/year, and as "water-scarce" if they fall below 1000 m^3/year (Damkjaer and Taylor, 2017). India supplies 4% of the global water resources and occupies 2.4% and represents 17% of the globe's total

 DOI: 10.1201/9781003432869-10

area and population, respectively, according to WRIS (2015). Given its large population, 1.34 billion, and water availability of 1427 m^3/capita/year, India is classified as a water-stressed region (FAO, 2016; Kakwani and Kalbar, 2020).

Already, some researchers have warned that, unless remedial features are employed, it may lead to depletion of various water resources and may reach water scarcity (Fitzhugh and Richter, 2004).

The hydrologic cycle is disturbed by various factors, of which climate change is one (Arnell, 1999; Oki and Kanae, 2006). The disturbance by global climate change is caused at the basin level (Gleick, 1986). The water demand increases exponentially with increasing temperature, an effect that cannot be balanced by available water resources; this increased demand is due to population increase and the rate of urbanization growth (Arnell, 1999; Falkenmark and Widstrand, 1992).

The huge demand for water resources is caused by several factors and it is very important to know, analyze, and identify which is involved (Kenway et al., 2011). Currently, the most challenging task is to guarantee enough water availability or distribution for the various sectors. To manage the water supply, the various sectors are considering the latest technology and well-organized models. It has been suggested that the circular economy (CE) provides a useful framework for sustainable water management. The CE is a good fit to assist with water supply management (Morseletto et al., 2022).

10.2 CONCEPT OF THE CIRCULAR ECONOMY

One strategy for sustainable development is the CE which lessens the demand on natural resources while increasing economic advantages. Pearce and Turner first presented the CE idea in their 1990 book *Economics of Natural Resources and the Environment* (Pearce and Turner, 1990). Subsequently, the Ellen MacArthur Foundation has made a substantial contribution to the advancement of CE practices since 2010 in a number of industries, including waste management, food production, sustainable building and design, indicator development, etc. (EMF, 2017). Hu et al. (2011) and Winans et al. (2017) state that the reduce, reuse, recycle, and recover (or CE) waste management principles form the foundation of the CE concept. Waste is viewed as a resource in the CE paradigm, and, given the issues the water industry has faced, there are some chances to manage wastewater sustainably from a CE standpoint.

The fundamental tenet of the CE concept is the alternative technique, or practice, that permits a change in emphasis from non-renewable resources to retrieving resources from trash, producing advantages for the economy, society, and environment (Lieder and Rashid, 2016). In order to preserve the environment and for individual products to continue to be useful after reaching the end of their useful lives, the valuable materials are recycled and used to create a new product, a practice known as the circular economy, or CE. The products of today can thus be transformed into resources for the future (Stahel, 1982). Water, energy, and materials can all be obtained from wastewater (Mo and Zhang, 2013; Verstraete et al., 2009). Recovered nutrients can be used in agriculture (Vaneeckhaute et al., 2018); wastewater can be used to recover water for use as consumable or non-potable water (Angelakis et al., 2018); and sludge can be used to recover energy (Rulkens, 2008;

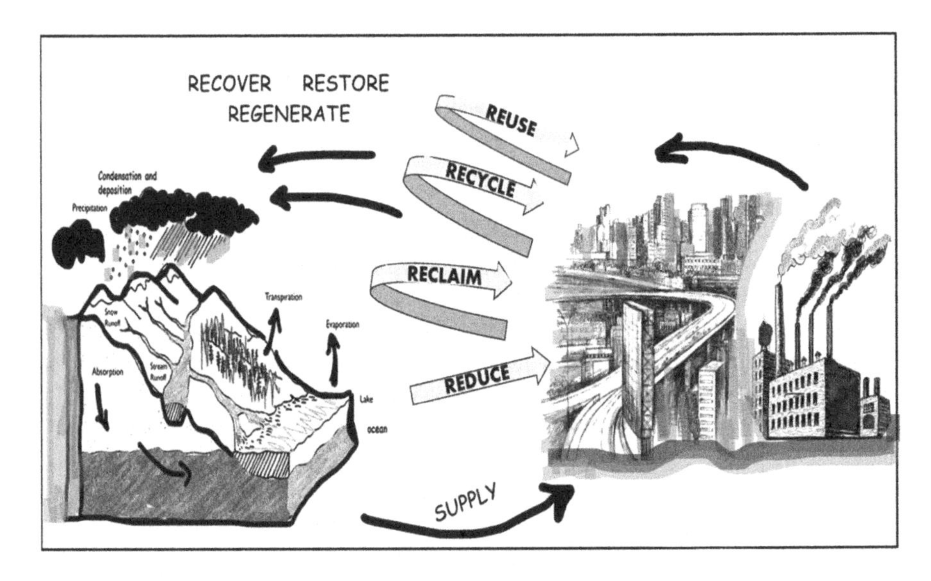

FIGURE 10.1 The circular economy approach in the water sector.

Batstone and Virdis, 2014). The 6Rs techniques (reduce, recycling, reuse, reclaim, recover, and restore) of CE must therefore be applied to the management of water and wastewater resources (McCarty et al., 2011). The circular economy approach to water is shown in Figure 10.1.

10.3 THE SIGNIFICANCE OF A CIRCULAR ECONOMY IN THE WATER SECTOR

The concepts of the CE provide an opportunity to recognize and utilize the full potential of water. The CE for water aims to maximize the value that is generated from water, procedures, and methods. To counteract the impact and utilization of extracting new water resources, wastewater treatment and renewable resource recovery are used. Different methodologies and multiple uses of the idea of CE were discussed (Prieto-Sandoval et al., 2018), and a consensus on the hypothesis of CE and its connection to eco-innovation was suggested. The CE strategy may be implemented, but its evaluation methodology and level of application in any given industry are not consistently developed. Based on six CE principles (BIS, 2017) offered thorough recommendations for implementing CE in organizations. However, it's still unclear how to monitor CE implementation techniques (Pauliuk, 2018). There is no agreement on a structure and metrics for evaluating the application of a CE. As a result, it may be advantageous to design sector-specific frameworks or indicators to encourage their implementation on a greater level (Saidani et al., 2019).

Almost everywhere in the world, significant work has been done on CE in a variety of fields (Prieto-Sandoval et al., 2018). Due to the irregularity in water supplies and the predominately linear method used on a large scale, CE in the water sector is currently receiving a lot of attention.

The wastewater 6Rs have been the subject of extensive research in a disaggregated form; nevertheless, the function of a CE in combining various methods calls for a multidimensional viewpoint. Additionally, there is uncertainty in the vocabulary used to describe the 6Rs, such as reuse and recycling (Blomsma and Tennant, 2020), and this is particularly noticeable in the water sector. The water sector, its implementation issues, potential from a CE perspective, and the monitoring system are not extensively addressed in previous CE research. Therefore, to close this gap, an evaluation of the CE in this sector is essential to gauge its potential.

The purpose of thisreview is to conduct a groundbreaking analysis of the global expansion of the CE idea in the water sector from an different economic, social, environmental, and technical standpoint. Understanding CE techniques that can facilitate successful execution, as well as the difficulties and chances associated with the water sector in India, will be helped by this review.

10.4 THE IMPORTANCE OF CE IN THE URBAN WATER ZONE

Water exchanges between different components of the water sector, including surface, groundwater, and urban water, are involved. These exchanges result in changes to the components' water content and quality. The fact that humans use water means that behavioral and management factors also take the front stage in the urban water sector. The 6Rs strategies can be used in the water sector. Although there is a lot of research done on wastewater in different aspects, it is necessary to contextualize these efforts from the standpoint of the CE. Urban water cycles are inherently complex due to the interactions and exchanges between these elements. As a result, the notion of a CE offers avenues for managing water resources sustainably. In order to attain water and wastewater resource circularity, it is necessary for the 6Rs for the water industry to adopt a CE.

The first strategy of the 6Rs is to minimize freshwater use, which will also result in less wastewater generation. Water-efficient equipment and source prevention/conservation can cut wastewater creation dramatically, lowering the demand for facilities that treat wastewater. (Arceivala and Asolekar, 2007). Consumer education and encouraging ethical behavior in businesses and organizations can help reduce sewage production. Installing water-saving plumbing fixtures and appliances, such as dual flush systems in favor of traditional flushing, can reduce water use (Bari et al., 2015). According to Liu et al. (2016), smart meter usage for requirement management can also help reduce water usage. Kalbar et al. (2014) reported that distribution network leaks in India account for 30–40% of water losses; consequently, eliminating leaks will also result in a decrease in freshwater demand. To lessen the demand for freshwater, reuse entails using spent water repeatedly available for different purposes while still in its raw form (Yang, 2012). Reusing wastewater is one way to fulfill the growing demand for water in areas with limited water supplies, but many parts of the world are still unaware of its enormous potential (Voulvoulis, 2018). As an example, wastewater from Hyderabad, a large city in India, is discharged into the Musi River; downstream of the city, it is used for agricultural purposes (Van Rooijen et al., 2005).

As a result, wastewater is used indirectly, i.e., partially treated sewage is reused for agricultural usage, increasing the nutrient supply for agriculture while decreasing freshwater requirement (Wintgens et al., 2016). However, because reusing directly leaves environment and human health hazards (Srinivasan and Reddy, 2009), it must be carefully considered before being implemented.

According to Jimenez and Asano (2008), recycling is the usage of treated wastewater for the same use for which it was created. The total decrease in freshwater consumption is largely due to efficient recycling. For instance, industrial effluent after treatment is utilised by the same industry for the same purpose. The rising demand for freshwater in India and many other countries has compelled towns to use treated water for other purposes, such as toilet flushing. The significance and recycling purposes and its representation for proper treatment are highlighted in research (Jefferson et al., 2004). The increasing demand for fresh water in India and many other nations has necessitated municipalities in these countries to use processed water for other uses such as restroom flushing. Jefferson et al. (2004) underlined the benefits of recycling as well as its characterization for optimal treatment. Wastewater reclamation entails treating wastewater and using it "outside of the loop" or process. Wastewater can be treated to a level that is comparable to drinkable water because of the quick invention and advancement of treatment technology (Asano et al., 2007). The level of wastewater pollution and the intended use of processed water determine the scope for reclamation. Multiple uses exist for wastewater that has been properly treated (Jimenez and Asano, 2008).

The process of recovering usable resources from wastewater, such as materials and energy, is commonly referred to as recovery. In an era of diminishing resources, source recovery from wastewater is a key step toward sustainability. Waste is a useful resource, just like wastewater is part of the CE paradigm (Puyol et al., 2017). Depending on the amount of pollutants present, wastewater can be a valuable source (Bracken et al., 2007).

To increase water availability during scarcity and to reserve water in the area, restoration entails the process of artificially refilling water sources, such as groundwater and aquifers. The water resources are replenished artificially using recharge strategies. For instance, rainwater harvesting, which involves pretreating rainwater, is frequently used to recharge groundwater and make it accessible during the dry season. Additionally, regulated aquifer recharge is used to restore rivers and lakes.

One such example is the work made to conserve the lakes in Bengaluru, where the restoration of bunds and spillways helped restore the groundwater. Similarly, effective wastewater restoration is used in the rejuvenation of Jaipur Mansagar Lake, which used the tertiary treatment of wastewater entering the lake or constructed wetland technology (Gowda and Sridhara, 2007; Asolekar et al., 2014). As a result, a water sector CE can significantly lower the demand for freshwater. Technology is important and plays an enormous role in recycling and revival (Villar and Merel, 2020).

Traditionally, the goal of wastewater treatment technology was to safeguard the environment by removing toxins.

A substantial intervention in the area of wastewater uses anaerobic treatment or low-energy wastewater treatment methods, principally to generate methane gas. Separation of diluted and concentrated streams by filtration processes, as well as anaerobic breakdown, need a major importance (Verstraete et al., 2009; McCarty et al., 2011). Furthermore, Batstone and Virdis (2014) stressed the importance of anaerobic digestion in a novel resource recovery paradigm that incorporates the partition release and recovery approach, as well as the low-energy mainstream strategy.

However, Batstone et al. (2015) examined the practical aspects of the suggestion made by researchers and continued to develop resource recovery programs with an emphasis on energy and economic considerations. Thus, a key component of wastewater management now involves technical advances that take resource recovery into account (Verstraete et al., 2009; McCarty et al., 2011).

Numerous obstacles to the significant technological advancement in the wastewater treatment sector are present. The CE strategy will assist in removing some of these obstacles, improving the water industry.

10.5 APPLICATION OF DIGITALIZATION IN THE WATER SECTOR

Digital technologies offer enormous benefits that are social, environmental, and economic as well. amplifying the benefits of their deployment even further. This is because they provide a diverse range of options, allowing for reduced environmental impacts and more efficient resource reallocation. The urban water sector is a strong practical example of the potential of digitalization, as the digital transformation of these facilities and pipes allows for improved management through constant activity control and monitoring. Because of the huge number of assets, the use of sensors is very important, offering real-time data on water quality and other parameters. A wide range of fixed and mobile sensors can be dispersed throughout the system to support daily operations by optimizing resource utilization, avoiding damage and giving relevant information for protection, maintenance, and a long-term plan.

10.6 OVERVIEW OF THE LITERATURE ON THE CIRCULAR ECONOMY (CE)

This section discusses research that uses the CE paradigm but focuses more on economic factors. In addition to agriculture, reclaimed wastewater is used for power plant cooling, which has economic advantages. However, depending on the location, it may become uneconomical to transport reclaimed wastewater from its source (a wastewater treatment facility, or WWTP), to its destination (a power plant). In comparison with conventional methods, a cost–benefit analysis was carried out, and the wastewater treatment, agricultural crop production, and energy-generating value chains all showed favorable results (Grundmann and Maa, 2017; Maa and Grundmann, 2016). The economy model (Okioga and Sireli, 2016) used a transportation optimization model to reduce the transport distance, optimize pipe cost, and close the loop of the water resource under the CE principle.

Although the aforementioned research primarily focused on financial investments, the cost-benefit analysis used to evaluate wastewater recycling efforts also took environmental concerns into account, offering helpful insights from a CE standpoint (Villar, 2018). The water and wastewater industry's financial and environmental costs and benefits were studied (Abu-Ghunmi et al., 2016). Jordan's centralized wastewater treatment facilities were the target of the study. A method for estimating water's market worth was given, and Jordan's current linear economic condition showed that not treating wastewater would result in a net loss. considering the financial benefits of treatment. In a different study, Kayal et al. (2019) concluded that technology is essential to the entire cost-benefit analysis of circularity and created a "Circonomics Index" to evaluate the level of circularity adopted in the WWTP. These studies indicate that the cost-benefit method is used to assess the economic benefits of the CE (Kalbar et al., 2012; Lorenzo-Toja et al., 2016).

10.6.1 Overview of Literature on Environmental and Technological Aspects and Addressing CE's Social Features

The CE employs reusing, sharing, restoring, reprocessing, and recycling to form a closed-loop system, decreasing resource inputs as well as waste. The research examining the environmental efficiency of the solutions suggested to attain CE falls under the heading of the environmental aspect, explaining how water, energy, and material resources were successfully implemented and exchanged in an eco-industrial place to reduce energy use and promote the protection of natural resources (Li and Ma, 2015). It is possible to accomplish balanced water source utilization in the farming and industrial schemes, and domestic purposes within the philosophy of the CE, and this is illustrated with the aid of case studies in China (Yang, 2012). An example in the same context is the circular utilization of water in various industrial plants (Dai et al., 2011). Similarly, there is the multi-objective optimization of cyclic routes and the symbiotic link between the pig industry, planting and sewage treatment, and the pig breeding industries (Liu and Xiao, 2015). It was recommended to use sludge from effluent treatment of the paper and pulp sector to create carbon adsorbents to attain a no-waste strategy for companies (Jaria et al., 2017). There are environmental benefits to producing fertilizer from fly ash waste from bioenergy generation and sludge from wastewater treatment from the forestry industry (Husgafvel et al., 2016). Sustainable resource management in the building sector involves using solar energy to generate electricity and reducing, reusing, and recycling water and wastewater resources while considering the concept of prosumers, i.e., consumers who are also producers (McLean and Roggema, 2019). A report focusing primarily on water resources and the working of the CE in the sectors of water allocation and sanitation at the federal level, described the problems, and the situation in Finland (Pimentel-Rodrigues and Siva-Afonso, 2019; Laitinen et al., 2020).

To increase an understanding of wastewater treatment and management among vineyard authorities, indicators for wastewater production and water use in the winery sector were created. Using CE indicators, the amount of waste that might be reused during the paper production process was also measured (Oliveira et al., 2019; Molina-Sânchez et al., 2018).

It is essential to assess social indicators and the ways in which circular economy practices can increase human well-being in society. Real-world case studies must be implemented to aid in the building of a circularity database. The application of these methodologies has implications for social concerns such as individual health, behavioral changes, safety, norms and guiding principles, recommendations, and governance. Addressing the issue of scarce water resources and management systems in the context of metropolitan areas, in accordance with the CE principle and policy importance, has been discussed in the literature (Da Rosa and Ramos, 2018; Molina-Giménez, 2018; Eneng et al., 2018). According to Villamar et al. (2018), it is essential to create regulations for safe reuse, recycling, and reclamation. To achieve CE application in the water sector, policies, rules, regulations, and laws must be developed. Additionally, the legal and governance framework must be strengthened. Additionally, from a sociological perspective, public acceptability and awareness are crucial for the successful implementation of a CE. The technology approach can treat wastewater to the point where it is safe for human consumption. Because this concept is still undesirable to the majority of people, reclaimed wastewater is most frequently used for industrial, urban, environmental, or recreational purposes. A brief discussion on different aspects, including the 6Rs strategies of the framework of CE, of the possibilities and applications of microbial technology in wastewater utilization and treatment is given by Nielsen (2017). Later, new approaches were analyzed for recovering useful materials from wastewater and potential advantages and difficulties were identified (Puyol et al., 2017). Akyol et al. (2020) studied techniques for treating anaerobic wastewater, recovering material and energy, and recycling wastewater inside the context of a biorefinery or water resource recovery facility. Gherghel et al. (2019) provided a review of the many kinds of sludge produced by wastewater treatment, along with strategies for managing sludge sustainably for effective recovery. Seco et al. (2018) claimed that resource recovery in wastewater treatment plants was feasible through sulfate-rich waste streams. In addition, sample biorefinery treatment flowsheets were included, along with the idea that a strategy combining industrialised ecology, cleaner production, and CE could lead to sustainability. The idea was proposed by the CAP Group to transform existing WWTPs in Italy into biorefineries with cutting-edge technologies and opportunities to recover valuable materials in an important step in the direction of a CE (Russo, 2018). A brief overview of anaerobic treatment technologies was given (Massara et al., 2017) in order to facilitate resource recovery and accomplish a CE, sustainable treatment, and water reuse. One such waste biorefinery example uses pyrolyzed heavy metal (lead) adsorbent made from anaerobically digested waste-activated sludge (Ho et al., 2017). Biorefineries thus support the successful application of the CE principle.

The membrane technique is employed in the recovery and treatment of agricultural and industrial wastes, with a similar approach to the removal of halogens for the goal of recycling (Castro-Muoz et al., 2018; Toth et al., 2018). The use of biofilms to treat wastewater has a long history, and extracellular polymeric substances are the source of the biofilms themselves (Seviour et al., 2019). However, the capabilities and advantages for wastewater treatment and recovery of extracellular polymeric substances are not yet well understood. A technical method was suggested for recovering gallium produced by the wafer fabrication sector (Jain et al., 2019). There was also a report on the water softening process application of the CE principle in drinking water treatment facilities (Schetters et al., 2015).

The water industry and information and communication technology (ICT) should be integrated, according to Kanakoudis and Tsitsifli (2019), to decrease water losses and enhance the management of water resources. The development of bioadsorbents using waste, energy-positive wastewater treatment plants, facilities for resource recovery or wastewater biological refineries, and natural treatment techniques are some of the key ideas addressed in this field.

10.7 WATER CIRCULAR ECONOMY STRATEGIES

A clear set of tactics represents an essential level of analysis for the circular economy of water (CEW). Reject, reconsider, reducing, reusing, recovery, restore, refurbish, reuse, recycling, and retrieve (also known as the R-strategies) are the most frequently used elements in the CE literature. minimizing, sustaining, and optimizing water consumption while ensuring environmental preservation and conservation through waste minimization, efficient use, and quality preservation.

10.8 PROSPECTS AND CHALLENGES FOR THE CIRCULAR ECONOMY OF WATER IN INDIA

Water scarcity in India is a year-round problem that affects hundreds of millions of people. The water industry is currently dealing with issues such as increasing water requirements, a system of intermittent water supply with increased non-revenue water, a large gap in the production and treatment of wastewater, and a linear approach method to water management. Because of the lack of the skilled personnel required to solve all of these challenges, dealing with technical, political, economic, and social obstacles are the most difficult elements of using and recycling wastewater treatment plant water. India relies on outside money and organizations to implement and operate water-related facilities. Geographic differences, for example, have a considerable impact on technological challenges. CEW has been successful in India in terms of social acceptance and reuse of treated water. Local governments should support and encourage initiatives to increase water reuse, particularly for agricultural and industrial purposes, across India. This includes the water geopolitical paradigm, water economy, water innovation, and water management and law in India, environmental and safety concerns regarding water reuse, and technological solutions for water recovery. The roles and duties for the creation and implementation of CEW initiatives are among the governance problems. It is critical to broaden understanding and acceptance of the CE by all people who use water, such as when water is recycled for many uses or used sequentially by different actors in cascading. Education, acceptance, and participation with regard to CEW tactics and solutions are critical for achieving this goal. A lack of enabling policies, easy access to fresh water, no restrictions on groundwater extraction, and no proper policies are all contributing factors to industries' delays in implementing CE. The proper functioning of CE in India faces a variety of obstacles in such a situation, which need rigorous analysis and teamwork. Here, some of the difficulties are described.

Technology is crucial for converting wastewater into a resource, and it has advanced to the point where conversion can produce drinkable water (Puyol et al., 2017). Regardless, existing technologies have a poor level of technological preparedness. To successfully implement CE policies of reclamation, recycling, and recovery, authorities must overcome several major obstacles, including the choice of suitable technology (Kalbar et al., 2012b). Currently, techniques are designed on a centralized level; although, as urbanization increases, the amount and quality of wastewater produced are constantly varying, and emerging pollutants pose novel difficulties. (Kalbar et al., 2018). Consequently, it is necessary to acknowledge the significance and function of natural therapy systems with decentralized treatment. Due to difficulties relating to the appropriateness of such technologies in Indian settings and high prices, onsite scaling-up is not done even after substantial research on lab-scale technological advances in nutrient recovery. Another major issue is the lack of a collection system. Problems with the centralized sewerage system, the centralized treatment facilities continue to be underutilized.

Economic issues include several things. According to customs, wastewater treatment costs between Rs. 6.5 and Rs. 9 per m^3 for recycling purposes and Rs. 3.5 to 6.5 per m^3 for annual operation and maintenance (Tare and Bose, 2009). Energy recovery is not used in typical wastewater treatment; as a result, the cost of energy for treatment predominates. The cost of energy as a whole could be considerably decreased if energy were produced at the WWTP. According to (Hastak et al., 2016). The use of recycled water for cultivation uses could lower the market value of the goods, which would lower overall earnings. Therefore, there is little incentive to employ recycled water in agriculture.

The proper operation of the infrastructure and, consequently, the implementation of the CE are severely hampered by a lack of understanding by the wastewater management authorities. There aren't enough specific policies in place to assist wastewater recycling and reclamation. Detailed procedures and rules are required. It is necessary to state precisely the amount of treatment required, the use of the water after treatment, its cost, the duties of the regulating agencies, etc. Policies encouraging the utilization of recovered resources are lacking. Government authorities are responsible for developing the operative methods and arrangements (such as centralized/decentralized systems) for the CEW. A successful transition to the CEW requires the participation of all essential stakeholders, especially when changing a system.

Given that stakeholders are accustomed to using water resources linearly, implementing the 6Rs is a significant issue. While long-term economic gains are harder to calculate and picture, they are easier to comprehend than the environmental benefits. Adoption of a decentralized treatment system on a large scale requires a fundamental shift in how society, business, and industry function, as well as support from the highest levels of government.

It is difficult to persuade the general population to accept the use of reclaimed water for any purpose. Regardless of whether recycled water is used for irrigation, certain micropollutants may still degrade plant growth or quality, affecting human health. As a result, thorough evaluations, that take into consideration local practices, are required. The concept of industrial ecology, or the lack of coordination and communication across industries, limits successful resource exchange. The potential for CE implementation represents a real opportunity but is currently underappreciated in underdeveloped nations like India. It is possible to take advantage of the lack of collection infrastructure while also advancing CE reclamation and recycling programs by developing decentralized facilities close together and reducing the load on centralized facilities. However, the trade-off between financial costs and environmental benefits must be evaluated. The treatment capacity of India's wastewater infrastructure is only about 37%, according to the Central Pollution Control Board (CPCB, 2016). As a result, the lack of infrastructure may present an opportunity to implement CE while constructing new wastewater management infrastructure. The vast mixed land-use pattern in India provides an opportunity to implement CE techniques. Adjacent business sectors or factories, for example, may use reclaimed water from a residential area (Guest et al., 2009). CE implementation also contributes to regional job creation (Lekshmi et al., 2020; Vanner et al., 2014). Because agriculture consumes 70% of India's water, using recycled water for agriculture can have a big positive influence on the country's overall economy. If urban lobal bodies (ULBs) effectively implement CE, it will give momentum to the development of long-lasting infrastructure in the water sector. For effective implementation, an assessment framework must be created, and indicators must be chosen. If the aforementioned issues are resolved and opportunities are taken advantage of by developing the necessary policies, the CE can help India make a successful transition to sustainable development.

10.9 CONCLUSION

A thorough understanding of the urban water cycle is necessary, given the current state of degraded and unbalanced water resources. There is now too little water and it is negatively affecting people's well-being. The management of water resources by all parties involved (consumers, authority agencies, ULBs, businesses, small- and medium-sized companies, and water and wastewater treatment plant authorities) can be extremely important in achieving sustainability. To advance the CE philosophy, the evaluation of water resources entering and leaving metropolitan areas, both qualitatively and quantitatively, needs careful consideration. The application of a CE also requires a structure for monitoring or assessment based on chosen indicators, including the six Rs. A thorough understanding of the urban water cycle is necessary given the current state of degraded and unbalanced water supplies.

REFERENCES

Abu-Ghunmi, D., Abu-Ghunmi, L., Kayal, B., Bino, A., 2016. Circular economy and the opportunity cost of not "closing the loop" of water industry: The case of Jordan. *J. Clean. Prod.* 131, 228–236. https://doi.org/10.1016/j.jclepro.2016.05.043.

Akyol, Ç., Foglia, A., Ozbayram, E.G., Frison, N., Katsou, E., Eusebi, A.L., Fatone, F., 2020. Validated innovative approaches for energy-efficient resource recovery and re-use from municipal wastewater: From anaerobic treatment systems to a biorefinery concept. *Crit. Rev. Environ. Sci. Technol.* 50, 869–902. https://doi.org/10.1080/10643389.2019.1634456.

Angelakis, A.N., Asano, T., Bahri, A., Jimenez, B.E., Tchobanoglous, G., 2018. Water reuse: From ancient to modern times and the future. *Front. Environ. Sci.* 6, 26. https://doi.org/10.3389/fenvs.2018.00026.

Arceivala, S.J., Asolekar, S.R., 2007. *Wastewater Treatment for Pollution Control and Reuse.* New Delhi: Tata McGraw-Hill.

Arnell, N., 1999. Climate change and global water resources. *Glob. Environ. Chang.* 9, S31–S49. https://doi.org/10.1016/S0959-3780(99)00017-5.

Asano, T., Burton, F.L., Leverenz, H.L., Tsuchihashi, R., Tchobanoglous, G., 2007. *Water Reuse: Issues, Technologies, and Applicaions.* New York: McGraw-Hill.

Asolekar, S.R., Kalbar, P.P., Chaturvedi, M.K.M., Maillacheruvu, K.Y., 2014. Rejuvenation of rivers and lakes in India: Balancing societal priorities with technological possibilities, in: *Comprehensive Water Quality and Purification.* Elsevier, pp. 181–229. https://doi.org/10.1016/B978-0-12-382182-9.00075-X.

Bari, M.A., Begum, R.A., Nesadurai, N., Pereira, J.J., 2015. Water consumption patterns in Greater Kuala Lumpur: Potential for reduction. *Asian J. Water, Environ. Pollut.* 12, 1–7. https://doi.org/10.3233/AJW-150001.

Batstone, D.J., Hülsen, T., Mehta, C.M., Keller, J., 2015. Platforms for energy and nutrient recovery from domestic wastewater: A review. *Chemosphere* 140, 2–11. https://doi.org/10.1016/j.chemosphere.2014.10.021.

Batstone, D.J., Virdis, B., 2014. The role of anaerobic digestion in the emerging energy economy. *Curr. Opin. Biotechnol.* 27, 142–149. https://doi.org/10.1016/j.copbio.2014.01.013.

Blomsma, F., Tennant, M., 2020. Circular economy: Preserving materials or products? Introducing the resource states framework. *Resour. Conserv. Recycl.* 156, 104698. https://doi.org/10.1016/j.resconrec.2020.104698.

Bracken, P., Wachtler, A., Panesar, A.R., Lange, J., 2007. The road not taken: How traditional excreta and greywater management may point the way to a sustainable future. *Water Supply* 7, 219–227. https://doi.org/10.2166/ws.2007.025.

Castro-Muñoz, R., Barragán-Huerta, B.E., Fíla, V., Denis, P.C., Ruby-Figueroa, R., 2018. Current role of membrane technology: From the treatment of agro-industrial by-products up to the valorization of valuable compounds. *Waste Biomass Valori.* 9, 513–529. https://doi.org/10.1007/s12649-017-0003-1.

CPCB, 2016. National status of waste water generation and treatment. CPCB Bulletin Volume I, July 2016, Updated on 6th December 2016. Retrieved from: http://www.sulabhenvis.nic.in/Database/STST_wastewater_2090.aspx (accessed January 2019).

Da Rosa, A.M., Ramos, A.L.S., 2018. Circular economy, human behavior and law, in: *WIT Transactions on the Built Environment*, pp. 35–44. https://doi.org/10.2495/UG180041.

Dai, Y., Li, T., Yang, R., Chen, D., Yan, L., Wang, X., Zhang, Y., 2011. Study on the demonstration of seawater desalination. *Adv. Mater. Res.* 183–185, 985–989. https://doi.org/10.4028/www.scientific.net/AMR.183-185.985.

Damkjaer, S., Taylor, R., 2017. The measurement of water scarcity: Defining a meaningful indicator. *Ambio* 46, 513–531. https://doi.org/10.1007/s13280-017-0912-z.

EMF (Ellen MacArthur Foundation), 2017. Our mission is to accelerate the transition to a circular economy. Retrieved from: https://www.ellenmacarthurfoundation.org (accessed 2019).

Eneng, R., Lulofs, K., Asdak, C., 2018. Towards a water balanced utilization through circular economy. *Manag. Res. Rev.* 41, 572–585. https://doi.org/10.1108/MRR-02-2018-0080.

Falkenmark, M., Widstrand, C., 1992. Population and water resources: A delicate balance. *Popul. Bull.* 47, 1–36.

FAO, 2016. AQUASTAT main database – Food and agriculture organization of the United Nations. Retrieved from: http://www.fao.org/nr/water/aquastat/data/query/index.html?lang=en (Last accessed June 2019).

Fitzhugh, T.W., Richter, B.D., 2004. Quenching urban thirst: Growing cities and their impacts on freshwater ecosystems. *Bioscience* 54, 741–754. https://doi.org/10.1641/0006-3568(2004)054[0741:QUTGCA]2.0.CO;2.

Gherghel, A., Teodosiu, C., De Gisi, S., 2019. A review on wastewater sludge valorisation and its challenges in the context of circular economy. *J. Clean. Prod.* 228, 244–263. https://doi.org/10.1016/j.jclepro.2019.04.240.

Gleick, P.H., 1986. Methods for evaluating the regional hydrologic impacts of global climatic changes. *J. Hydrol.* 88, 97–116. https://doi.org/10.1016/0022-1694(86)90199-X.

Gowda, K., Sridhara, M.V., 2007. Conservation of tanks/lakes in the Bangalore metropolitan area. *Manag. Environ. Qual. An Int. J.* 18, 137–151. https://doi.org/10.1108/14777830710725812.

Grundmann, P., Maaß, O., 2017. Wastewater reuse to cope with water and nutrient scarcity in agriculture-A case study for Braunschweig in Germany, in: *Competition for Water Resources: Experiences and Management Approaches in the US and Europe.* Elsevier Inc., pp. 352–365. https://doi.org/10.1016/B978-0-12-803237-4.00020-3.

Guest, J.S., Skerlos, S.J., Barnard, J.L., Beck, M.B., Daigger, G.T., Hilger, H., Jackson, S.J., Karvazy, K., Kelly, L., Macpherson, L., Mihelcic, J.R., Pramanik, A., Raskin, L., Van Loosdrecht, M.C.M., Yeh, D., Love, N.G., 2009. A new planning and design paradigm to achieve sustainable resource recovery from wastewater. *Environ. Sci. Technol.* 43, 6126–6130. https://doi.org/10.1021/es9010515.

Hastak, S., Labhasetwar, P., Kundley, P., Gupta, R., 2016. Changing from intermittent to continuous water supply and its influence on service level benchmarks: A case study in the demonstration zone of Nagpur, India. *Urban Water J.* 14, 768–772. https://doi.org/10.1080/1573062X.2016.1240808.

Ho, S.-H., Chen, Y., Yang, Z., Nagarajan, D., Chang, J.-S., Ren, N., 2017. High-efficiency removal of lead from wastewater by biochar derived from anaerobic digestion sludge. *Bioresour. Technol.* 246, 142–149. https://doi.org/10.1016/j.biortech.2017.08.025.

Hu, J., Xiao, Z., Zhou, R., Deng, W., Wang, M., Ma, S., 2011. Ecological utilization of leather tannery waste with circular economy model. *J. Clean. Prod.* 19, 221–228. https://doi.org/10.1016/j.jclepro.2010.09.018.

Husgafvel, R., Karjalainen, E., Linkosalmi, L., Dahl, O., 2016. Recycling industrial residue streams into a potential new symbiosis product – The case of soil amelioration granules. *J. Clean. Prod.* 135, 90–96. https://doi.org/10.1016/j.jclepro.2016.06.092.

Jain, R., Fan, S., Kaden, P., Tsushima, S., Foerstendorf, H., Barthen, R., Lehmann, F., Pollmann, K., 2019. Recovery of gallium from wafer fabrication industry wastewaters by Desferrioxamine B and E using reversed-phase chromatography approach. *Water Res.* 158, 203–212. https://doi.org/10.1016/j.watres.2019.04.005.

Jaria, G., Silva, C.P., Ferreira, C.I.A., Otero, M., Calisto, V., 2017. Sludge from paper mill effluent treatment as raw material to produce carbon adsorbents: An alternative waste management strategy. *J. Environ. Manage.* 188, 203–211. https://doi.org/10.1016/j.jenvman.2016.12.004.

Jefferson, B., Palmer, A., Jeffrey, P., Stuetz, R., Judd, S., 2004. Grey water characterisation and its impact on the selection and operation of technologies for urban reuse. *Water Sci. Technol.* 50, 157–164. https://doi.org/10.2166/wst.2004.0113.

Jimenez, B., Asano, T., 2008. *Water Reuse: An International Survey of Current Practice, Issues and Needs.* IWA.

Kakwani, N.S., Kalbar, P.P., 2020. Review of circular economy in urban water sector: Challenges and opportunities in India. *J. Environ. Manage.* 271, 111010.

Kalbar, P.P., Karmakar, S., Asolekar, S.R., 2012a. Selection of an appropriate wastewater treatment technology: A scenario-based multiple-attribute decision-making approach. *J. Environ. Manage.* 113, 158–169. https://doi.org/10.1016/j.jenvman.2012.08.025.

Kalbar, P.P., Karmakar, S., Asolekar, S.R., 2012b. Technology assessment for wastewater treatment using multiple-attribute decision-making. *Technol. Soc.* 34, 295–302. https://doi.org/10.1016/j.techsoc.2012.10.001.

Kalbar, P.P., Kulkarni, V., Gokhale, P., 2014. Role of hydraulic modeling in development of road map for achieving 24x7 continuous water supply, in: *Proceedings of 46th IWWA Annual Event Bangalore, 17th–19th January 2014*, pp. 175–182.

Kalbar, P.P., Muñoz, I., Birkved, M., 2018. WW LCI v2: A second-generation life cycle inventory model for chemicals discharged to wastewater systems. *Sci. Total Environ.* 622–623, 1649–1657. https://doi.org/10.1016/j.scitotenv.2017.10.051.

Kanakoudis, V., Tsitsifli, S., 2019. Water networks management: New perspectives. *Water* 11, 1–4. https://doi.org/10.3390/w11020239.

Kayal, B., Abu-Ghunmi, D., Abu-Ghunmi, L., Archenti, A., Nicolescu, M., Larkin, C., Corbet, S., 2019. An economic index for measuring firm's circularity: The case of water industry. *J. Behav. Exp. Financ.* 21, 123–129. https://doi.org/10.1016/j.jbef.2018.11.007.

Kenway, S., Gregory, A., McMahon, J., 2011. Urban water mass balance analysis. *J. Ind. Ecol.* 15, 693–706. https://doi.org/10.1111/j.1530-9290.2011.00357.x.

Laitinen, J., Antikainen, R., Hukka, J.J., Katko, T.S., 2020. Water supply and sanitation in a green economy society: The case of Finland. *Public Work. Manag. Policy*, 1–18. https://doi.org/10.1177/1087724X19847211.

Lekshmi, B., Sharma, S., Sutar, R.S., Parikh, Y.J., Ranade, D.R., Asolekar, S.R., 2020. Circular economy approach to women empowerment through reusing treated rural wastewater using constructed wetlands, in: Ghosh, S. (Ed.), *Waste Management as Economic Industry Towards Circular Economy*. Springer Singapore, pp. 1–10. https://doi.org/10.1007/978-981-15-1620-7_1.

Li, Y., Ma, C., 2015. Circular economy of a papermaking park in China: A case study. *J. Clean. Prod.* 92, 65–74. https://doi.org/10.1016/j.jclepro.2014.12.098.

Lieder, M., Rashid, A., 2016. Towards circular economy implementation: A comprehensive review in context of manufacturing industry. *J. Clean. Prod.* 115, 36–51. https://doi.org/10.1016/j.jclepro.2015.12.042.

Liu, A., Giurco, D., Mukheibir, P., 2016. Urban water conservation through customised water and end-use information. *J. Clean. Prod.* 112, 3164–3175. https://doi.org/10.1016/j.jclepro.2015.10.002.

Liu, X., Xiao, X., 2015. The optimization of cyclic links of live pig-industry chain based on circular economics. *Sustainability* 8, 26. https://doi.org/10.3390/su8010026.

Lorenzo-Toja, Y., Vázquez-Rowe, I., Amores, M.J., Termes-Rifé, M., Marín-Navarro, D., Moreira, M.T., Feijoo, G., 2016. Benchmarking wastewater treatment plants under an eco-efficiency perspective. *Sci. Total Environ.* 566–567, 468–479. https://doi.org/10.1016/j.scitotenv.2016.05.110.

Maaß, O., Grundmann, P., 2016. Added-value from linking the value chains of wastewater treatment, crop production and bioenergy production: A case study on reusing wastewater and sludge in crop production in Braunschweig (Germany). *Resour. Conserv. Recycl.* 107, 195–211. https://doi.org/10.1016/j.resconrec.2016.01.002.

Massara, T.M., Komesli, O.T., Sozudogru, O., Komesli, S., Katsou, E., 2017. A mini review of the techno-environmental sustainability of biological processes for the treatment of high organic content industrial wastewater streams. *Waste Biomass Valori* 8, 1665–1678. https://doi.org/10.1007/s12649-017-0022-y.

McCarty, P.L., Bae, J., Kim, J., 2011. Domestic wastewater treatment as a net energy producer–Can this be achieved? *Environ. Sci. Technol.* 45, 7100–7106. https://doi.org/10.1021/es2014264.

McLean, L., Roggema, R., 2019. Planning for a Prosumer future: The case of Central Park, Sydney. *Urban Plan.* 4, 172–186. https://doi.org/10.17645/up.v4i1.1746.

Mo, W., Zhang, Q., 2013. Energy–nutrients–water nexus: Integrated resource recovery in municipal wastewater treatment plants. *J. Environ. Manage.* 127, 255–267. https://doi.org/10.1016/j.jenvman.2013.05.007.

Molina-Giménez, A., 2018. Water governance in the smart city, in: *WIT Transactions on the Built Environment*, pp. 13–22. https://doi.org/10.2495/UG180021.

Molina-Sánchez, E., Leyva-Díaz, J., Cortés-García, F., Molina-Moreno, V., 2018. Proposal of sustainability indicators for the waste management from the paper industry within the circular economy model. *Water* 10, 1014. https://doi.org/10.3390/w10081014.

Morseletto, P., Mooren, C.E., Munaretto, S., 2022. Circular economy of water: Definition, strategies and challenges. *Circ. Econ. Sust.* 2, 1463–1477. https://doi.org/10.1007/s43615-022-00165-x.

Nielsen, P.H., 2017. Microbial biotechnology and circular economy in wastewater treatment. *Microb. Biotechnol.* 10, 1102–1105. https://doi.org/10.1111/1751-7915.12821.

Oki, T., Kanae, S., 2006. Global hydrological cycles and world water resources. *Science* 313, 1068–1072. https://doi.org/10.1126/science.1128845.

Okioga, I.T., Sireli, Y., 2016. The circular economy of wastewater: Supply prioritization of reclaimed water for power plant cooling. *Proc. Int. Annu. Conf. Am. Soc. Eng. Manag.*, 1–11.

Oliveira, M., Costa, J.M., Fragoso, R., Duarte, E., 2019. Challenges for modern wine production in dry areas: Dedicated indicators to preview wastewater flows. *Water Supply* 19, 653–661. https://doi.org/10.2166/ws.2018.171.

Pauliuk, S., 2018. Critical appraisal of the circular economy standard BS 8001:2017 and a dashboard of quantitative system indicators for its implementation in organizations. *Resour. Conserv. Recycl.* 129, 81–92. https://doi.org/10.1016/j.resconrec.2017.10.019.

Pearce, D.W., Turner, R.K., 1990. *Economics of Natural Resources and the Environment.* Baltimore: Johns Hopkins University Press.

Pimentel-Rodrigues, C., Siva-Afonso, A., 2019. Reuse of resources in the use phase of buildings. Solutions for water. *IOP Conf. Ser. Earth Environ. Sci.* 225, 1–8. https://doi.org/10.1088/1755-1315/225/1/012050.

Prieto-Sandoval, V., Jaca, C., Ormazabal, M., 2018. Towards a consensus on the circular economy. *J. Clean. Prod.* 179, 605–615. https://doi.org/10.1016/j.jclepro.2017.12.224.

Puyol, D., Batstone, D.J., Hülsen, T., Astals, S., Peces, M., Krömer, J.O., 2017. Resource recovery from wastewater by biological technologies: Opportunities, challenges, and prospects. *Front. Microbiol.* 7. https://doi.org/10.3389/fmicb.2016.02106.

Rulkens, W., 2008. Sewage sludge as a biomass resource for the production of energy: Overview and assessment of the various options. *Energy & Fuels* 22, 9–15. https://doi.org/10.1021/ef700267m.

Russo, A., 2018. Innovation and circular economy in water sector: The CAP Group, in: Gilardoni, A. (Ed.), *The Italian Water Industry.* Cham: Springer International Publishing, pp. 215–224. https://doi.org/10.1007/978-3-319-71336-6_15.

Saidani, M., Yannou, B., Leroy, Y., Cluzel, F., Kendall, A., 2019. A taxonomy of circular economy indicators. *J. Clean. Prod.* 207, 542–559. https://doi.org/10.1016/j.jclepro.2018.10.014.

Schetters, M.J.A., van der Hoek, J.P., Kramer, O.J.I., Kors, L.J., Palmen, L.J., Hofs, B., Koppers, H., 2015. Circular economy in drinking water treatment: Reuse of ground pellets as seeding material in the pellet softening process. *Water Sci. Technol.* 71, 479–486. https://doi.org/10.2166/wst.2014.494.

Seco, A., Aparicio, S., González-Camejo, J., Jiménez-Benítez, A., Mateo, O., Mora, J.F., Noriega-Hevia, G., Sanchis-Perucho, P., Serna-García, R., Zamorano-López, N., Giménez, J.B., Ruiz-Martínez, A., Aguado, D., Barat, R., Borrás, L., Bouzas, A., Martí, N., Pachés, M., Ribes, J., Robles, A., Ruano, M.V., Serralta, J., Ferrer, J., 2018. Resource recovery from sulphate-rich sewage through an innovative anaerobic-based water resource recovery Page 35 of 37 facility (WRRF). *Water Sci. Technol.* 78, 1925–1936. https://doi.org/10.2166/wst.2018.492.

Seviour, T., Derlon, N., Dueholm, M.S., Flemming, H.-C., Girbal-Neuhauser, E., Horn, H., Kjelleberg, S., van Loosdrecht, M.C.M., Lotti, T., Malpei, M.F., Nerenberg, R., Neu, T.R., Paul, E., Yu, H., Lin, Y., 2019. Extracellular polymeric substances of biofilms: Suffering from an identity crisis. *Water Res.* 151, 1–7. https://doi.org/10.1016/j.watres.2018.11.020.

Srinivasan, J.T., Reddy, V.R., 2009. Impact of irrigation water quality on human health: A case study in India. *Ecol. Econ.* 68, 2800–2807. https://doi.org/10.1016/j.ecolecon.2009.04.019.

Stahel, W.R., 1982. The product-life factor, in: Orr, S.G. (Ed.), *An Inquiry into the Nature of Sustainable Societies: The Role of the Private Sector.* Houston: Houston Area Research Center (Series: 1982 Mitchell Prize Papers), NARC, pp. 72–105.

Swilling, M., Hajer, M., Baynes, T., Bergesen, J., Labbé, F., Musango, J.K., Ramaswami, A., Robinson, B., Salat, S., Suh, S., Currie, P., Fang, A., Hanson, A., Kruit, K., Reiner, M., Smit, S., Tabory, S., 2018. *The Weight of Cities: Resource Requirements of Future Urbanization.* UN Environment – International Resource Panel.

Tare, V., Bose, P., 2009. *Compendium of Sewage Treatment Technologies National River Conservation Directorate.* Ministry of Environment and Forests, Government of India.

Toth, A.J., Haaz, E., Nagy, T., Tarjani, A.J., Fozer, D., Andre, A., Valentinyi, N., Mizsey, P., 2018. Novel method for the removal of organic halogens from process wastewaters enabling water reuse. *Desalin. Water Treat.* 130, 54–62. https://doi.org/10.5004/dwt.2018.22987.

UN-DESA (United Nations Department of Economic and Social Affairs), 2018. World urbanization prospects: The 2018 revision, Highlights, United Nations, New York. Retrieved from: https://population.un.org/wup/Publications/Files/WUP2018-KeyFacts.pdf (accessed July 2019).

Van Rooijen, D.J., Turral, H., Wade Biggs, T., 2005. Sponge city: Water balance of mega-city water use and wastewater use in Hyderabad, India. *Irrig. Drain.* 54, S81–S91. https://doi.org/10.1002/ird.188.

Vaneeckhaute, C., Belia, E., Meers, E., Tack, F.M.G., Vanrolleghem, P.A., 2018. Nutrient recovery from digested waste: Towards a generic roadmap for setting up an optimal treatment train. *Waste Manag.* 78, 385–392. https://doi.org/10.1016/j.wasman.2018.05.047.

Vanner, R., Bicket, M., Withana, S., Brink, P.T., Razzini, P., Van Dijl, E., Watkins, E., Hestin, M., Tan, A., Guilche, S., Hudson, C., 2014. Scoping study to identify potential circular economy actions, priority sectors, material flows and value chains. European Commission. https://doi.org/10.2779/29525.

Verstraete, W., Van de Caveye, P., Diamantis, V., 2009. Maximum use of resources present in domestic "used water." *Bioresour. Technol.* 100, 5537–5545. https://doi.org/10.1016/j.biortech.2009.05.047.

Villamar, C.A., Vera-Puerto, I., Rivera, D., De la Hoz, F.D., 2018. Reuse and recycling of livestock and municipal wastewater in Chilean agriculture: A preliminary assessment. *Water* 10, 1–22. https://doi.org/10.3390/w10060817.

Villar Del, A., 2018. The reuse of reclaimed water in urban areas: A cost–benefit analysis, in: *WIT Transactions on the Built Environment*, pp. 323–332. https://doi.org/10.2495/UG180301.

Villarín, M.C., Merel, S., 2020. Paradigm shifts and current challenges in wastewater management. *J. Hazard. Mater.* 390, 1–21. https://doi.org/10.1016/j.jhazmat.2020.122139.

Vörösmarty, C.J., Sahagian, D., 2000. Anthropogenic disturbance of the terrestrial water cycle. *Bioscience* 50, 753–765. https://doi.org/10.1641/0006-3568.
Voulvoulis, N., 2018. Water reuse from a circular economy perspective and potential risks from an unregulated approach. *Curr. Opin. Environ. Sci. Heal.* 2, 32–45. https://doi.org/10.1016/j.coesh.2018.01.005.
Winans, K., Kendall, A., Deng, H., 2017. The history and current applications of the circular economy concept. *Renew. Sustain. Energy Rev.* 68, 825–833. https://doi.org/10.1016/j.rser.2016.09.123.
Wintgens, T., Nattorp, A., Elango, L., Asolekar, S.R., 2016. *Natural Water Treatment Systems for Safe and Sustainable Water Supply in the Indian Context: Saph Pani*. London: IWA Publishing. https://doi.org/10.2166/9781780408392.
WRIS (Water Resource Information System of India), 2015. India's water wealth. Retrieved from: http://www.india-wris.nrsc.gov.in/wrpinfo/index.php?title=India%27s_Water_Wealth (accessed June 2019).
Yang, Z., 2012. Study on the circular utilization of water resource in three industries, in: *2012 2nd International Conference on Remote Sensing, Environment and Transportation Engineering*. IEEE, pp. 1–4. https://doi.org/10.1109/RSETE.2012.6260753.

11 Circular Economy on Utilization of Wastewater for Energy Recovery

Shiv Prasad
Division of Environment Sciences, ICAR- Indian Agricultural Research Institute, New Delhi, India

Vikas Chandra Gupta
Department of Biotechnology, College of Engineering and Technology-IILM University, Greater Noida, Uttar Pradesh, India

11.1 INTRODUCTION

Increasing volumes of untreated wastewater or raw sewage from domestic and various non-domestic sources and its management have become a significant global issue. Worldwide, untreated raw sewage is dumped into waterbodies, increasing water pollution and adversely affecting drinking water quality and human health (Sinduja et al., 2023). Sewage is untreated wastewater that needs proper treatment before discharge into waterbodies or its use for various purposes. It contains mainly water and different organic and inorganic substances, nutrients, and toxic heavy metals, including pathogenic microbes (Maćerak et al., 2023). If mishandled, it may cause severe soil, water, and air pollution. However, sewage sludge (SS) is a semi-solid waste produced during wastewater treatment and industrial waste processing. SS is considered a valuable bioresource. If managed efficiently by recycling to recover energy under a circular economy, it would help to maximize SS usage and maintain a sustainable future (Ambay et al., 2023). A schematic sewage treatment, with energy recovery from wastewater, is illustrated in Figure 11.1.

As shown in Figure 11.1, sewage/wastewater treatment plants (STPs) include various stages: primary treatment, where solids, oils, fats, and organic matter are removed physically, and water is prepared to be sent to secondary treatment. The secondary treatment combines biochemical processes to eliminate organic matter and suspended solids, including various toxic pollutants, through activated sludge, sorption and decomposition, and oxidation of organic compounds (Dai et al., 2023). Further, tertiary treatments or advanced processes are performed by using disinfection agents like chlorination, ozonation, and UV-radiation before discharge.

DOI: 10.1201/9781003432869-11

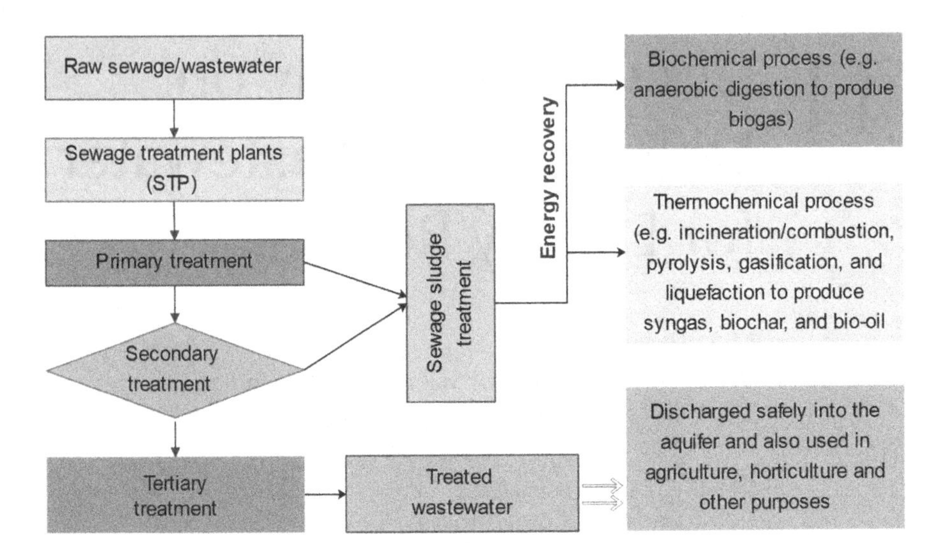

FIGURE 11.1 A flow scheme of a sewage treatment plant and recovery of energy.

Worldwide, treated liquid effluent is reported to be suitable for disposal in the natural environment. Treated water from STPs is discharged safely into aquifers and used for agriculture, horticulture, and other purposes.

Sewage sludge generation is increasing faster as STPs are being installed. Vast amounts of SS generation and SS management are becoming challenging. Scientists, industrial, and government organizations are trying to use it wisely due to its potential as a bioresource. Sewage sludge from wastewater treatment plants is considered a critical bioresource for generating heat and energy, e.g., biogas, fuel gas, syngas, biochar, and bio-oil through the biological and biochemical conversion processes (Ameri et al., 2023). These processes have now been considered to be a sustainable and promising alternative to sewage sludge management/treatment/volume reduction and energy recovery (Syed-Hassan et al., 2017; Singh et al., 2020). This chapter discusses the circular economy with respect to utilization of wastewater for energy recovery, and challenges and solutions for economic sewage sludge utilization.

11.2 WASTEWATER (SEWAGE) GENERATION IN INDIA

Wastewater (sewage) is the polluted water generated by various human activities. It is typically categorized by how it is generated, specifically domestic, industrial, or storm sewage (stormwater). According to CPCB Annual Report 2020–21, sewage treatment capacity has been enhanced by 50% since 2014. However, total sewage generation increased by 16% (72,368 megaliters/day (MLD)) compared to 2014, which was around 62,000 MLD. The current developed and operationalized wastewater treatment capacity is 26,869 MLD. State-wise sewage generation and treatment capacity are given in Table 11.1.

According to the UN-WWAP report (2017), high-income countries treat 70% of their wastewater, upper-middle-income countries 38%, lower-middle-income

TABLE 11.1
State-wise Sewage Generation and Treatment Capacity of Urban Centers in India

States /UTs	Sewage Generation (in MLD)	Installed Capacity (in MLD)	Proposed Capacity (in MLD)	Treatment Capacity (in MLD) Planned/Proposed	Operational Treatment Capacity (in MLD)
Andaman & Nicobar Islands	23	0	0	0	0
Andhra Pradesh	2,882	833	20	853	443
Arunachal Pradesh	62	0	0	0	0
Assam	809	0	0	0	0
Bihar	2,276	10	621	631	0
Chandigarh	188	293	0	293	271
Chhattisgarh	1,203	73	0	73	73
Dadra & Nagar Haveli	67	24	0	24	24
Goa	176	66	38	104	44
Gujarat	5,013	3,378	0	3,378	3,358
Haryana	1,816	1,880	0	1,880	1,880
Himachal Pradesh	116	136	19	155	99
Jammu & Kashmir	665	218	4	222	93
Jharkhand	1,510	22	617	639	22
Karnataka	4,458	2,712	0	2,712	1,922
Kerala	4,256	120	0	120	114
Lakshadweep	13	0	0	0	0
Madhya Pradesh	3,646	1,839	85	1,924	684
Maharashtra	9,107	6,890	2,929	9,819	6,366
Manipur	168	0	0	0	0
Meghalaya	112	0	0	0	0

(Continued)

TABLE 11.1 (CONTINUED)
State-wise Sewage Generation and Treatment Capacity of Urban Centers in India

States /UTs	Sewage Generation (in MLD)	Installed Capacity (in MLD)	Proposed Capacity (in MLD)	Treatment Capacity (in MLD) Planned/Proposed	Operational Treatment Capacity (in MLD)
Mizoram	103	10	0	10	0
Nagaland	135	0	0	0	0
NCT of Delhi	3,330	2,896	0	2,896	2,715
Orissa	1,282	378	0	378	55
Pondicherry	161	56	3	59	56
Punjab	1,889	1,781	0	1,781	1,601
Rajasthan	3,185	1,086	109	1,195	783
Sikkim	52	20	10	30	18
Tamil Nadu	6,421	1,492	0	1,492	1,492
Telangana	2,660	901	0	901	842
Tripura	237	8	0	8	8
Uttar Pradesh	8,263	3,374	0	3,374	3,224
Uttarakhand	627	448	67	515	345
West Bengal	5,457	897	305	1,202	337
Total	**72,368**	**31,841**	**4,827**	**36,668**	**26,869**

Source: CPCB Annual Report 2020–21

countries 28%, and low-income countries 8%. Globally, on average, 20% of wastewater is being treated. However, according to the CPCB Annual Report 2020–21, India's wastewater treatment capacity was 27.3%, and sewage treatment capacity was 18.6%, with another 5.2% capacity added per year. India's sewage treatment capacity is above the global average of 20%. However, considering the enormity of the problem, it needs to be adequately addressed. There is a need for swift measures to increase sewage treatment plants (STPs) to increase sewage treatment capacity, achieve efficient sludge management (Kosiński et al., 2023), and harness the potential of untapped resources for energy recovery.

11.3 CHARACTERISTICS OF WASTEWATER

Wastewater contains domestic sewage and industrial effluents. It is characterized based on suspended solids, organic matter (Biochemical Oxygen Demand [BOD] and Chemical Oxygen Demand [COD]), temperature, pH, and other parameters. Usually, wastewater is considered to have suspended solids concentrations less than 1000 mg/L with negligible amounts of grit (inorganic non-soluble solids), which can be removed mainly by simple pretreatment. In addition, wastewater can be classified as low, medium, or high strength based on BOD (or degradable matter) concentration. BOD measures the amount of O_2 required to remove waste organic matter from water in the decomposition process by aerobic bacteria (those bacteria that live only in an environment containing oxygen). In contrast, COD is the amount of O_2 required to break down the organic material via oxidation (Purwaningrum et al., 2023). Table 11.2 indicates the range of BOD concentrations associated with this classification and provides examples of wastewater sources.

According to Singh et al. (2020), in India, dewatered sludge has a total solid content of 25–45%, organics of 23–40%, and a calorific value of 8–21 MJ/kg. They reported an energy recovery potential of 555–1068 kWh/tonne by incineration and 315–608 kWh/tonne for AD sludge on a dry matter basis. However, Syed-Hassan et al. (2017) assessed sludge's heating value to be higher at 15–20 MJ/kg for dry sludge rich in organic content. Thus, it is considered a suitable substrate for waste-to-energy and fuel production under a circular economy. Sludge treatment is cost-intensive; it

TABLE 11.2
Classification of Wastewater Strength and Examples

Wastewater Strength	BOD Range (mg/L)	Examples of Sources
Low	<1,000	Municipal, agricultural (including flushed manures), pulp and paper
Medium	1,000–10,000	Food processing, canning, citrus processing, dairy processing, juice processing, brewery
High	10,000–200,000	Ethanol production, distilleries, biodiesel production, petrochemical, slaughterhouse

Source: https://biogas.ifas.ufl.edu/wastewater.asp

accounts for around 50% of the total cost of wastewater treatment and contributes 40% of total greenhouse gas (GHG) emissions associated with the STP (Liu et al., 2013). If dewatered sludge is used for energy generation, it can also help to mitigate pollution and climate change problems.

11.4 ENERGY RECOVERY FROM WASTEWATER SLUDGE UNDER A CIRCULAR ECONOMY

Energy recovery from wastewater sludge is a vital part of the circular economy (CE). CE is being realized where wastewater treatment is evolving in the circular economy by combining reusable water with energy production and resource recovery (Ali et al., 2022). Waste-to-energy from wastewater sludge through recycling is a fundamental step toward a CE (Bora et al., 2020). The key benefit of adopting CE principles in wastewater treatment is that recovering valuable resources and achieving their reuse can transform costlier sanitation into a self-sustaining service. It can also add value to the local economy (Rodriguez et al., 2020).

Currently, various technologies are used for energy recovery from wastewater sludge. Biochemical processes (e.g., anaerobic digestion) and thermochemical processes (e.g., incineration/combustion, pyrolysis, gasification, hydrothermal liquefaction [HTL], supercritical water gasification [SCWG], and supercritical water oxidation [SCWO]) are traditional methods to recover energy and valuable resources (Skaggs et al., 2018; Bora et al., 2020; Prasad et al., 2020) from sludge and biosolids. However, understanding the unrestricted energy in the wastewater sludge is critical to harnessing the power and reducing the wastewater treatment plant's energy needs. It is also being considered to combat pollution, minimize fossil fuel requirements, and create new revenue by using it to produce energy on-site or sell it to the local electricity grid. Energy recovery from wastewater sludge via biochemical and thermal conversion technologies is shown in Figure 11.2.

Wastewater treatment plants (WSTs) play a crucial role in purifying wastewater. A considerable amount of sewage sludge (SS) is also generated during this treatment process, a potential energy source (Song et al., 2014). SS generation is expected to increase continuously. However, it contains toxic heavy metals, and SS decomposition emits greenhouse gases (Dume et al., 2021), which need proper treatment in an environmentally friendly manner. Since sewage treatment plants (STP) are representative energy-consuming facilities (Cho et al., 2014), biogas plants can play an important role in considerably reducing wastewater treatment plant's energy needs and are an essential part of a wastewater treatment plant facility; due to this, this approach is being promoted globally. The AD process is being used to generate heat and electricity at STPs. Moreover, the AD process can reduce the organic matter content of sludge by 40 to 50% (Pettpain et al., (2006), significantly reducing the final sludge volume, stabilizing sewage sludge, and reducing pathogens and odor emissions.

Adaptation of energy-self-sufficient systems in wastewater treatment plants is now becoming important worldwide due to techno-economic feasibility and the success of the co-digestion process. Furthermore, it helps to keep waste materials, products, and sewage sludge (SS) economically viable by promoting recycling and reuse. According to Ghimire et al. (2021), two major routes can be adapted to

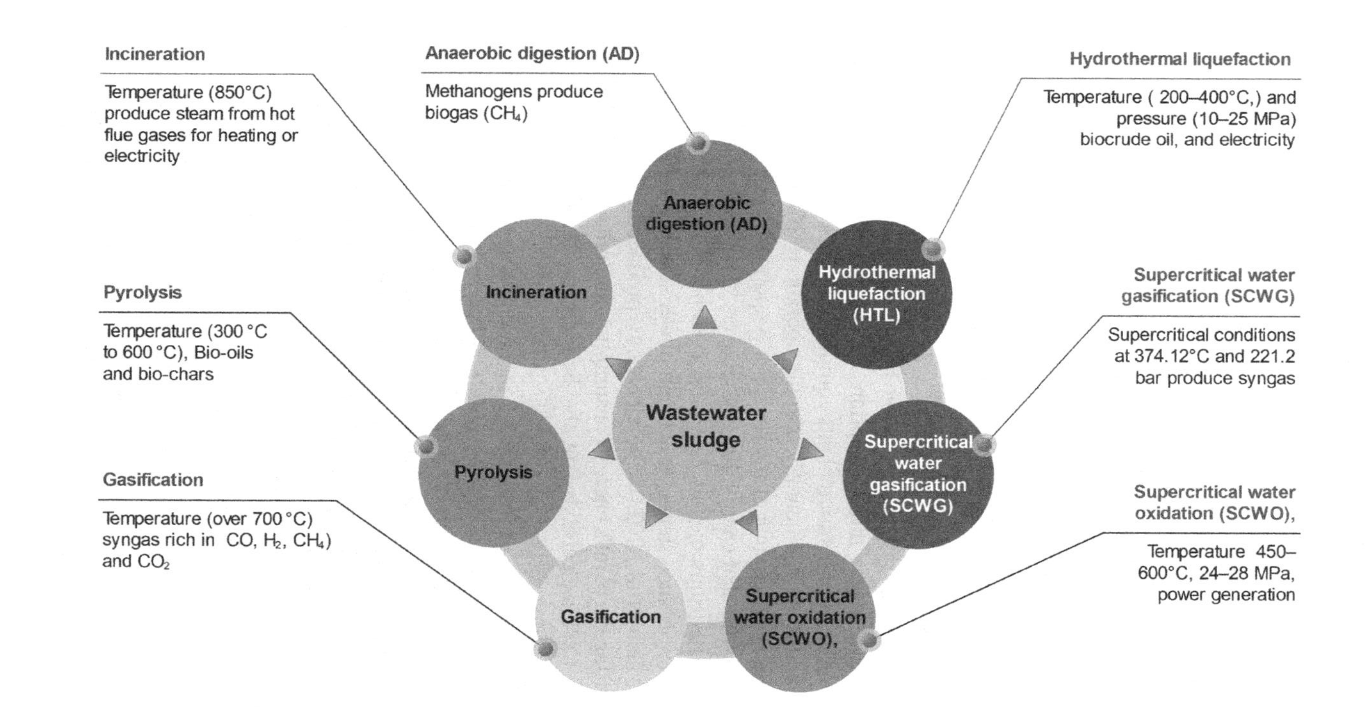

FIGURE 11.2 Energy recovery from wastewater sludge via biothermal conversion technologies.

the development of energy self-sufficiency in wastewater treatment: (i) to enhance C-capture (sludge) via anaerobic co-digestion to recover energy schemes, and (ii) to innovate and integrate energy and resource-efficient anaerobic treatment, thermal energy recovery and bioelectrochemical systems (BES); details are presented in the following subsections.

11.4.1 Thermochemical Energy Recovery Technology

Thermochemical energy recovery technology is promising for sewage sludge reduction, energy recovery, and the effective destruction of pathogens (Hu et al., 2021). Although many advanced thermochemical energy conversion technologies, such as pyrolysis, gasification, hydrothermal liquefaction, supercritical water gasification, and supercritical water oxidation, have been invented, there are still many technical limitations and challenges in their commercialization (Oladejo et al., 2019; Hu et al., 2021).

11.4.1.1 Incineration

Incineration involves the controlled combustion of sludge at high temperatures, converting the organic matter into appropriate forms of energy. The process typically occurs in specialized incineration plants equipped with state-of-the-art equipment to ensure efficient and environmentally friendly operations (Yang et al., 2023). During incineration, the water content of the sludge is evaporated, and the remaining solids are burned, releasing heat energy. Energy recovery from sludge incineration can be achieved through various means. The most common approach involves using the heat generated to produce steam, which can power steam turbines to generate electricity (Abdelsalam et al., 2023). The electricity produced can be used within the WWTP, offsetting the plant's energy consumption and reducing reliance on external power sources. Excess electricity can also be fed back into the grid, contributing to the local energy supply.

The heat energy obtained from sludge incineration can also be utilized for other purposes. It can be used for heating buildings, preheating water for treatment, or even for industrial applications such as drying or space heating (Chen et al., 2022). By effectively utilizing the energy released during incineration, WWTPs can achieve greater energy efficiency and reduce their carbon footprint.

Energy recovery from sludge incineration offers multiple benefits to wastewater treatment plants (WWTPs) and the environment. Firstly, it contributes to producing renewable energy, reducing reliance on coal and petroleum, and mitigating greenhouse gas (GHG) emissions (Yang et al., 2023). Additionally, incineration significantly reduces sludge volume by converting it into ash and gases, minimizing the need for extensive transport, disposal, and land use while reducing environmental impacts (Shabani et al., 2020). Moreover, energy recovery from sludge incineration provides cost savings for WWTPs by offsetting energy consumption and potentially generating revenue through electricity sales (Peltola et al., 2023). Furthermore, the incineration process allows for the recovery of valuable mineral resources, including phosphorus and metals, from the ash, promoting resource conservation and sustainable practices through recycling and reuse (Al-Ghouti et al., 2021). Thus, energy recovery from sludge incineration simultaneously contributes to renewable energy generation, waste reduction, cost savings, and resource conservation, making it a comprehensive and sustainable solution for sludge management in WWTPs.

Energy recovery from wastewater sludge through incineration technology presents a promising solution for managing sludge generated by WWTPs. These plants can generate renewable energy, reduce waste volume, and achieve greater resource efficiency by harnessing the heat energy released during incineration. Incorporating incineration technology into wastewater treatment systems is a more sustainable and eco-friendly approach to sludge management.

11.4.1.2 Pyrolysis

Wastewater treatment plants (WWTPs) generate significant amounts of wastewater sludge as a by-product of the treatment process. Traditional disposal methods for sludge present challenges due to its high water content and complex composition (Xiao et al., 2022). However, technological advancements have led to the development of pyrolysis as a promising method for energy recovery from wastewater sludge. Pyrolysis involves the thermal decomposition of organic materials, such as sludge, in the absence of oxygen, resulting in the production of gases, biochar, and bio-oil (Aboughaly and Fattah, 2023) This process can be carried out in specialized pyrolysis reactors, which provide controlled heating and conditions for the optimal conversion of the sludge. Pyrolysis offers several advantages, including its ability to handle different types of sludge and the potential for producing valuable by-products.

Energy recovery from sludge through pyrolysis technology has plenty of benefits. It produces renewable energy by converting sludge into biofuels and biogas, reducing reliance on fossil fuels and mitigating greenhouse gas emissions (Hasan et al., 2021). Also, pyrolysis significantly reduces sludge volume by transforming it into biochar, bio-oil, and gases, minimizing the need for extensive sludge disposal and land use and reducing environmental impacts (Wang et al., 2019). Additionally, pyrolysis enables the recovery of valuable resources from sludge. The biochar can be used as a soil amendment, enhancing soil quality and nutrient retention, while the bio-oil and gases can be processed into valuable fuels, promoting resource efficiency (Tarpani and Azapagic, 2023).

Furthermore, energy recovery from sludge pyrolysis offers cost savings for wastewater treatment plants (WWTPs) as the generation of valuable by-products, such as biofuels, can create additional revenue streams, and the reduction in sludge volume reduces disposal and transportation costs (Shanmugam et al., 2022). Therefore, the energy recovery from sludge through pyrolysis technology presents a comprehensive solution contributing to renewable energy generation, waste reduction, resource recovery, and cost-effectiveness. This makes it a promising approach for sustainable sludge management in WWTPs.

11.4.1.3 Gasification

Wastewater treatment plants (WWTPs) face the challenge of managing wastewater sludge generated during wastewater treatment and gasification has emerged as a promising method for energy recovery from sludge (Liu et al., 2020). Gasification involves converting organic materials, such as sludge, into a gaseous fuel known as synthesis gas or syngas (Okolie et al., 2019). This process occurs at high temperatures in an oxygen-limited environment, leading to the breakdown of complex organic compounds. Gasification can be performed using various gasification systems, such as fluidized bed gasifiers or entrained flow gasifiers, which offer efficient and controlled conditions for sludge conversion. Gasification technology enables the recovery of energy from sludge in different forms. The produced syngas can be utilized in combined heat and power (CHP) systems, generating electricity and heat for internal plant use or supplying energy to the grid (Brachi et al., 2022). The syngas can also be further processed to remove impurities and converted into substitute natural gas (SNG) for use in heating or as a renewable fuel for transportation (Gupta et al., 2021).

Energy recovery from sludge gasification is advantageous mainly because it contributes to renewable energy production by converting sludge into syngas, reducing reliance on fossil fuels, and promoting environmental sustainability (Tezer et al., 2022). The gasification significantly reduces sludge volume, minimizing the need for extensive sludge disposal and reducing the environmental impact of traditional methods (Huang et al., 2022). Also, gasification enables the recovery of valuable resources, such as phosphorus and metals, from the ash produced during the process, promoting resource conservation (Kwapinski et al., 2021). In addition, the energy recovery from sludge gasification can lead to cost savings for WWTPs through electricity and heat generation, offsetting energy consumption costs and potentially creating additional revenue streams. Thus, sludge gasification offers advantages of renewable energy generation, waste reduction, resource conservation, and cost-effectiveness for sustainable sludge management. By embracing gasification technology, WWTPs can contribute to a more sustainable and environmentally friendly approach to sludge treatment and energy production.

11.4.1.4 Hydrothermal Technologies

Hydrothermal technologies have gained attention as a viable method for energy recovery from wastewater sludge. Hydrothermal technologies involve subjecting sludge to high-temperature and high-pressure conditions in water or steam, leading to the decomposition of organic matter and the production of energy-rich products (Ghadge et al., 2022). These technologies include hydrothermal carbonization (HTC) and hydrothermal liquefaction (HTL), and energy recovery options from hydrothermal technologies are diverse. HTC produces hydrochar, a solid product that can be used as a renewable energy source through combustion or gasification. HTL, on the other hand, generates bio-oil, which can be refined and used as a substitute for fossil fuels. Additionally, the aqueous phase produced during hydrothermal processes can be treated further for nutrient recovery or anaerobic digestion to generate biogas (Sharma et al., 2020).

The benefits of energy recovery from sludge using hydrothermal technologies are significant as these methods offer a sustainable solution for sludge management by converting sludge into energy-rich products, that reduce the reliance on fossil fuels and help to mitigate greenhouse gas emissions. At the same time, hydrothermal technologies can minimize waste by reducing sludge volume, minimizing the need for extensive disposal. Furthermore, the recovery of valuable energy products, such as hydrochar and bio-oil, provides opportunities for resource conservation and potential revenue generation (Ipiales et al., 2021). Thus, the energy recovery from wastewater sludge through hydrothermal technologies offers a promising approach for both sludge management and renewable energy production. The ability to convert sludge into valuable energy products, such as hydrochar and bio-oil, highlights the potential of hydrothermal technologies in promoting sustainability and resource efficiency in wastewater treatment processes.

11.4.1.5 Other Thermochemical Technologies

Energy recovery from wastewater treatment plants (WWTPs) is crucial to sustainable wastewater management. Thermochemical technologies have attracted increasing attention as promising methods for converting wastewater and wastewater sludge into valuable energy resources. Emerging thermochemical technologies offer profitable solutions for energy recovery from wastewater treatment plants (WWTPs) while enhancing overall sustainability and wastewater treatment capabilities. Hydrothermal liquefaction (HTL) utilizes high temperature and pressure to convert wastewater sludge into bio-oil, leading to the breakdown of organic matter into liquid fuel. HTL offers multiple benefits, including increased energy recovery efficiency, sludge volume reduction, and the potential to produce valuable co-products like biochar (De la Torre Bayo et al., 2022; Yaqoob et al., 2021). Electrochemical oxidation (EO) involves using electrical energy to break down organic compounds in wastewater and sludge, offering high energy recovery potential and assisting in removing pollutants and pathogens. EO has shown promising results regarding energy generation and wastewater treatment efficiency, although further research is required to optimize performance and address electrode fouling and maintenance issues (Yap et al., 2023).

Supercritical water gasification (SCWG) is an advanced thermochemical technology that utilizes high-pressure water to convert organic matter into synthesis gas (syngas), a mixture of carbon monoxide (CO) and hydrogen (H_2) gases. SCWG offers numerous advantages, including high energy recovery efficiency, reduced greenhouse gas emissions, and the potential for nutrient recovery from wastewater. However, character design and catalyst selection challenges must be overcome for successful implementation at WWTPs (Okolie et al., 2019). Therefore, emerging thermochemical technologies present exciting opportunities for energy recovery from WWTPs, offering benefits such as increased energy conversion efficiency, reduced environmental impacts, and improved wastewater treatment capabilities. Nonetheless, further research and development are necessary to optimize these technologies, address technical challenges, and ensure their economic viability and scalability in practical applications.

11.4.2 Biochemical Energy Recovery Technology

Biochemical energy recovery technology from wastewater and sludge mainly involves microbial processes such as biodegradation, which makes chemical changes in the organic substrate. Anaerobic digestion (AD) is a typical example of energy recovery from organic waste to biogas. Bioelectrochemical systems (BESs) are emerging technologies which convert chemical energy into electrical energy by employing microorganisms as catalysts.

11.4.2.1 Biogas Technology for Recovery of Energy

The first digestion plant, built in 1859 near Mumbai, marked the initial exploration of biogas technology in India for sewage treatment (Nguyen, 2020). Traditionally, biogas plants primarily utilized animal dung as feedstock. However, the literature suggests that organic waste material can yield more biogas and address the dung shortage issue in countries like India. Biogas, also known as marsh gas, landfill gas, or digester gas, is predominantly composed of methane (CH_4), carbon dioxide (CO_2), and trace amounts of other gases such as hydrogen sulfide (H_2S). Its composition varies depending on the substrate used for methanogenesis (Prasad et al., 2017). Biogas technology involves the anaerobic digestion of organic matter, including sewage sludge, municipal solid waste, and other biodegradable feedstocks, by a consortium of microorganisms, producing biogas (Mulu et al., 2021).

Primary and secondary water treatment systems produce massive amounts of carbon-containing sewage sludge (SS). C-removal, by capturing this potential sludge, and its anaerobic digestion (AD) can enhance biogas (CH_4) production with higher energy recovery. Wastewater treatment plants increase their AD capacity to process organic wastes to generate additional revenues. They transform organic waste materials, products (fats, oil, grease, food waste), and sewage sludge (SS) into high-value beneficial products such as biogas (CH_4) and biosolids under a circular economy (Ghimire et al., 2021). Direct electricity is an attractive option through wastewater treatment facilities with large capacities (>50 million gallons/day (MGD)) that may give energy recovery efficiency of up to 40% (based on CH_4 generation). Microbial systems that can convert chemical energy (in the form of organic compounds) into electrical power are known as bioelectrochemical systems, and they show potential for higher energy recovery of up to 65%. (Rittmann, 2006). An integrated approach of biogas and bioelectricity production from sewage sludge is shown in Figure 11.3.

The energy recovery potential from wastewater through biogas technology is substantial. By capturing and utilizing the biogas generated during anaerobic digestion, wastewater treatment plants (WWTPs) can produce renewable energy through heat and electricity. This energy can be used on-site to power various plant operations or be fed into the local grid, contributing to the overall energy supply and generating income. Biogas technology not only facilitates energy recovery but also offers environmental benefits and improves livelihoods and sustainability (Prasad et al., 2022). By capturing methane emissions from wastewater, biogas technology helps mitigate climate change by reducing the release of this potent greenhouse gas (Sarpong et al., 2020). Furthermore, the anaerobic digestion process reduces the organic load and pathogens in wastewater, improving effluent quality and environmental safety.

In addition to energy recovery, biogas technology enables the recovery of resources and nutrients. The by-product of anaerobic digestion ("digestate") is a nutrient-rich fertilizer (Chozhavendhan et al., 2022). The nutrients in wastewater, such as nitrogen, phosphorus, and potassium, are effectively recycled back into the agricultural

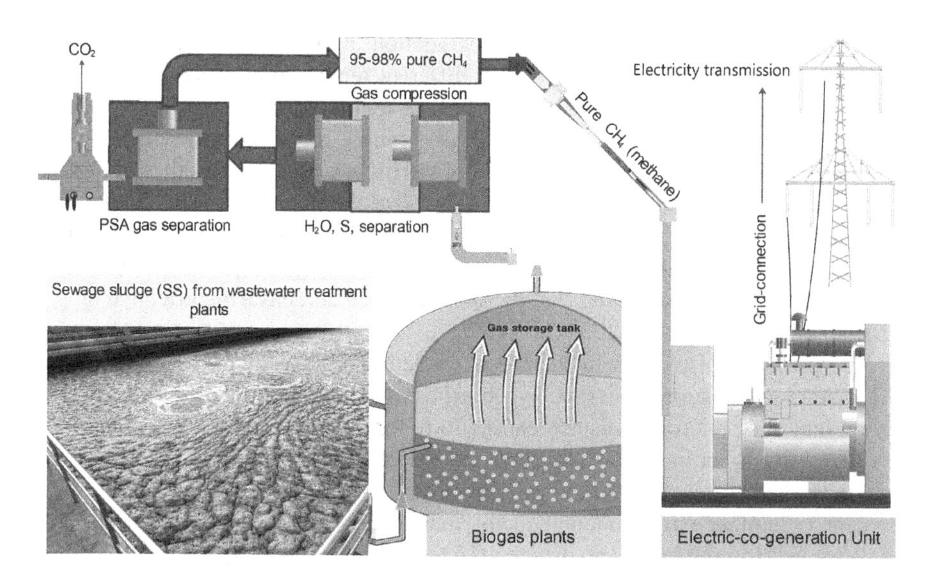

FIGURE 11.3 Integrated approach of biogas and bioelectric production from sewage sludge.

or horticultural sectors, reducing reliance on synthetic fertilizers and promoting a closed-loop nutrient system.

Using biogas technology for energy recovery from wastewater also offers economic advantages. The biogas produced can be converted into valuable energy products such as heat and electricity or even upgraded to biomethane for injection into the natural gas grid (Riley et al., 2020). These energy products can be monetized, providing a revenue stream for WWTPs. Additionally, biogas technology reduces operating costs associated with fossil fuel usage and can lead to potential savings for sludge disposal. Using biogas technology for energy recovery from wastewater aligns with the circular economy principles, promoting resource efficiency, environmental sustainability, and economic viability (Zvimba et al., 2021). By converting organic waste into a valuable energy resource, biogas technology contributes to reduced greenhouse gas emissions and fosters a more sustainable energy ecosystem. Furthermore, the resource and nutrient recovery potential of biogas technology supports the circularity of the wastewater management system. However, further research, technological advancements, and supportive policies are necessary to optimize the efficiency and scalability of biogas technology within the context of circular wastewater management.

Environmental factors play a crucial role in the successful implementation of biogas technology for the recovery of energy. These factors can influence the efficiency, effectiveness, and environmental sustainability of the biogas production processes. Understanding and managing these ecological factors are essential for maximizing the benefits and minimizing the potential negative impacts of biogas technology (Chrispim et al., 2021). Temperature is a critical environmental factor significantly affecting microbial activity and biodegradation rates in anaerobic digestion. Different types of microorganisms have different temperature optima for optimal biogas production. Maintaining the appropriate temperature range for the specific microbial consortia in the biogas system is essential to ensure efficient digestion and maximize biogas yield. The pH level of the biogas system is another important environmental factor that influences the activity of the microorganisms involved in anaerobic digestion (Yang et al., 2022). The optimal pH range for biogas production typically lies between 6 and 8. Monitoring and controlling the pH level within this range is crucial to maintaining the microbial balance and ensuring effective biogas production (Admasu et al., 2022).

The presence of oxygen in the biogas system can have detrimental effects on the anaerobic digestion process. Oxygen exposure can inhibit the growth and activity of anaerobic microorganisms, leading to reduced biogas production. Therefore, it is essential to maintain an anaerobic environment and prevent oxygen intrusion into the biogas system through proper system design and gas management. Moisture content is another environmental factor that can impact biogas production. The moisture level in the biogas feedstock affects water availability for microbial activity. Appropriate moisture control is necessary to ensure sufficient water content for microbial metabolism without causing excessive liquid accumulation or dryness in the system. Nutrient availability, particularly the carbon-to-nitrogen (C/N) ratio, is crucial for the microbial community in the biogas system. An optimal C/N ratio in the feedstock promotes microbial growth and activity, producing biogas efficiently. Monitoring and adjusting the C/N ratio through appropriate feedstock mixing or supplementation can help maintain optimal nutrient conditions.

In addition to these factors, relevant waste management practices, such as waste segregation, can minimize contaminants in the feedstock and ensure the quality of biogas production. Adequate mixing and agitation in the anaerobic digestion system are also essential to facilitate the breakdown of organic matter and maximize biogas yield. Considering and managing these environmental factors are vital for optimizing biogas production processes and maximizing energy recovery. By maintaining suitable temperature and pH conditions, preventing oxygen intrusion, managing moisture content, and ensuring nutrient availability, the environmental sustainability and efficiency of biogas technology can be enhanced, contributing to a cleaner and more sustainable energy future.

Operational factors play a critical role in the successful implementation of biogas technology for the recovery of energy. These factors encompass various aspects of the biogas production process, including feedstock management, system design, operational parameters, and maintenance. Understanding and addressing these operational factors are essential for optimizing biogas production, ensuring system reliability, and maximizing energy recovery (Yadav et al., 2022). Proper feedstock management is crucial for efficient biogas production. Factors such as feedstock type, composition, and availability must be considered. The feedstock should have a suitable carbon-to-nitrogen (C/N) ratio and moisture content to support microbial activity and biogas generation. Regularly monitoring feedstock characteristics and proper handling, storage, and pretreatment techniques are essential for maintaining consistent and high-quality feedstock for the biogas system. System design plays a significant role in the operational efficiency of biogas technology. The design should consider factors such as reactor type, size, configuration, gas collection and storage systems, and safety measures. Well-designed systems facilitate optimal mixing, efficient gas capture, and adequate heat management, leading to higher biogas yields and energy recovery.

Operational parameters, such as hydraulic retention time, organic loading rate, and temperature, must be carefully monitored. These parameters impact microbial activity, biogas production rate, and system stability. Regular monitoring and adjusting operational parameters ensure optimal microbial digestion and biogas production conditions, preventing process upsets and maximizing energy recovery. Maintenance and regular system inspections are essential to ensure the smooth operation of biogas technology. Proper maintenance activities include routine cleaning, repair of equipment, and preventive measures to avoid system failures or leaks. Regular inspections help in promptly identifying and addressing any operational issues, ensuring uninterrupted biogas production and energy recovery. Effective gas management is crucial for biogas technology.

Proper collection, storage, and utilization of biogas require efficient gas handling systems, including gas collection covers, piping, and storage tanks. Ensuring gas quality and preventing leaks are essential for safety and efficient energy recovery. Operator training and knowledge transfer are vital operational factors. Well-trained operators understand the biogas production process, operational requirements, and safety protocols. They can effectively manage the system, monitor performance, and troubleshoot issues, ensuring the optimal operation of biogas technology. Addressing operational factors is crucial for successfully implementing biogas technology for energy recovery. Proper feedstock management, system design, operational parameter control, maintenance, gas management, and operator training contribute to biogas production's efficiency, reliability, and safety. By optimizing these operational factors, biogas technology can maximize energy recovery, reduce environmental impacts, and contribute to a sustainable energy future

11.4.2.2 Integrated Approach of Energy Recovery from Sewage Sludge Using BES

The integrated approach to energy recovery from sewage sludge (SS) using bioelectrochemical systems (BESs) has now gained importance due to its eco-friendly nature and efficient pollutant reduction capability. BESs are generally described as systems that involve electrode reactions using microorganisms as biocatalysts (Ambaye et al., 2023) Microbial fuel and electrolysis cells are prominent examples of BESs as fast-growing biotechnologies (Abourached et al., 2014). Different types of other BESs have also been reported, such as biofuel cells, microbial chemical cells, microbial desalination cells, microbial electrolysis cells, microbial electrochemical snorkels, microbial fuel cells, and sediment fuel cells (Jin and Fallgren, 2022). Using wastewater for energy recovery based on microbial fuel cells (MFCs) holds significant promise in the circular economy of using wastewater for sustainable energy production. MFCs are a bioelectrochemical system that harnesses the metabolic activity of microorganisms to generate electricity while treating wastewater. An integrated approach to energy recovery from sewage sludge (SS) using BESs is shown in Figure 11.4.

This innovative approach offers numerous benefits, including efficient wastewater treatment, renewable energy production, and resource recovery (Elhenawy et al., 2022). In the circular economy framework, wastewater is considered a valuable resource rather than a waste product. Traditional wastewater treatment processes consume significant energy and generate sludge as a by-product. However, MFCs provide an alternative solution by simultaneously treating wastewater and generating electricity, transforming wastewater treatment into a resource-recovery process (Kaszycki et al., 2021). MFCs employ microorganisms, typically bacteria, that can transfer electrons to and from solid electrodes. The microorganisms oxidize organic matter present in the wastewater, releasing electrons that are then captured by the anode electrode. The electrons flow through an external circuit, creating an electrical current that can be utilized as electricity (Figure 11.4). Meanwhile, a cathode electrode, typically exposed to oxygen, facilitates the reduction reaction by consuming protons and electrons, forming water (Zhou et al., 2022).

The utilization of wastewater and sewage sludge for energy recovery based on MFCs offers several key advantages.

Efficient Wastewater Treatment. MFCs provide effective and sustainable wastewater treatment. The microbial activity in the MFCs degrades organic pollutants, reducing the wastewater's chemical oxygen demand (COD) and biological oxygen demand (BOD). This leads to cleaner effluent discharge, helping to protect natural water bodies and ecosystems (Ardakani and Gholikandi, 2020).

Renewable Energy Generation. MFCs offer a renewable energy source through electricity production. The electrons generated by the microorganisms during wastewater treatment are harnessed to produce electrical current. Although the power output of individual MFCs is relatively low, the scalability potential of this technology

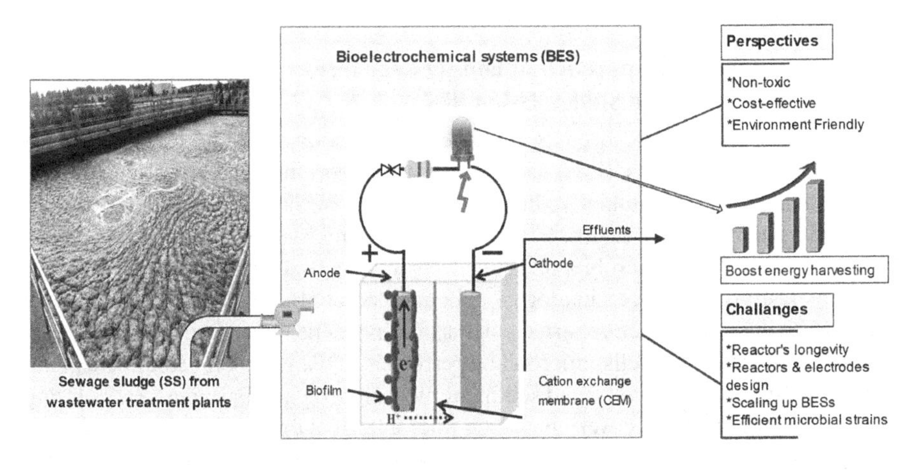

FIGURE 11.4 An integrated approach to energy recovery from sewage sludge using BESs.

allows for deploying multiple MFCs in series or parallel configurations, increasing the overall energy production (Shabani et al., 2020).

Resource Recovery. MFCs enable the recovery of valuable resources from wastewater. The treatment process can promote phosphorus precipitation, allowing its recovery as struvite, a valuable fertilizer. Additionally, nitrogen compounds can be recovered as ammonium salts, which can be used in agricultural or industrial applications (Munoz-Cupa et al., 2021).

Carbon Footprint Reduction. The utilization of MFCs for wastewater treatment and energy recovery contributes to reducing carbon emissions. By replacing traditional energy-intensive wastewater treatment methods, MFCs lessen the overall carbon footprint associated with wastewater management. Furthermore, the renewable electricity generated by MFCs can displace fossil fuel-based energy sources, further reducing carbon emissions (Bhatia et al., 2023).

Water–Energy Nexus. Integrating MFC technology into wastewater treatment creates synergies between the water and energy sectors. The energy generated by MFCs can power various processes within the wastewater treatment plant, reducing the reliance on external energy sources. This integration enhances the overall efficiency and sustainability of the water and energy systems (Kurniawan et al., 2022).

11.5 CONCLUSION

Utilizing wastewater for energy recovery offers multiple benefits within the circular economy context. Energy recovery from wastewater sludge through biochemical, bioelectrical, or thermal technology presents a promising energy production and waste management solution. Current technologies convert solid sludge to energy, providing efficient wastewater treatment, renewable energy generation, resource recovery, and carbon footprint reduction – the other benefits of energy recovery from sludge waste reduction and potential cost savings. Integrating biochemical, bioelectrical or thermal technology in wastewater treatment plants creates synergies between the water and energy sectors, contributing to a more eco-friendly approach to sustainable sludge management and a more resilient future.

REFERENCES

Abdelsalam, M. Y., Friedrich, K., Mohamed, S., Chebeir, J., Lakhian, V., Sullivan, B., Abdalla, A., Van Ryn, J., Girard, J., & Lightstone, M. F. (2023). Integrated community energy and harvesting systems: A climate action strategy for cold climates. *Applied Energy*, *346*, 121291.

Aboughaly, M., & Fattah, I. M. R. (2023). *Production of Biochar from Biomass Pyrolysis for Removal of PFAS from Wastewater and Biosolids: A Critical Review*. https://doi.org/10.20944/preprints202304.0309.v1

Abourached, C., Catal, T., & Liu, H. (2014). Efficacy of single-chamber microbial fuel cells for removal of cadmium and zinc with simultaneous electricity production. *Water Research*, *51*, 228–233.

Admasu, A., Bogale, W., & Mekonnen, Y. S. (2022). Experimental and simulation analysis of biogas production from beverage wastewater sludge for electricity generation. *Scientific Reports*, *12*(1). https://doi.org/10.1038/s41598-022-12811-3

Al-Ghouti, M. A., Khan, M., Nasser, M. S., Al-Saad, K., & Heng, O. E. (2021). Recent advances and applications of municipal solid wastes bottom and fly ashes: Insights into sustainable management and conservation of resources. *Environmental Technology & Innovation*, *21*, 101267.

Ali, M., Hong, P. Y., Mishra, H., Vrouwenvelder, J., & Saikaly, P. E. (2022). Adopting the circular model: Opportunities and challenges of transforming wastewater treatment plants into resource recovery factories in Saudi Arabia. *Water Reuse*, *12*(3), 346–365.

Ambaye, T. G., Vaccari, M., Franzetti, A., Prasad, S., Formicola, F., Rosatelli, A., Hassani, A., Aminabhavi, T. M., & Rtimi, S. (2023). Microbial electrochemical bioremediation of petroleum hydrocarbons (PHCs) pollution: Recent advances and outlook. *Chemical Engineering Journal*, *452*, 139372.

Ameri, B., & Hanini, S. (2023). Analogous study of cumulative biogas production by anaerobic digestion of sewage treatment plant sludge, the proposal of universal dimensionless models. *Energy Science & Engineering*, *11*(7), 2366–2384.

Ardakani, M. N., & Gholikandi, G. B. (2020). Microbial fuel cells (MFCs) in integration with anaerobic treatment processes (AnTPs) and membrane bioreactors (MBRs) for simultaneous efficient wastewater/sludge treatment and energy recovery-A state-of-the-art review. *Biomass and Bioenergy*, *141*, 105726.

Bhatia, L., Jha, H., Sarkar, T., & Sarangi, P. K. (2023). Food waste utilization for reducing carbon footprints towards the sustainable and cleaner environment: A review. *International Journal of Environmental Research and Public Health*, *20*(3), 2318.

Bora, R. R., Richardson, R. E., & You, F. (2020). Resource recovery and waste-to-energy from wastewater sludge via thermochemical conversion technologies in support of circular economy: A comprehensive review. *BMC Chemical Engineering*, *2*(1), 1–16.

Brachi, P., Di Fraia, S., Massarotti, N., & Vanoli, L. (2022). Combined heat and power production based on sewage sludge gasification: An energy-efficient solution for wastewater treatment plants. *Energy Conversion and Management: X*, *13*, 100171.

Chen, Z., Hou, Y., Liu, M., Zhang, G., Zhang, K., Zhang, D., Yang, L., Kong, Y., & Du, X. (2022). Thermodynamic and economic analyses of sewage sludge resource utilization systems integrating drying, incineration, and power generation processes. *Applied Energy*, *327*, 120093.

Cho, I. H., Ko, I. B., & Kim, J. T. (2014). Technology trend on the increase of biogas production and sludge reduction in wastewater treatment plants: Sludge pretreatment techniques. *Korean Chemical Engineering Research*, *52*, 413–424.

Chozhavendhan, S., Karthigadevi, G., Bharathiraja, B., Kumar, R. P., Abo, L. D., Prabhu, S. V., Balachandar, R., & Jayakumar, M. (2022). Current and prognostic overview on the strategic exploitation of anaerobic digestion and digestate: A review. *Environmental Research*, *216*, 114526.

Chrispim, M. C., Scholz, M., & Nolasco, M. A. (2021). Biogas recovery for sustainable cities: A critical review of enhancement techniques and key local conditions for implementation. *Sustainable Cities and Society*, *72*, 103033.

Dai, W., Pang, J., Ding, J., Wang, Y., Zhang, L., Ren, N., & Yang, S. (2023). Study on the removal characteristics and degradation pathways of highly toxic and refractory organic pollutants in real pharmaceutical factory wastewater treated by a pilot-scale integrated process. *Frontiers in Microbiology*, *14*. https://doi.org/10.3389/fmicb.2023.1128233

De la Torre Bayo, J. J., Martín Pascual, J., Torres Rojo, J. C., & Zamorano Toro, M. (2022). Waste to energy from municipal wastewater treatment plants: A science mapping. *Sustainability*, *14*(24), 16871.

Dume, B., Hanč, A., Švehla, P., Chane, A., & Nigussie, A. (2021). Carbon Dioxide and Methane Emissions during the composting and vermicomposting of sewage sludge under the effect of different proportions of Straw Pellets. *Environmental Sciences Proceedings*, *8*, 7.

Elhenawy, S., Khraisheh, M., AlMomani, F., Al-Ghouti, M., & Hassan, M. K. (2022). From waste to watts: Updates on key applications of microbial fuel cells in wastewater treatment and energy production. *Sustainability*, *14*(2), 955.

Ghadge, R., Nagwani, N., Saxena, N., Dasgupta, S., & Sapre, A. (2022). Design and scale-up challenges in hydrothermal liquefaction process for biocrude production and its upgradation. *Energy Conversion and Management: X*, *14*, 100223.

Ghimire, U., Sarpong, G., & Gude, V. G. (2021). Transitioning wastewater treatment plants toward circular economy and energy sustainability. *ACS Omega*, *6*(18), 11794–11803.

Gupta, P. K., Kumar, V., & Maity, S. (2021). Renewable fuels from different carbonaceous feedstocks: A sustainable route through Fischer–Tropsch synthesis. *Journal of Chemical Technology & Biotechnology*, *96*(4), 853–868.

Haghighat, M., Majidian, N., & Hallajisani, A. (2020). Production of bio-oil from sewage sludge: A review on the thermal and catalytic conversion by pyrolysis. *Sustainable Energy Technologies and Assessments*, *42*, 100870.

Hasan, M. M., Rasul, M. G., Khan, M. M. K., Ashwath, N., & Jahirul, M. I. (2021). Energy recovery from municipal solid waste using pyrolysis technology: A review on current status and developments. *Renewable and Sustainable Energy Reviews*, *145*, 111073.

Hu, M., Ye, Z., Zhang, H., Chen, B., Pan, Z., & Wang, J. (2021). Thermochemical conversion of sewage sludge for energy and resource recovery: Technical challenges and prospects. *Environmental Pollutants and Bioavailability*, *33*(1), 145–163.

Huang, C., Mohamed, B. A., & Li, L. Y. (2022). Comparative life-cycle assessment of pyrolysis processes for producing bio-oil, biochar, and activated carbon from sewage sludge. *Resources, Conservation and Recycling*, *181*, 106273.

Ipiales, R. P., de La Rubia, M. A., Diaz, E., Mohedano, A. F., & Rodriguez, J. J. (2021). Integration of hydrothermal carbonization and anaerobic digestion for energy recovery of biomass waste: An overview. *Energy & Fuels*, *35*(21), 17032–17050.

Jin, S., & Fallgren, P. H. (2022). Feasibility of using bioelectrochemical systems for bioremediation. In S. Das (Ed.), *Microbial biodegradation and bioremediation* (pp. 493–507). Elsevier.

Kaszycki, P., Głodniok, M., & Petryszak, P. (2021). Towards a bio-based circular economy in organic waste management and wastewater treatment–The Polish perspective. *New Biotechnology*, *61*, 80–89.

Kosiński, P., Kask, B., Franus, M., Piłat-Rożek, M., Szulżyk-Cieplak, J., & Łagód, G. (2023). The possibility of using sewage sludge pellets as thermal insulation. *Advances in Science and Technology Research Journal*, *17*(2), 161–172.

Kurniawan, T. A., Othman, M. H. D., Liang, X., Ayub, M., Goh, H. H., Kusworo, T. D., Mohyuddin, A., & Chew, K. W. (2022). Microbial fuel cells (MFC): A potential game-changer in renewable energy development. *Sustainability*, *14*(24), 16847.

Kwapinski, W., Kolinovic, I., & Leahy, J. J. (2021). Sewage sludge thermal treatment technologies with a focus on phosphorus recovery: A review. *Waste and Biomass Valorization*, *12*(11), 5837–5852. https://doi.org/10.1007/s12649-020-01280-2

Liu, B. B., Wei, Q., Zhang, B., & Bi, J. (2013). Life cycle GHG emissions of sewage sludge treatment and disposal options in Tai Lake Watershed, China. *Science of The Total Environment*, *447*, 361–369.

Liu, Z., Mayer, B. K., Venkiteshwaran, K., Seyedi, S., Raju, A. S. K., Zitomer, D., & McNamara, P. J. (2020). The state of technologies and research for energy recovery from municipal wastewater sludge and biosolids. *Current Opinion in Environmental Science & Health*, *14*, 31–36.

Maćerak, A. L., Mandić, A. K., Pešić, V., Pilipović, D. T., Bečelić-Tomin, M., & Kerkez, Đ. (2023). "Green" nzvi-biochar as fenton catalyst: Perspective of closing-the-loop in wastewater treatment. *Molecules*, *28*(3), 1425.

Mulu, E., M'Arimi, M. M., & Ramkat, R. C. (2021). A review of recent developments in application of low cost natural materials in purification and upgrade of biogas. *Renewable and Sustainable Energy Reviews*, *145*, 111081.

Munoz-Cupa, C., Hu, Y., Xu, C., & Bassi, A. (2021). An overview of microbial fuel cell usage in wastewater treatment, resource recovery and energy production. *Science of The Total Environment*, *754*, 142429.

Nguyen, T. V. A. (2020). *Comparison of carbon footprint of different waste treatment systems in Hanoi-Vietnam*. Hanoi, Vietnam: Lappeenranta-Lahti University of Technology, M.Sc. Thesis.

Okolie, J. A., Rana, R., Nanda, S., Dalai, A. K., & Kozinski, J. A. (2019). Supercritical water gasification of biomass: A state-of-the-art review of process parameters, reaction mechanisms and catalysis. *Sustainable Energy & Fuels*, *3*(3), 578–598.

Oladejo, J., Shi, K. Q., Luo, X., Yang, G., & Wu, T. (2019). A review of sludge-to-energy recovery methods. *Energies*, *12*(1), 60.

Peltola, P., Ruottu, L., Larkimo, M., Laasonen, A., & Myöhänen, K. (2023). A novel dual circulating fluidized bed technology for thermal treatment of municipal sewage sludge with recovery of nutrients and energy. *Waste Management*, *155*, 329–337.

Pettpain, F. (2006). Municipal sludge digestion attracts interest as energy prices rise. *Water World Magazine*, *21*. https://www.waterworld.com/wastewater/article/16200704

Prasad, S., Rathore, D., & Singh, A. (2017). Recent advances in biogas production. *Chem Eng Process Tech*, *3*(2), 1038.

Prasad, S., Singh, A., Dhanya, M. S., Rathore, D., & Rakshit, A. (2022). Biogas technology for livelihoods and sustainability. In A. Rakshit, S. Chakraborty, M. Parihar, V. S. Meena, P. K. Mishra, & H. B. Singh (Eds.), *Innovation in small-farm agriculture: Improving livelihoods and sustainability* (p. 341). CRC Press. ISBN:9781000574142. https://doi.org/10.1201/9781003164968

Prasad, S., Singh, A., Korres, N. E., Rathore, D., Sevda, S., & Pant, D. (2020). Sustainable utilization of crop residues for energy generation: A life cycle assessment (LCA) perspective. *Bioresource Technology*, *303*, 122964.

Purwaningrum, S. I., Syarifuddin, H., Nizori, A., & Wibowo, Y. G. (2023). Wastewater treatment plant design for batik wastewater with off-site system method in Ulu Gedong sub-district, Jambi City. *Jurnal Presipitasi: Media Komunikasi Dan Pengembangan Teknik Lingkungan*, *20*(1), 153–164. https://doi.org/10.14710/presipitasi.v20i1.153-164

Riley, D. M., Tian, J., Güngör-Demirci, G., Phelan, P., Villalobos, J. R., & Milcarek, R. J. (2020). Techno-economic assessment of chp systems in wastewater treatment plants. *Environments*, *7*(10), 74.

Rittmann, B. E. (2006). Microbial ecology to manage processes in environmental biotechnology. *Trends in Biotechnology*, *24*(6), 261–266. https://doi.org/10.1016/j.tibtech.2006.04.003

Rodriguez, D. J., Serrano, H. A., Delgado, A., Nolasco, D., & Saltiel, G. (2020). *From waste to resource: Shifting paradigms for smarter wastewater interventions in Latin America and the Caribbean*. World Bank.

Sarpong, G., Gude, V. G., Magbanua, B. S., & Truax, D. D. (2020). Evaluation of energy recovery potential in wastewater treatment based on codigestion and combined heat and power schemes. *Energy Conversion and Management*, *222*, 113147.

Shabani, M., Younesi, H., Pontié, M., Rahimpour, A., Rahimnejad, M., & Zinatizadeh, A. A. (2020). A critical review on recent proton exchange membranes applied in microbial fuel cells for renewable energy recovery. *Journal of Cleaner Production*, *264*, 121446.

Shanmugam, K., Gadhamshetty, V., Tysklind, M., Bhattacharyya, D., & Upadhyayula, V. K. K. (2022). A sustainable performance assessment framework for circular management of municipal wastewater treatment plants. *Journal of Cleaner Production*, *339*, 130657.

Sharma, H. B., Sarmah, A. K., & Dubey, B. (2020). Hydrothermal carbonization of renewable waste biomass for solid biofuel production: A discussion on process mechanism, the influence of process parameters, environmental performance and fuel properties of hydrochar. *Renewable and Sustainable Energy Reviews*, *123*, 109761.

Sinduja, M., Sathya, V., Maheswari, M., Dinesh, G. K., Prasad, S., & Kalpana, P. (2023). Groundwater quality assessment for agricultural purposes at Vellore District of Southern India: A geospatial based study. *Urban Climate*, *47*, 101368.

Singh, V., Phuleria, H. C., & Chandel, M. K. (2020). Estimation of energy recovery potential of sewage sludge in India: Waste to watt approach. *Journal of Cleaner Production*, *276*, 122538.

Skaggs, R. L., Coleman, A. M., Seiple, T. E., & Milbrandt, A. R. (2018). Waste-to-energy biofuel production potential for selected feedstocks in the conterminous United States. *Renewable and Sustainable Energy Reviews*, *82*, 2640– 2651.

Song, H. W., Han, S. K., Kim, C. G., & Shin, H. G. (2014). A study on the viscosity characteristics of dewatered sewage sludge according to thermal hydrolysis reaction. *Journal of the Korea Organic Resource Recycling Association*, *22*, 27–34.

Syed-Hassan, S. S. A., Wang, Y., Hu, S., Su, S., & Xiang, J. (2017). Thermochemical processing of sewage sludge to energy and fuel: Fundamentals, challenges and considerations. *Renewable and Sustainable Energy Reviews*, *80*, 888–913.

Tarpani, R. R. Z., & Azapagic, A. (2023). Life cycle sustainability assessment of advanced treatment techniques for urban wastewater reuse and sewage sludge resource recovery. *Science of The Total Environment*, *869*, 161771.

Tezer, Ö., Karabağ, N., Öngen, A., Çolpan, C. Ö., & Ayol, A. (2022). Biomass gasification for sustainable energy production: A review. *International Journal of Hydrogen Energy*, *47*(34), 15419–15433. https://doi.org/10.1016/j.ijhydene.2022.02.158

UNESCO & UN-WWAP. (2017). *The United Nations world water development report, 2017: Wastewater: The untapped resource*. UNESCO.

Wang, Z., Liu, K., Xie, L., Zhu, H., Ji, S., & Shu, X. (2019). Effects of residence time on characteristics of biochars prepared via co-pyrolysis of sewage sludge and cotton stalks. *Journal of Analytical and Applied Pyrolysis*, *142*, 104659.

Xiao, Y., Raheem, A., Ding, L., Chen, W.-H., Chen, X., Wang, F., & Lin, S.-L. (2022). Pretreatment, modification and applications of sewage sludge-derived biochar for resource recovery: A review. *Chemosphere*, *287*, 131969.

Yadav, P., Yadav, S., Singh, D., & Giri, B. S. (2022). Sustainable rural waste management using biogas technology: An analytical hierarchy process decision framework. *Chemosphere*, *301*, 134737.

Yang, F., Lei, F., & Zhen, X. (2022). Influence of organic loading rate and temperature fluctuation caused by solar energy heating on food waste anaerobic digestion. *Waste Management & Research: The Journal for a Sustainable Circular Economy*, *40*(9), 1440–1449.

Yang, H., Guo, Y., Fang, N., & Dong, B. (2023). Life cycle assessment of greenhouse gas emissions of typical sewage sludge incineration treatment route based on two case studies in China. *Environmental Research, 231*, 115959.

Yap, B. J. T., Heng, G. C., Ng, C. A., Bashir, M. J. K., & Lock, S. S. M. (2023). Enhancement of electrochemical–anaerobic digested palm oil mill effluent waste activated sludge in solids minimization and biogas production: Bench–scale verification. *Processes, 11*(6), 1609.

Yaqoob, A. A., Ibrahim, M. N. M., Umar, K., Parveen, T., Ahmad, A., Lokhat, D., & Setapar, S. H. M. (2021). A glimpse into the microbial fuel cells for wastewater treatment with energy generation. *Desalination Water Treat, 214*, 379–389.

Zhou, E., Lekbach, Y., Gu, T., & Xu, D. (2022). Bioenergetics and extracellular electron transfer in microbial fuel cells and microbial corrosion. *Current Opinion in Electrochemistry, 31*, 100830.

Zvimba, J. N., Musvoto, E. V, Nhamo, L., Mabhaudhi, T., Nyambiya, I., Chapungu, L., & Sawunyama, L. (2021). Energy pathway for transitioning to a circular economy within wastewater services. *Case Studies in Chemical and Environmental Engineering, 4*, 100144.

12 Circular Economy on Resource Recovery from Industrial Wastewater

Sumarlin Shangdiar
Institute of Environmental Engineering, National Sun Yat-Sen University, Kaohsiung, Taiwan
Center for Emerging Contaminants Research, National Sun Yat-Sen University, Kaohsiung, Taiwan

Kassian T.T. Amesho
Institute of Environmental Engineering, National Sun Yat-Sen University, Kaohsiung, Taiwan
Center for Emerging Contaminants Research, National Sun Yat-Sen University, Kaohsiung, Taiwan
Tshwane School for Business and Society, Faculty of Management of Sciences, Tshwane University of Technology, Pretoria, South Africa
The International University of Management, Centre for Environmental Studies, Main Campus, Dorado Park Ext 1, Windhoek, Namibia
Destinies Biomass Energy and Farming Pty Ltd, P. O. Box 7387, Swakopmund, Namibia
Regent Business School, Durban, South Africa

Abner Kukeyinge Shopati
Namibia Business School (NBS), Faculty of Commerce, Management and Law, University of Namibia, Private Bag 13301, Main Campus, Windhoek, Namibia

Timoteus Kadhila
School of Education, Department of Higher Education and Lifelong Learning, University of Namibia, Private Bag 13301, Windhoek, Namibia

DOI: 10.1201/9781003432869-12

Sioni Iikela
The International University of Management,
Centre for Environmental Studies, Main Campus,
Dorado Park Ext 1, Windhoek, Namibia

E.I. Edoun
Tshwane School for Business and Society, Faculty
of Management of Sciences, Tshwane University
of Technology, Pretoria, South Africa

Nastassia Thandiwe Sithole
Department of Chemical Engineering, Faculty
of Engineering and the Built Environment at the
University of Johannesburg, South Africa

S. Murugapoopathi
Department of Mechanical Engineering, PSNA College of
Engineering and Technology, Dindigul, Tamil Nadu, India

S. Surendarnath
Department of Mechanical Engineering, DVR
& Dr. HS MIC College of Technology (A),
Vijayawada, Andhra Pradesh, India

12.1 INTRODUCTION

The concept of a circular economy has emerged as a promising paradigm shift in the realm of sustainable resource management and environmental stewardship (Nayak and Bhushan, 2019; Roy et al., 2023). Circular economy principles have gained increasing recognition for their potential to address the complex challenges posed by industrial wastewater, a significant by-product of various industrial processes (Castellet-Viciano et al., 2022; Völker et al., 2020). The contemporary industrial landscape has witnessed remarkable advances in technology and production methods, yet it has also led to escalating concerns about the depletion of natural resources, pollution, and environmental degradation (Allwood, 2014; Dustin et al., 2017). In this context, the application of circular economy practices to industrial wastewater management holds the promise of not only mitigating environmental harm but also unlocking valuable resources that can contribute to a more sustainable and resilient future.

Circular economy, in essence, embodies a regenerative approach that seeks to minimize waste generation while maximizing resource efficiency (Willi et al., 2015). Unlike the traditional linear model of "take–make–dispose", a circular economy promotes the notion of closing the loop through strategies such as recycling, reusing, and recovering resources (Peter and Rutqvist, 2016). This approach aligns well with the imperative of addressing the challenges posed by industrial wastewater, which often contains valuable materials that can be reclaimed and reused through innovative processes (Matthias and Meisch, 2015; Andreas et al., 2018).

The significance of circular economy practices for resource recovery from industrial wastewater is underscored by the pressing need to transition toward more sustainable industrial practices (Petit-Boix and Leipold 2018; Bocken et al., 2017a). Industrial processes generate substantial volumes of wastewater laden with various pollutants and contaminants, including heavy metals, organic compounds, and other harmful substances (BSI, 2017; Atabaki et al., 2020). Disposing of such wastewater without proper treatment poses a considerable risk to aquatic ecosystems and public health (Eurostat, 2018). Moreover, the traditional linear approach to wastewater management results in the squandering of potentially valuable resources embedded within these effluents (Huang et al., 2019; Abhishek et al., 2020).

The purpose of this chapter is to provide a comprehensive review of circular economy practices in the context of resource recovery from industrial wastewater. By elucidating the multifaceted dimensions of circular economy strategies, this chapter aims to shed light on the transformative potential of these approaches in tackling the challenges of industrial wastewater management. Through an in-depth examination of relevant literature and case studies, we seek to unravel the intricate interplay between circular economy principles and the recovery of resources from industrial wastewater (Adeleye et al., 2023; Ambaye et al., 2021). By delving into the mechanisms, benefits, challenges, and future prospects of these practices, this chapter seeks to contribute to a nuanced understanding of how circular economy principles can be harnessed to drive sustainable resource recovery from industrial wastewater.

In the following sections, we will embark on a journey that traverses the landscape of circular economy practices, unraveling their significance for industrial wastewater management. We will delve into the various dimensions of circular economy strategies, exploring their potential to not only mitigate environmental harm but also foster economic growth and social well-being. Through a synthesis of existing knowledge, empirical evidence, and innovative insights, this chapter seeks to provide a comprehensive foundation for understanding and implementing circular economy practices for resource recovery from industrial wastewater. In doing so, we aspire to contribute to the advancement of knowledge in this critical field and inspire transformative action toward a more sustainable and circular industrial future.

12.2 CIRCULAR ECONOMY PRACTICES FOR RESOURCE RECOVERY FROM INDUSTRIAL WASTEWATER

The application of circular economy principles to industrial wastewater management holds significant promise for addressing the dual challenges of environmental sustainability and resource scarcity (Kurniawan et al., 2023; Hu et al., 2021). Circular economy practices emphasize the creation of closed-loop systems that minimize waste generation and maximize resource utilization, aligning with the regenerative approach to environmental stewardship (Abinandan et al., 2018; Al-Jabri et al., 2020). In the context of industrial wastewater, these practices offer a strategic pathway toward not only treating wastewater but also recovering valuable resources embedded within these effluents.

12.2.1 Overview of Circular Economy Practices for Resource Recovery

Circular economy practices encompass a spectrum of strategies that prioritize the optimization of resource flows, emphasizing the reduction of resource consumption and waste generation (Amorim de Carvalho et al., 2021; Ambulkar and Nathanson, 2022). Resource recovery from industrial wastewater aligns with these principles, aiming to extract valuable materials, energy, and water from effluents that would otherwise be discarded (Han et al., 2021; Huang et al., 2022). The recovery of resources, such as metals, nutrients, and organic matter, not only mitigates the environmental impact of wastewater discharge but also presents opportunities for generating economic value and promoting circularity (Karakas et al., 2020; Lee and Lei, 2022).

12.2.2 Biological and Physicochemical Treatment Technologies

A pivotal aspect of circular economy practices for resource recovery from industrial wastewater involves the utilization of advanced treatment technologies that facilitate the extraction and purification of valuable resources (Rufí-Salís et al., 2022; Oliveira et al., 2020). Biological and physicochemical treatment methods have emerged as key approaches in this regard, offering versatile solutions to address diverse wastewater streams and contaminants.

12.2.2.1 Biological Treatment Technologies

Biological treatment technologies harness the metabolic capabilities of microorganisms to degrade organic pollutants and facilitate resource recovery (Zhang et al., 2021; Wang et al., 2022). Microbial processes, such as aerobic and anaerobic digestion, biofiltration, and constructed wetlands, play a pivotal role in removing organic matter and nutrients from wastewater while concurrently generating biogas, valuable biomass, and nutrient-rich sludge (Schambeck et al., 2020b; Schellenberg et al., 2020). These recovered resources can subsequently be utilized for energy generation or soil amendment, or even as feedstocks for bioproducts.

12.2.2.2 Physicochemical Treatment Technologies

Physicochemical treatment technologies encompass a range of processes that leverage physical and chemical mechanisms to separate, concentrate, and recover valuable resources from wastewater (Pahunang et al., 2021; Chaudhary et al., 2020). Techniques such as precipitation, adsorption, membrane filtration, and electrochemical methods enable the removal of contaminants, heavy metals, and other pollutants from wastewater, resulting in concentrated streams that can be further processed to recover valuable materials (Pan et al., 2022; Ali et al., 2020). The materials recovered can include metals, rare earth elements, and even clean water suitable for reuse.

In summary, circular economy practices hold tremendous potential for advancing resource recovery from industrial wastewater, contributing to both environmental sustainability and economic viability. The application of biological and

TABLE 12.1
Various Biological Treatment Technologies for the Extraction and Purification of Valuable Resources in the Context of Circular Economy Practices

Biological Treatment Technology	Description	Merits	Limitations
Anaerobic digestion	Microbial degradation of organic matter in the absence of oxygen, producing biogas (methane-rich) and a nutrient-rich digestate.	• Biogas production for energy. • Nutrient-rich digestate for soil amendment. • Reduction in organic load.	• Requires proper feedstock composition. • Longer processing time. • Potential odor issues. • May not remove all contaminants.
Microalgae cultivation	Growing microalgae using wastewater nutrients for biomass production, which can be used as a biofuel or nutrient-rich supplements.	• Efficient nutrient removal. • High growth rate. • CO_2 sequestration. • Versatile end-products.	• Intensive cultivation requirements. • Harvesting and processing challenges. • Susceptibility to contamination.
Phytoremediation	Use of plants to remove, sequester, or transform contaminants from wastewater while accumulating valuable elements.	• Low-cost. • Natural treatment. • Diverse plant species. • Potential for biomass and metal recovery.	• Species selection is crucial. • Slower treatment. • May require large land area. • Limited to specific contaminants.
Bioleaching	Microbial processes to solubilize metals from solid waste, enabling metal recovery via bioaccumulation.	• Selective metal extraction. • Reduced reliance on chemical leaching. • Potential for low-grade ores.	• Process optimization needed. • Slow kinetics, pH- and temperature- sensitive. • Effluent treatment required.
Microbial fuel cells	Microbes generate electricity during wastewater treatment, allowing for simultaneous resource recovery.	• Simultaneous energy and nutrient recovery. • Low operational costs. • Suitable for low-strength wastewaters.	• Energy-intensive initial setup. • Low power output. • Limited scale, complex operation and maintenance.
Aerobic granular sludge	Dense microbial aggregates capable of simultaneous nutrient removal and accumulation of valuable compounds.	• High treatment efficiency. • Compact reactor design. • Good settling properties. • Potential for nutrient recovery.	• Long start-up time. • Complex granule formation. • Limited information on specific applications.

physicochemical treatment technologies underscores the versatility of circular solutions in managing wastewater while harnessing the value inherent in these effluents. As the next sections of this chapter delve deeper into the mechanisms and applications of these strategies, we will gain a comprehensive understanding of how circular economy practices drive the transformation of industrial wastewater management toward a more sustainable and resource-efficient paradigm.

12.3 BENEFITS OF CIRCULAR ECONOMY PRACTICES FOR RESOURCE RECOVERY FROM INDUSTRIAL WASTEWATER

The adoption of circular economy practices for resource recovery from industrial wastewater offers multifaceted benefits that extend beyond mere waste management, contributing to enhanced sustainability, economic gains, and improved environmental stewardship. This section elucidates the diverse advantages associated with circular economy strategies, underscoring their pivotal role in transforming industrial wastewater management into a regenerative and resource-efficient paradigm.

12.3.1 Reduction in Water Consumption and Wastewater Generation

One of the paramount benefits of integrating circular economy practices into industrial wastewater management is the notable reduction in water consumption and wastewater generation. Traditional linear approaches to wastewater management often result in significant water wastage and the generation of large volumes of effluents (Ferreira dos Santos et al., 2022). In contrast, circular economy practices emphasize the optimization of water utilization through closed-loop systems that prioritize the recycling and reuse of treated water (Gherghel et al., 2019). This approach not only conserves scarce freshwater resources but also minimizes the discharge of pollutants into natural water bodies, thus mitigating the environmental impact of industrial activities (Gupta et al., 2019).

12.3.2 Increased Revenue from Recovered Resources

Circular economy practices for resource recovery from industrial wastewater yield substantial economic benefits, primarily through the extraction and valorization of valuable materials and substances (Bacelo et al., 2020). By implementing advanced treatment technologies that enable the recovery of metals, nutrients, and energy from wastewater, industries can capitalize on the creation of new revenue streams (Kizito et al., 2017). The recovered resources, such as metals and biogas, can be sold or utilized internally, reducing the need for raw material extraction and thereby curbing the associated environmental burdens (Ji et al., 2023).

12.3.3 Reduced Energy Costs

Circular economy practices offer a compelling avenue for curbing energy costs associated with industrial wastewater treatment (Le Corre et al., 2009). Traditional wastewater treatment methods often necessitate energy-intensive processes, such

TABLE 12.2
Various Physicochemical Treatment Technologies for the Extraction and Purification of Valuable Resources in the Context of Circular Economy Practices

Physicochemical Treatment Technology	Description	Merits	Limitations
Precipitation	Chemical reactions are used to form insoluble precipitates that capture and separate target metals or compounds from wastewater.	• Effective for metal removal. • Simple operation. • Potential for metal recovery. • Produces solid waste.	• pH control required. • May lead to sludge generation. • Limited to specific contaminants. • Chemical costs.
Adsorption	Pollutants adhere to solid surfaces, such as activated carbon or other adsorbents, facilitating their removal from wastewater.	• Versatile and effective for various contaminants. • High removal efficiency. • Potential for adsorbent regeneration.	• Adsorbent replacement. • Saturation limits. • Possible secondary waste generation. • Cost of adsorbent materials.
Ion exchange	Ion-exchange resins selectively remove target ions from wastewater and release other ions in exchange.	• High removal efficiency. • Suitable for metal recovery. • Regenerable resins. • Effective for low concentrations.	• Resin replacement or regeneration. • Limited resin capacity. • Brine waste generation, requires chemicals.
Membrane filtration	Semi-permeable membranes separate pollutants from water based on size or charge, enabling concentration or recovery.	• High removal efficiency. • Versatile for various contaminants. • Potential for concentrate recovery.	• Membrane fouling. • Energy intensive. • Limited to certain pollutants. • Operational and maintenance costs.
Electrocoagulation	Electrochemical reactions generate coagulants that help remove contaminants through aggregation and settling.	• Effective for turbidity and metal removal. • Reduced chemical use. • Potential for metal recovery.	• Energy intensive. • Electrode consumption. • Requires monitoring and control. • Potential sludge handling.

(*Continued*)

TABLE 12.2 (CONTINUED)
Various Physicochemical Treatment Technologies for the Extraction and Purification of Valuable Resources in the Context of Circular Economy Practices

Physicochemical Treatment Technology	Description	Merits	Limitations
Flocculation and sedimentation	Coagulants or flocculants are added to wastewater, promoting particle aggregation and settling in a clarifier.	• Simple and cost-effective. • Effective for solids removal. • Potential for reuse or recovery of settled solids.	• Requires chemical addition. • Sludge handling issue. • Settling may be slow. • Potential for effluent quality issues.
Supercritical fluid extraction	Solvent in a supercritical state selectively extracts target compounds from wastewater, leaving behind purified solvent.	• High selectivity. • Potential for recovery of high-value compounds. • Low environmental impact.	• Energy-intensive use. • Specialized equipment needed. • Limited applications. • High operating and capital costs.

TABLE 12.3
Challenges and Barriers to Implementing Circular Economy Practices for Resource Recovery from Industrial Wastewater

Challenges and Barriers	Description	Examples
Regulatory and policy	Complex and evolving regulations can hinder the adoption of circular practices.	• Stringent discharge limits. • Lack of clear guidelines for recovered materials. • Conflicting policies.
Economic and financial	High initial costs, uncertain returns on investment, and lack of economic incentives.	• Expensive technology implementation. • Limited market demand for recovered products. • Inadequate funding.
Technical and technological	Limited availability of suitable technologies, technical complexities, and integration difficulties.	• Inadequate wastewater treatment infrastructure. • Lack of expertise in resource recovery. • Compatibility issues with existing processes.
Organizational and cultural	Resistance to change, lack of awareness, and organizational inertia.	• Employee reluctance to adopt new practices. • Lack of training, cultural resistance to recycling and reuse. • Inadequate internal support for innovation.
Knowledge and information	Lack of data, research gaps, and limited information sharing.	• Insufficient understanding of resource recovery technologies. • Absence of comprehensive data on recovered materials. • Limited knowledge dissemination.
Collaboration and partnerships	Inadequate cross-sector collaboration and stakeholder engagement.	• Limited cooperation between industries and regulators. • Absence of collaborative networks. • Insufficient engagement with local communities.

as aeration and chemical dosing, for pollutant removal and effluent purification (Muhmood et al., 2019). Circular economy strategies, particularly those involving biological treatment technologies, capitalize on microbial metabolic processes that can generate energy-rich by-products, such as biogas, during resource recovery (Li et al., 2019). The utilization of such by-products for on-site energy generation not only reduces operational energy expenses but also contributes to a more sustainable and self-sufficient wastewater treatment system (Li et al., 2020).

12.3.4 Improved Environmental Performance

The integration of circular economy practices into industrial wastewater management aligns seamlessly with the broader goals of environmental sustainability (Li et al., 2020). By diverting valuable resources from being discarded into waste streams, circular economy strategies significantly diminish the environmental impact associated with waste disposal and the extraction of virgin resources (Lin et al., 2021). Moreover, the reduction in wastewater discharges resulting from efficient resource recovery translates into fewer pollutants entering natural ecosystems, leading to improved water quality and ecosystem health (Mehta et al., 2015).

In summary, the adoption of circular economy practices for resource recovery from industrial wastewater engenders a host of benefits that resonate across environmental, economic, and societal dimensions. From water conservation and revenue generation to energy savings and enhanced environmental performance, circular economy strategies hold immense potential for ushering in a transformative shift in industrial wastewater management practices.

12.4 CHALLENGES AND BARRIERS TO IMPLEMENTING CIRCULAR ECONOMY PRACTICES FOR RESOURCE RECOVERY FROM INDUSTRIAL WASTEWATER

While the integration of circular economy practices into industrial wastewater management holds significant promise, it is not without its share of challenges and barriers. This section elucidates the multifaceted impediments that stakeholders may encounter when endeavoring to implement circular economy strategies for resource recovery from industrial wastewater. These challenges span regulatory, technical, economic, and organizational dimensions, underscoring the need for comprehensive and innovative approaches to surmount these barriers and facilitate the transition toward sustainable wastewater management.

12.4.1 Regulatory and Technical Barriers

Regulatory frameworks play a pivotal role in shaping the landscape of industrial wastewater management and resource recovery practices. However, certain regulatory hurdles can inadvertently hinder the adoption of circular economy strategies. Often, existing regulations and policies may be tailored to linear wastewater treatment approaches, thereby failing to accommodate innovative circular practices (Zhang et al., 2021). Ambiguities in regulatory guidelines, permitting processes, and compliance requirements can pose challenges to industries seeking to implement novel resource recovery techniques (Zhang et al., 2021). Moreover, the absence of clear regulatory incentives for circular economy practices may deter organizations from investing in the necessary infrastructure and technologies (Sanchez et al., 2021).

Technical complexities represent another formidable challenge. The successful implementation of circular economy strategies necessitates the deployment of sophisticated treatment technologies that enable resource recovery while ensuring effective pollutant removal (Sharma et al., 2020a). However, integrating these technologies into existing wastewater treatment systems can be intricate, requiring careful consideration of compatibility, operability, and maintenance (Alcalde-Calonge et al., 2022; Zhang et al., 2021). The intricate nature of some resource recovery processes, such as the extraction of valuable metals, demands specialized technical expertise and equipment, potentially impeding widespread adoption (Sharma et al., 2021).

12.4.2 Economic and Organizational Barriers

Economic considerations represent a critical factor influencing the adoption of circular economy practices in industrial wastewater management. While resource recovery has the potential to generate revenue and reduce costs in the long run, initial investments required for infrastructure upgrades and technology implementation can be substantial (Zhang et al., 2021). Industries may encounter financial barriers when attempting to justify these upfront expenditures to stakeholders or investors. Additionally, uncertainties surrounding the market value of recovered resources, such as fluctuating commodity prices, can pose risks to the economic viability of circular economy initiatives (Li et al., 2020).

Organizational barriers further complicate the adoption of circular economy practices. Resistance to change, entrenched operational routines, and a lack of awareness among personnel about circular economy concepts can impede the successful implementation of resource recovery strategies (Gupta et al., 2019). Collaborative efforts across departments and functions may be required to fully embrace circularity, necessitating effective communication and coordination (Bacelo et al., 2020). Moreover, industries operating in diverse sectors may encounter sector-specific challenges that require tailored solutions, underscoring the need for customized approaches to circular economy integration (Ji et al., 2023).

In conclusion, the journey toward implementing circular economy practices for resource recovery from industrial wastewater is accompanied by a constellation of challenges and barriers. Regulatory complexities, technical intricacies, economic considerations, and organizational dynamics collectively influence the success of circular economy initiatives. Addressing these barriers requires a concerted effort from policymakers, researchers, industries, and stakeholders, with a focus on regulatory reform, technological innovation, economic incentives, and organizational transformation.

12.5 CASE STUDIES AND EXAMPLES

As the principles of the circular economy continue to gain momentum, industries across diverse sectors have embarked on pioneering initiatives to integrate circularity into their wastewater management strategies. This section investigates selected case studies that spotlight the successful implementation of circular economy practices for resource recovery from industrial wastewater. These examples underscore the practical feasibility of circular approaches and offer insights into the diverse strategies employed by industries to achieve sustainable resource utilization and environmental stewardship (Figure 12.1).

12.5.1 Industrial Symbiosis in the Textile Sector

One illustrative case emerges from the textile sector, an industry traditionally associated with substantial water consumption and pollutant discharge. A textile dyeing and finishing plant in Denmark undertook a transformative approach by engaging in industrial symbiosis with neighboring companies. This collaborative endeavor enabled the plant to recover heat and wastewater from adjacent industries, utilizing the recovered resources for dyeing processes and reducing freshwater consumption

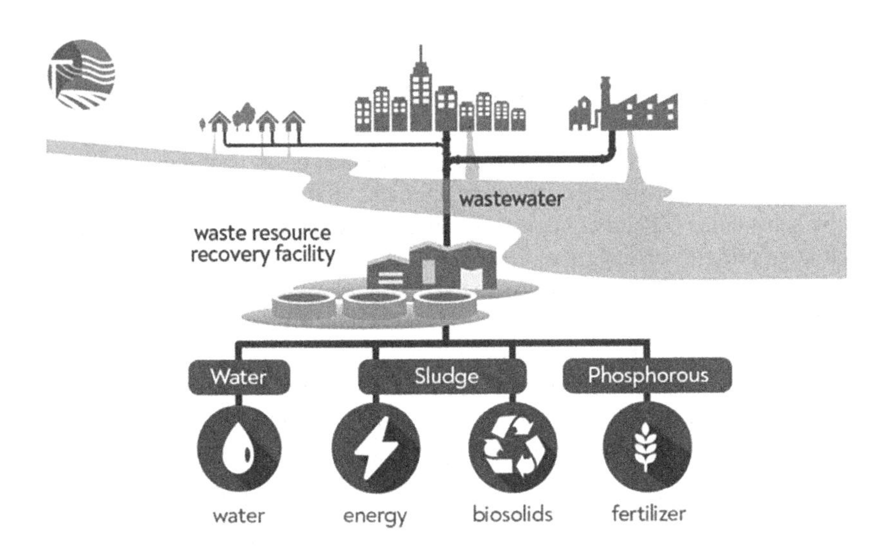

FIGURE 12.1 Wastewater? From waste to resource. Source: https://www.worldbank.org/en/topic/water/publication/wastewater-initiative

by 40%. The implementation of this closed-loop system not only mitigated the environmental impact of wastewater discharge but also led to considerable energy savings and cost reduction. Figure 12.2 shows industrial symbiosis for ensuring the implementation of circular economy practices.

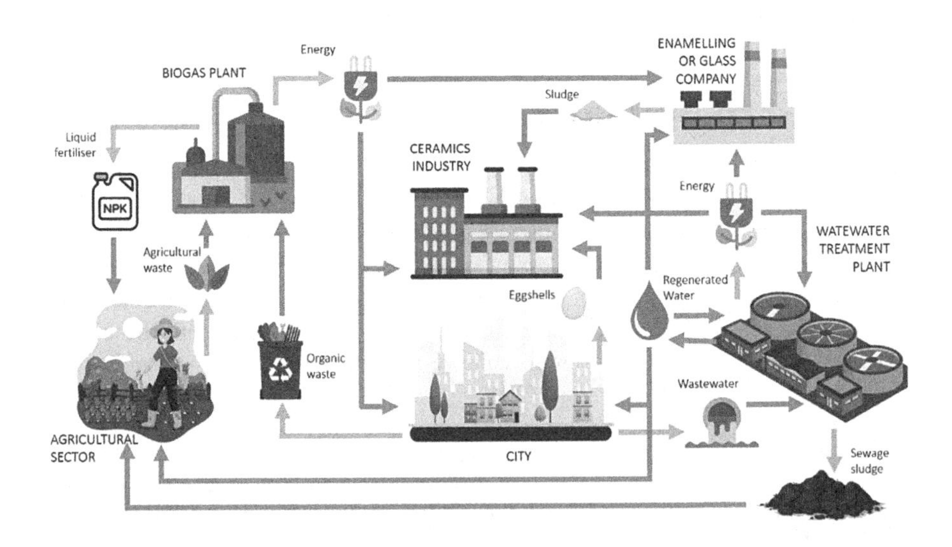

FIGURE 12.2 Industrial symbiosis: Ensuring the implementation of circular economy practices. Source: Castellet-Viciano et al. (2022)

12.5.2 Metal Recovery from Electroplating Effluents

Electroplating industries have long grappled with the challenge of treating effluents rich in heavy metals. An electroplating facility in South Korea adopted an innovative resource recovery strategy by implementing a combination of electrochemical and ion-exchange technologies (Chaudhary et al., 2020). Through this integrated approach, the facility achieved efficient removal and recovery of metal contaminants from wastewater streams, resulting in the extraction of valuable metals for potential reuse (Schellenberg et al., 2020). This case exemplifies the potential for circular economy practices to transform wastewater treatment from a burden to an opportunity for resource extraction and revenue generation.

12.5.3 Agricultural Waste Valorization

The agricultural sector, a significant contributor to water pollution through runoff and nutrient leaching, has also witnessed the integration of circular economy principles. An agricultural cooperative in the Netherlands embraced a circular approach by implementing an on-site biogas plant to digest livestock manure and crop residues (Ali et al., 2020). The biogas produced served dual purposes, generating renewable energy for on-farm operations and producing nutrient-rich digestate that could be returned to the fields as an organic fertilizer (Pan et al., 2022). This circular solution not only curtailed the negative environmental impact of agricultural waste but also aligned with broader sustainability objectives. Figure 12.3 shows an overview of the valorization agro-food wastes into high-value products.

12.5.4 Brewery Effluent Upcycling

The beverage industry has also leveraged circular economy principles to enhance wastewater management. A brewery in Belgium innovatively harnessed the potential of wastewater-derived resources by employing a microalga-based treatment system (Ferreira dos Santos et al., 2022). This system effectively removed nutrients from brewery effluent while concurrently cultivating a microalgal biomass rich in proteins and lipids (Schambeck et al., 2020b). The harvested biomass was subsequently valorized as a feedstock for animal feed, exemplifying the potential for circular strategies to close material loops and generate value from wastewater by-products.

In conclusion, the presented case studies serve as illuminating examples of circular economy practices applied to industrial wastewater management. These instances highlight the versatility and adaptability of circular strategies, with industries adopting tailored approaches to suit their specific contexts. By embracing circularity, these industries have not only achieved resource recovery but also unlocked economic, environmental, and societal benefits. These examples underscore the potential of circular economy principles to revolutionize the way industries perceive and manage wastewater, propelling them toward more sustainable and resilient practices.

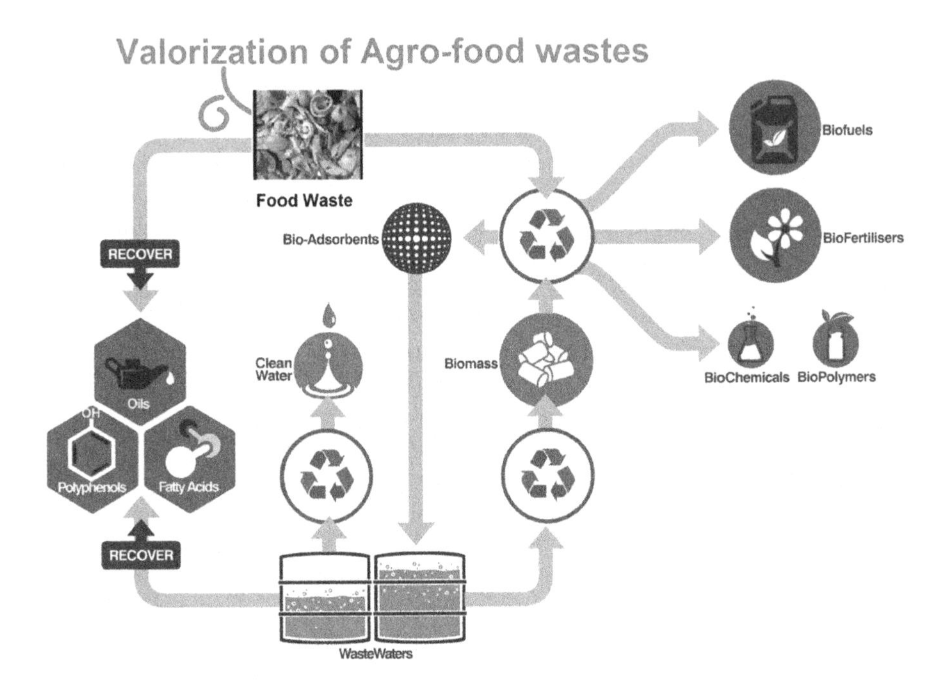

FIGURE 12.3 Overview of the valorization of agro-food wastes into high-value products (obtained with permission from Nayak & Bhushan, 2019)

12.6 FUTURE DIRECTIONS AND RESEARCH NEEDS

The integration of circular economy practices into industrial wastewater management presents a dynamic and evolving landscape that holds significant promise for sustainable resource recovery. As industries continue to recognize the intrinsic value of circular approaches, there exists a compelling imperative to advance research efforts and drive innovation toward more efficient, effective, and comprehensive strategies. This section delves into the future directions of circular economy practices for resource recovery from industrial wastewater and highlights critical research needs that will shape the trajectory of this field.

12.6.1 Optimization of Circular Economy Strategies

While the presented case studies exemplify successful implementations of circular economy practices, there remains a need for in-depth investigations to optimize the performance and applicability of these strategies across diverse industrial sectors. The development of robust and adaptable circular solutions requires a nuanced understanding of the intricate interplay between technological, economic, and environmental factors. Future research endeavors should focus on refining and tailoring circular economy practices to specific industrial contexts, encompassing considerations such as wastewater characteristics, process compatibility, and resource potential.

12.6.2 Technological Advancements

Advances in technology continue to be a driving force behind the evolution of circular economy practices. As researchers explore innovative treatment technologies, novel avenues for resource recovery are being revealed. One notable frontier lies in the application of emerging technologies such as nanotechnology, advanced membrane processes, and bioremediation techniques. These technologies hold the potential to enhance the efficiency of resource extraction, streamline treatment processes, and expand the range of recoverable resources. Consequently, future research should prioritize the exploration of cutting-edge technologies and their integration within circular wastewater management frameworks.

12.6.3 Multi-sector Collaborations and Knowledge Exchange

The realization of circular economy practices necessitates collaborative efforts that transcend traditional boundaries. Cross-sector collaborations among industries, academia, government agencies, and research institutions can facilitate knowledge exchange, promote best practices, and catalyze the adoption of circular solutions. Interdisciplinary partnerships offer a platform for holistic problem-solving, enabling the identification of synergies, co-benefits, and innovative approaches that extend beyond individual industrial domains. Future research should emphasize the establishment of collaborative networks that foster a culture of shared expertise, fostering a collective journey toward sustainable resource recovery.

12.6.4 Lifecycle Assessment and Circular Metrics

The comprehensive assessment of circular economy practices requires robust methodologies that capture the environmental, economic, and societal implications of resource recovery from industrial wastewater. Lifecycle assessment (LCA) and circular metrics emerge as indispensable tools for quantifying the net benefits of circular strategies and evaluating their potential trade-offs. Future research endeavors should focus on refining LCA methodologies specific to wastewater treatment, considering variables such as resource depletion, energy consumption, greenhouse gas emissions, and socioeconomic impacts. The integration of circular metrics can provide a standardized framework for benchmarking circular performance, enabling informed decision-making and policy formulation.

In conclusion, the future of circular economy practices for resource recovery from industrial wastewater is marked by a dynamic interplay of optimization, technological innovation, collaborative synergy, and robust assessment frameworks. The pursuit of these future directions, coupled with a commitment to rigorous research and knowledge dissemination, holds the promise of ushering in a new era of sustainable and resilient industrial wastewater management. By addressing research needs and embracing emerging developments, the field is poised to revolutionize resource recovery, minimize environmental impact, and contribute to a circular and regenerative economy.

12.7 CONCLUSION

The adoption of circular economy practices for resource recovery from industrial wastewater represents a paradigm shift that holds significant promise for sustainable industrial development. This chapter has navigated through the intricate landscape of circular economy strategies, highlighting their role in reshaping industrial wastewater management toward a regenerative and resource-efficient approach. As we conclude this investigation, it is imperative to summarize the key insights, underscore their implications, and outline the imperative for industries to embark on this transformative journey.

Throughout this chapter, we have examined the multifaceted realm of circular economy practices for resource recovery from industrial wastewater. We delved into the fundamental principles of circularity, encompassing the cradle-to-cradle approach, waste valorization, and closed-loop systems. The significance of circular design and innovative treatment technologies were underscored as pivotal enablers of efficient resource extraction and utilization. The comprehensive benefits of circular practices, ranging from reduced water consumption to enhanced economic viability, were revealed through illustrative case studies. The inherent challenges and barriers that industries must overcome were identified, ranging from regulatory hurdles to organizational constraints.

12.7.1 Nurturing the Transition: A Call to Action

As industries stand at the precipice of a circular revolution, the time has come to heed the call for action. The evidence presented throughout this chapter underscores the transformative potential of circular economy practices. It is incumbent upon industries to embark on a journey of recalibrating their wastewater management strategies, harnessing the latent potential of by-products and effluents that were once considered liabilities. By embracing circularity, industries can not only optimize resource recovery but also forge a path toward resilience, reduced environmental impact, and enhanced profitability.

12.7.2 Charting the Path Forward: Future Directions and Research Imperatives

The journey toward a circular economy for industrial wastewater management is not without its challenges. The realization of circularity necessitates concerted efforts, interdisciplinary collaborations, and continual research advancements. The future directions of this field should be characterized by a commitment to optimizing circular strategies, harnessing cutting-edge technologies, fostering cross-sector partnerships, and robustly assessing circular performance through comprehensive lifecycle assessment. Furthermore, the identification of innovative financing mechanisms, policy incentives, and regulatory frameworks will play a pivotal role in expediting the adoption of circular practices across diverse industrial sectors.

12.7.3 Embracing a Sustainable Future

In summary, the concept of circular economy practices for resource recovery from industrial wastewater is more than a theoretical construct – it is a pragmatic and actionable approach that promises to redefine the way industries engage with their waste streams. This chapter serves as a clarion call for industries to embrace circularity as a guiding principle, thereby harnessing the inherent potential of industrial wastewater for sustained value creation. By adopting circular practices, industries can simultaneously achieve economic prosperity, environmental stewardship, and social responsibility. The path forward requires determination, innovation, and an unwavering commitment – a collective endeavor to steer industries toward a more sustainable and circular future.

ACKNOWLEDGMENTS

We would like to express our sincere gratitude to Dr. Chingakham Chinglenthoiba (John) for his invaluable assistance in proofreading and providing constructive feedback on our manuscript. His expertise and meticulous attention to detail greatly enhanced the quality and clarity of our work. We are truly grateful for his time and dedication in helping us improve this research.

DECLARATION OF COMPETING INTERESTS

The authors declare that they have no known competing financial interests or personal relationships that could have appeared to influence the work reported in this paper.

DECLARATION OF GENERATIVE AI AND AI-ASSISTED TECHNOLOGIES IN THE WRITING PROCESS

Statement: During the preparation of this work, the authors used ChatGPT as AI-assisted technology in order to improve readability and language. After using this tool, the authors reviewed and edited the content as necessary and take full responsibility for the content of the publication.

REFERENCES

Abhishek, P., Bishnu, A., Aitazaz, F., 2020. Biochar-assisted wastewater treatment and waste valorization. In A.A. Ahmed & H.H.A. Mohammed (Eds.), *Applications of Biochar for Environmental Safety* (pp. Ch. 14). IntechOpen. https://doi.org/10.5772/intechopen.92288.

Abinandan, S., Subashchandrabose, S.R., Venkateswarlu, K., Megharaj, M., 2018. Nutrient removal and biomass production: Advances in microalgal biotechnology for wastewater treatment. *Crit. Rev. Biotechnol.* 38 (8), 1244–1260. https://doi.org/10.1080/07388551.2018.1472066.

Adeleye, A.T., Bahar, M.M., Megharaj, M., Rahman, M.M., 2023. Recent developments and mechanistic insights on adsorption technology for micro- and nanoplastics removal in aquatic environments. *J. Water Process Eng.* 53, 103777. https://doi.org/10.1016/j.jwpe.2023.103777.

Alcalde-Calonge, A., Sáez-Martínez, F.J., Ruiz-Palomino, P., 2022. Evolution of research on circular economy and related trends and topics: A thirteen-year review. *Ecol. Inform.* 70, 101716.

Ali, J., Rasheed, T., Afreen, M., Anwar, M.T., Nawaz, Z., Anwar, H., Rizwan, K., 2020. Modalities for conversion of waste to energy—Challenges and perspectives. *Sci. Total Environ.* 727, 138610. https://doi.org/10.1016/J.SCITOTENV.2020.138610.

Al-Jabri, H., Das, P., Khan, S., Thaher, M., Abdulquadir, M., 2020. Treatment of wastewaters by microalgae and the potential applications of the produced biomass—A review. *Water.* 13 (1), 27. https://doi.org/10.3390/W13010027.

Allwood, J.M., 2014. Squaring the circular economy: The role of recycling within a hierarchy of material management strategies. In E. Worell & M.A. Reuter (Eds.), *Handbook of Recycling* (pp. 445–477). Amsterdam: Elsevier.

Ambaye, T.G., Vaccari, M., van Hullebusch, E.D., Amrane, A., Rtimi, S., 2021. Mechanisms and adsorption capacities of biochar for the removal of organic and inorganic pollutants from industrial wastewater. *Int. J. Environ. Sci. Technol.* 18 (10), 3273–3294. https://doi.org/10.1007/s13762-020-03060-w.

Ambulkar, A., Nathanson, J.A., 2022. Wastewater treatment. Encyclopedia Britannica. 5 Sep. 2022. https://www.britannica.com/technology/wastewater-treatment. (Accessed 31 October 2022).

Amorim de Carvalho, C., de, Ferreira dos Santos, A., Tavares Ferreira, T.J., Sousa Aguiar Lira, V.N., Mendes Barros, A.R., Bezerra dos Santos, A., 2021. Resource recovery in aerobic granular sludge systems: Is it feasible or still a long way to go? *Chemosphere.* 274, 129881. https://doi.org/10.1016/j.chemosphere.2021.129881.

Andreas, M., Haas, W., Wiedenhofer, D., Krausmann, F., Philip Nuss, P., Blengini, G.A., 2018. Measuring progress towards a circular economy: A Monitoring framework for economy-wide material loop closing in the EU28. *J. Ind. Ecol.* 23 (1), 62–76. https://doi.org/10.1111/jiec.12809.

Atabaki, M.S., Mohammadi, M., Naderi, B., 2020. New robust optimization models for closed-loop supply chain of durable products: Towards a circular economy. *Comput. Ind. Eng.* 146, 106520. https://doi.org/10.1016/j.cie.2020.106520.

Bacelo, H., Pintor, A.M.A., Santos, S.C.R., Boaventura, R.A.R., Botelho, C.M.S., 2020. Performance and prospects of different adsorbents for phosphorus uptake and recovery from water. *Chem. Eng. J.* 381, 122566. https://doi.org/10.1016/j.cej.2019.122566.

Bocken, N.M.P., Olivetti, E.A., Cullen, J.M., Potting, J., Lifset, R., 2017. Taking the circularity to the next level: A special issue on the circular economy. *J. Ind. Ecol.* 21, 476–482.

BSI, 2017. *BS 8001:2017.* Framework for *I*mplementing the *P*rinciples of the *Circular Economy* in *Organizations*—Guide. London: The British Standards Institution. www.bsigroup.com/en-GB/standards/benefits-of-using-standards/becoming-more-sustainable-with-standards/BS8001-Circular-Economy/. Accessed April 9, 2018.

Castellet-Viciano, L., Hernández-Chover, V., BellverDomingo, Á., Hernández-Sancho, F., 2022. Industrial symbiosis: A mechanism to guarantee the implementation of circular economy practices. *Sustainability.* 14, 15872. https://doi.org/10.3390/su142315872.

Chaudhary, R., Tong, Y.W., Dikshit, A.K., 2020. Kinetic study of nutrients removal from municipal wastewater by Chlorella vulgaris in photobioreactor supplied with CO2–enriched air. *Environ. Technol.* 41, 617–626. https://doi.org/10.1080/09593330.2018.1508250.

Dustin, B., Hazell, J., Hill, H., 2017. *The Guide to the Circular Economy: Capturing Value and Managing Material Risk.* London: Routledge.

Eurostat, 2018. Eurostat—statistics explained. Circular economy. http://ec.europa.eu/eurostat/web/circular-economy/material-flow-diagram. Accessed August 14, 2018.

Ferreira dos Santos, A., Amancio Frutuoso, F.K., de Amorim de Carvalho, C., Sousa Aguiar Lira, V.N., Mendes Barros, A.R., Bezerra dos Santos, A., 2022. Carbon source affects the resource recovery in aerobic granular sludge systems treating wastewater. *Bioresour. Technol.* 357, 127355. https://doi.org/10.1016/j.biortech.2022.127355.

Gherghel, A., Teodosiu, C., De Gisi, S., 2019. A review on wastewater sludge valorisation and its challenges in the context of circular economy. *J. Clean. Prod.* 228, 244–263. https://doi.org/10.1016/j.jclepro.2019.04.240.

Gupta, S., Pawar, S.B., Pandey, R.A., 2019. Current practices and challenges in using microalgae for treatment of nutrient rich wastewater from agro-based industries. *Sci. Total Environ.* 687, 1107–1126. https://doi.org/10.1016/j.scitotenv.2019.06.115.

Han, W., Jin, P., Chen, D., Liu, X., Jin, H., Wang, R., Liu, Y., 2021. Resource reclamation of municipal sewage sludge based on local conditions: A case study in Xi'an, China. *J. Clean. Prod.* 316, 128189. https://doi.org/10.1016/J.JCLEPRO.2021.128189.

Hu, Q., Jung, J., Chen, D., Leong, K., Song, S., Li, F., Mohan, B.C., Yao, Z., Prabhakar, A.K., Lin, X.H., Lim, E.Y., Zhang, L., Souradeep, G., Ok, Y.S., Kua, H.W., Li, S.F.Y., Tan, H.T.W., Dai, Y., Tong, Y.W., Peng, Y., Joseph, S., Wang, C.-H., 2021. Biochar industry to circular economy. *Sci. Total Environ.* 757, 143820. https://doi.org/10.1016/j.scitotenv.2020.143820.

Huang, M., Wang, Z., Chen, T., 2019. Analysis on the theory and practice of industrial symbiosis based on bibliometrics and social network analysis. *J. Clean. Prod.* 213, 956–967.

Huang, R., Xu, J., Xie, L., Wang, H., Ni, X., 2022. Energy neutrality potential of wastewater treatment plants: A novel evaluation framework integrating energy efficiency and recovery. *Front. Environ. Sci. Eng.* 16 (9), 117. https://doi.org/10.1007/S11783-022-1549-0.

Ji, L., Sun, Y., Liu, J., Chiu, Y., 2023. Analysis of the circular economy efficiency of China's industrial wastewater and solid waste - Based on a comparison before and after the 13th five-year plan. *Sci. Total Environ.* 881, 63435. http://dx.doi.org/10.1016/j.scitotenv.2023.163435.

Karakas, I., Sam, S.B., Cetin, E., Dulekgurgen, E., Yilmaz, G., 2020. Resource recovery from an aerobic granular sludge process treating domestic wastewater. *J. Water Process Eng.* 34, 101148. https://doi.org/10.1016/j.jwpe.2020.101148.

Kizito, S., Luo, H., Wu, S., Ajmal, Z., Lv, T., Dong, R., 2017. Phosphate recovery from liquid fraction of anaerobic digestate using four slow pyrolyzed biochars: Dynamics of adsorption, desorption and regeneration. *J. Environ. Manag.* 201, 260–267. https://doi.org/10.1016/j.jenvman.2017.06.057.

Kurniawan, T.A., Othman, M.H.D., Liang, X., Goh, H.H., Gikas, P., Chong, K.-K., Chew, K.W., 2023. Challenges and opportunities for biochar to promote circular economy and carbon neutrality. *J. Environ. Manage.* 332, 117429. https://doi.org/10.1016/j.jenvman.2023.117429.

Le Corre, K.S., Valsami-Jones, E., Hobbs, P., Parsons, S.A., 2009. Phosphorus recovery from wastewater by struvite crystallization: A review. *Crit. Rev. Environ. Sci. Technol.* 39 (6), 433–477. https://doi.org/10.1080/10643380701640573.

Lee, Y.-J., Lei, Z., 2022. Wastewater treatment using microalgal-bacterial aggregate process at zero-aeration scenario: Most recent research focuses and perspectives. *Bioresour. Technol. Rep.* 17, 100943. https://doi.org/10.1016/j.biteb.2021.100943.

Li, K., Liu, Q., Fang, F., Luo, R., Lu, Q., Zhou, W., Huo, S., Cheng, P., Liu, J., Addy, M., Chen, P., Chen, D., Ruan, R., 2019. Microalgae-based wastewater treatment for nutrients recovery: A review. *Bioresour. Technol.* 291, 121934. https://doi.org/10.1016/j.biortech.2019.121934.

Li, S., Zeng, W., Wang, B., Xu, H., Peng, Y., 2020. Obtaining three cleaner products under an integrated municipal sludge resources scheme: Struvite, short-chain fatty acids and biological activated carbon. *Chem. Eng. J.* 380, 122567. https://doi.org/10.1016/J.CEJ.2019.122567.

Lin, R., O'Shea, R., Deng, C., Wu, B., Murphy, J.D., 2021. A perspective on the efficacy of green gas production via integration of technologies in novel cascading circular biosystems. *Renew. Sustain. Energy Rev.* 150, 111427. https://doi.org/10.1016/J.RSER.2021.111427.

Matthias, L., Meisch, S., 2015. Securitising Sustainability? Questioning the 'water, energy and food-security nexus. *Water Alternatives.* 8 (1), 695–709.

Mehta, C.M., Khunjar, W.O., Nguyen, V., Tait, S., Batstone, D.J., 2015. Technologies to recover nutrients from waste streams: A critical review. *Crit. Rev. Environ. Sci. Technol.* 45, 385–427. https://doi.org/10.1080/10643389.2013.866621.

Muhmood, A., Lu, J., Dong, R., Wu, S., 2019. Formation of struvite from agricultural wastewaters and its reuse on farmlands: Status and hindrances to closing the nutrient loop. *J. Environ. Manag.* 230, 1–13. https://doi.org/10.1016/j.jenvman.2018.09.030.

Nayak, A., Bhushan, B., 2019. An overview of the recent trends on the waste valorization techniques for food wastes. *J. Environ. Manag.* 233, 352–370. https://doi.org/10.1016/j.jenvman.2018.12.041.

Oliveira, A.S., Amorim, C.L., Ramos, M.A., Mesquita, D.P., Inoc^encio, P., Ferreira, E.C., van Loosdrecht, M., Castro, P.M.L., 2020. Variability in the composition of extracellular polymeric substances from a full-scale aerobic granular sludge reactor treating urban wastewater. *J. Environ. Chem. Eng.* 8 (5), 104156. https://doi.org/10.1016/j.jece.2020.104156.

Pahunang, R.R., Buonerba, A., Senatore, V., Oliva, G., Ouda, M., Zarra, T., Munoz, R., Pahunang, R.R., Buonerba, A., Senatore, V., Oliva, G., Ouda, M., Zarra, T., Muñoz, R., Puig, S., Ballesteros, F.C., Li, C.-W., Hasan, S.W., Belgiorno, V., Naddeo, V., 2021. Advances in technological control of greenhouse gas emissions from wastewater in the context of circular economy. *Sci. Total Environ.* 792, 148479. https://doi.org/10.1016/j.scitotenv.2021.148479.

Pan, Z., Qiu, C., Yang, Q., Wei, H., Pan, J., Sheng, J., Li, J., 2022. Adding waste iron shavings in a pilot-scale two-stage SBRs to develop aerobic granular sludge treating real wastewater. *J. Water Process Eng.* 47, 102811. https://doi.org/10.1016/J.JWPE.2022.102811.

Peter, L., Rutqvist, J., 2016. *Waste to Wealth: The Circular Economy Advantage.* London: Palgrave Macmillan.

Petit-Boix, A., Leipold, S., 2018. Circular economy in cities: Reviewing how environmental research aligns with local practices. *J. Clean. Prod.* 195, 1270–1281. https://doi.org/10.1016/j.jclepro.2018.05.281.

Roy, P., Mohanty, A.K., Dick, P., Misra, M., 2023. A review on the challenges and choices for food waste valorization: Environmental and economic impacts. *ACS Environ. Au.* 3 (2), 58–75. https://doi.org/10.1021/acsenvironau.2c00050.

Rufí-Salís, M., Petit-Boix, A., Leipold, S., Villalba, G., Rieradevall, J., Molin´e, E., Gabarrell, X., Carrera, J., Suarez-Ojeda, M.E., 2022. Increasing resource circularity in wastewater treatment: environmental implications of technological upgrades. *Sci. Total Environ.* 838, 156422. https://doi.org/10.1016/j.scitotenv.2022.156422.

Sanchez, A.S., Martins, G., 2021. Nutrient recovery in wastewater treatment plants: Comparative assessment of different technological options for the metropolitan region of Buenos Aires. *J. Water Process Eng.* 41, 102076. https://doi.org/10.1016/j.jwpe.2021.102076.

Schambeck, C.M., Magnus, B.S., de Souza, L.C.R., Leite, W.R.M., Derlon, N., Guimaraes, L.B., da Costa, R.H.R., 2020. Biopolymers recovery: dynamics and characterization of alginate-like exopolymers in an aerobic granular sludge system treating municipal wastewater without sludge inoculum. *J. Environ. Manag.* 263, 110394. https://doi.org/10.1016/j.jenvman.2020.110394.

Schellenberg, T., Subramanian, V., Ganeshan, G., Tompkins, D., Pradeep, R., 2020. Wastewater discharge standards in the evolving context of urban sustainability–the case of India. *Front. Environ. Sci.* 8, 30. https://doi.org/10.3389/fenvs.2020.00030.

Sharma, H.B., Vanapalli, K.R., Samal, B., Sharma, H.B., Vanapalli, K.R., Samal, B., Cheela, V.R.S., Dubey, B.K., Bhattacharya, J., 2021. Circular economy approach in solid waste management system to achieve UN-SDGs: Solutions for post-COVID recovery. *Sci. Total Environ.* 800, 149605.

Sharma, J., Kumar, S.S., Kumar, V., Malyan, S.K., Mathimani, T., Bishnoi, N.R., Pugazhendhi, A., 2020. Upgrading of microalgal consortia with CO_2 from fermentation of wheat straw for the phycoremediation of domestic wastewater. *Bioresour. Technol.* 305, 123063. https://doi.org/10.1016/J. BIORTECH.2020.123063.

Völker, T., Kovacic, Z., Strand, R., 2020. Indicator development as a site of collective imagination? The case of European Commission policies on the circular economy. *Cult. Organ.* 26, 103–120. https://doi.org/10.1080/14759551.2019.1699092.

Wang, Q., Li, H., Shen, Q., Wang, J., Chen, X., Zhang, Z., Lei, Z., Yuan, T., Shimizu, K., Liu, Y., Lee, D.-J., 2022. Biogranulation process facilitates cost-efficient resources recovery from microalgae-based wastewater treatment systems and the creation of a circular bioeconomy. *Sci. Total Environ.* 828, 154471. https://doi.org/10.1016/j.scitotenv.2022.154471.

Willi, H., Krausmann, F., Wiedenhofer, D., Heinz, M., 2015. How circular is the global economy? An assessment of material flows, waste production, and recycling in the European Union and the World in 2005. *J Ind. Ecol.* 19 (5), 765–777. https://doi.org/10.1111/jiec.12244.

Zhang, M., Ji, B., Liu, Y., 2021. Microalgal-bacterial granular sludge process: A game changer of future municipal wastewater treatment? *Sci. Total Environ.* 752, 141957. https://doi.org/10.1016/j.scitotenv.2020.141957.

13 Nanomaterials from Waste and Their Application for Wastewater Treatment

Mohamed N. Abd El-Ghany
Botany and Microbiology Department,
Faculty of Science, Cairo University, Giza, Egypt

Fatma M. Shahat
Chemistry/Microbiology Department,
Faculty of Science, Cairo University, Giza, Egypt

Mariam M. Hassan
Chemistry/Biotechnology Department,
Faculty of Science, Cairo University, Giza, Egypt

13.1 SOURCES OF WASTEWATER

Humans produce several types of waste but wastewater is one of the wastes produced most frequently. The composition of water differs as a result of different sources of contaminant (Bhatia et al., 2021; You et al., 2022). The pollutants threaten human health and biodiversity (Saravanan et al., 2021).

13.1.1 Domestic Wastewater

Domestic wastewater is released from houses, bathrooms, kitchens, etc. It has received attention due to the increasing levels because of overpopulation and urbanization (Chiu et al., 2015). The concentration of contaminants is increasing continuously because of the different usage of chemical products, such as personal care products and pharmaceuticals (Xu et al., 2020). Domestic water can be classified into two types: gray water and black water. The difference between gray and black water is that gray water included all the washing water except that produced from toilets (Jefferson et al., 2000).

DOI: 10.1201/9781003432869-13

13.1.2 Agricultural Wastewater

Agricultural wastewater is the water that results from farming, agricultural products, and animal manure wastewater (Liu and Hong, 2021). Furthermore, this water contains different types of synthetic pesticides, herbicides, and fertilizers.

13.1.3 Industrial Wastewater

Industrial wastewater is the net result of industrial processes which could be produced as a by-product or intermediate product. In addition to the different pollutants that are produced from the industrial process itself (Maurya et al., 2022), industrial wastewater has low biodegradability due to the presence of various toxicants such as heavy metals, antibiotics, and oils. In addition, industrial wastewater has a concentration of organic compounds (Liu and Hong, 2021; Udaiyappan et al., 2017).

13.2 THE EFFECT OF CONTAMINATED WATER ON THE ECOSYSTEM

All sectors require clean water, such as industry, agriculture, daily domestic life, and even the ecosystem needs clean water to maintain balance and allow every organism to survive (Huang et al., 2022). The presence of azo dyes cause:

- Reduction in the transparency which decreases the light flow for photosynthetic organisms.
- Decrease in dissolved oxygen (Corso and Maganha de Almeida, 2009).
- Toxicity to the aquatic environment because they contain aromatic hydrocarbons, metal salts, and chlorides (Alhassani et al., 2007; Daneshvar et al., 2007).

Wastewater is considered to be a harmful environment because it contains a lot of microorganisms (including pathogens) and protozoa surrounded by organic compounds such as carbohydrates, proteins, vitamins, and fats. In addition to the presence of inorganic substances, such as sulfur, sodium, arsenic, and ammonium salts (Wen et al., 2020), the pathogens included in wastewater cause several diseases such as typhoid, hepatitis, and cholera (Wen et al., 2020).

Exposure to wastewater can cause several diseases, such as those causing renal, cardiovascular, and neurological disorders. Oil is one of the common pollutants of wastewater. Oil contains hydrocarbons which affect marine life. In addition, oil prevents the passage of sunlight and activates microbial growth which decreases the level of oxygen in the water. Furthermore, oil activates anaerobic conditions, so that the wastewater content of hazardous substances such as ammonia and sulfide increases. The presence of oil in water causes the death of marine animals because it blocks their respiratory system. Furthermore, the sticking of oil to the feathers of aquatic birds directly damages the birds and impacts negatively on the food chain (Chen et al., 2019; Saravanan et al., 2020). Heavy metals are not biodegradable substances.

Therefore, they accumulate in the environment which causes an imbalance in the ecosystem. Heavy metals become harmful to humans and the ecosystem through absorption and consumption. Uptake of heavy metals through the roots of plants result in the accumulation of heavy metals in plants and hence in animals that feed on the plants, affecting the food chain (Joseph et al., 2019; Cao et al., 2020). For humans, the consumption of drinking water containing heavy metals results in bioaccumulation of the heavy metals. Heavy metals activate oxidative stress, causing genetic disorders and imbalanced expression of genes (Jacob et al., 2018; Yin et al., 2019). The presence of microplastics in water, without proper management or disposal, causes the accumulation of these plastics. Marine organisms feed on the plastics and, as a result, marine organisms suffer from the inability to digest food due to the accumulation of plastics which ultimately causes the death of these organisms (Anjana et al., 2020).

The presence of high levels of pharmaceuticals or their metabolites in wastewater at concentrations greater than the allowable limit cause increases in antimicrobial resistance, feminizing fish levels, and increased fish predation (Kidd et al., 2007; Wellington et al., 2013; Brodin et al., 2013; Horký et al., 2021). There is potential to accumulate viruses in the ecosystem through the usage of contaminated wastewater in different fields, such as agriculture (Lahrich et al., 2021). Therefore, pharmaceutical contamination threatens global health (Wilkinson et al., 2022).

Generally, the transformation events that occur chemically or biologically in the wastewater cause adverse consequences to human health and the ecosystem (Saravanan et al., 2021).

13.3 TYPES OF CONTAMINANT IN WASTEWATER

Wastewaters differ in terms of contaminants according to the sector from which they are produced. These contaminants could be an insecticide, dyes, polymers, disinfection by-products, polychlorinated bisphenols, polycyclic aromatic hydrocarbons, etc. (Huang et al., 2022). Water contamination is reflected in the changes in the chemical and physical characteristics of the wastewater (Ensafi and Karimi-Maleh, 2010). Therefore, from this point, one can classify the contaminants into three types: organic, inorganic, and pathogenic.

13.3.1 Organic Contaminants

One of the most significant contaminants in wastewater is oil, with 13% of oil pollution caused by travel on surface water (Ullmann et al., 2013). Pharmaceutical ingredients appear in wastewater due to their daily usage, disposal, or excretion, and the high level of their manufacture (Kidd et al., 2007; Wellington et al., 2013; Brodin et al., 2013; Horký et al., 2021). A study made across 258 of the world's rivers in different continents showed that the most active pharmaceuticals that appeared in samples were carbamazepine, metformin, and caffeine, the latter being widely used on a daily level. In addition, at least any one of these compounds occurred in 25.7% of the samples that have been collected. Moreover, it is present in excess of allowable level that aquatic organisms can live in, which can cause antimicrobial resistance (Wilkinson et al., 2022).

TABLE 13.1
Major Environmental Pollutants and their Toxic Effects (Reprinted with Permission from Saravanan et al., 2021, BMC)

Major Pollutants	Sources	Toxic Effect	
		On Humans	**On the Ecosystem**
Heavy metals	Industrial effluents, pesticides, mining, sewage sludge	Carcinogenic diseases, organ impairment	Oxidative stress in plants, bioaccumulation
Dyes	Industrial effluents from textile, painting, paper, printing and tanning industries	Carcinogenic, mutagenic, organ dysfunction	Reduced photosynthetic activity, increases BOD and COD, overall suppression of plant growth
Oil	Transportation oil spills, industrial effluents	Respiratory problems, carcinogenic, neurological, eyes and nose irritation	Destroys aquatic habitat, reduces dissolved oxygen (DO) in water
Plastics	Disposal of packaging material and industrial wastes into ocean	Liver dysfunction, lung problems, deafness, impaired immune function	Blocks respiratory system in aquatic organisms
Pesticides and herbicides	Agricultural practices	Endocrine system disruption, organ impairment	Biomagnification, decreases biodiversity
Pathogenic microorganisms	Domestic discharge, sewage, hospital wastes	Serious effect on human metabolism	Reduction in DO content

13.3.2 Inorganic Contaminants

Heavy metals are metals that have a density greater than 4 g/cm^3. The industries that use heavy metals are the paper, electroplating, and fertilizer manufacture sectors (Manikandan and Saravanan, 2018). Azo dyes and their derivatives are part of the most dangerous pollutants due to their toxicity, mutagenicity, persistency, and nonbiodegradability. In addition, azo dyes are frequently used in several types of industries such as the plastic and leather industries (Selvaraj et al., 2021). Although azo dyes represent a challenging pollutant that is difficult to eliminate, they can be removed either using physicochemical or biological techniques (Selvaraj et al., 2021; Saratale et al., 2009a; Saratale et al., 2009b).

13.3.3 Pathogens

Viruses are abundant pathogens in wastewater (Lahrich et al., 2021), in addition to bacteria, fungi, protozoan, and helminths, and each of them has its own detrimental effects on human health and the ecosystem around them.

13.4 NANOMATERIALS

13.4.1 Definition and Synthesis

Nanoparticles are materials that are synthesized in the range of nanoscale of 1–100 nm. They can be classified according to their dimensions, namely zero-dimensional (0D), one-dimensional (1D), and two-dimensional (2D) nanomaterials (Khan and Malik, 2019). Nanoparticles have a wide range of applications in several sectors due to their physical, chemical, and surface properties which differ from their bulkier counterparts. The synthesis of nanomaterials can occur either naturally, as in volcanoes, or in industry, using specific physic-chemical characterization methods (Awan et al., 2020). There are two approaches to the synthesis of nanomaterials, either top-down and bottom-up. Top-down synthesis requires the breakage of van der Waals forces and this approach is easy to perform. Bottom-up synthesis requires formation of a bond, either ionic or covalent (Vickers, 2017).

Using the physical and chemical synthesis methods may require reagents that are toxic to the environment and harm human health. In addition, nanoparticle size is controlled by varying several parameters, such as temperature, pH, reaction time, etc. (Gan and Li, 2012).

13.4.2 Green Synthesis

This is the ideal method that can be used for the synthesis of nanoparticles due to the following reasons. It is eco-friendly, cost-effective, and the reagents used are biodegradable. Green synthesis can be carried out using microorganisms, plants, or biomolecules extracted from microorganisms or plants. The systemic metabolism varies between microorganisms.

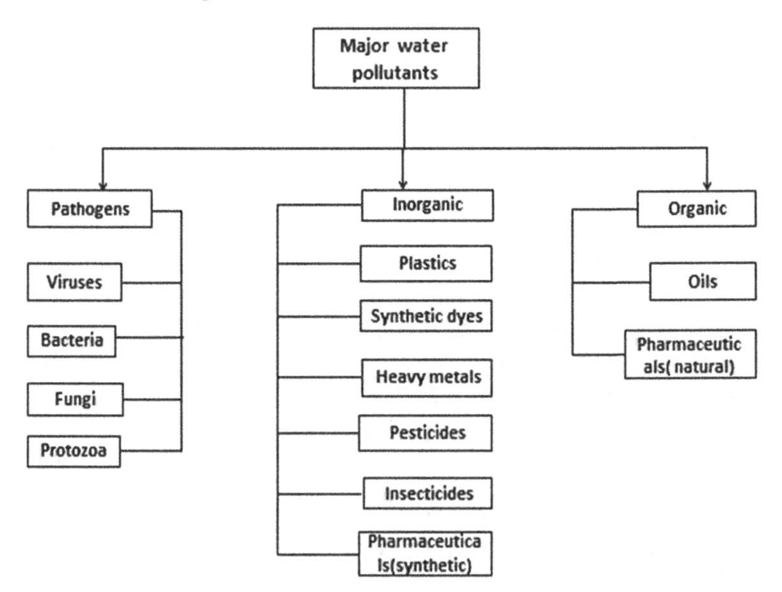

FIGURE 13.1 The major water pollutants.

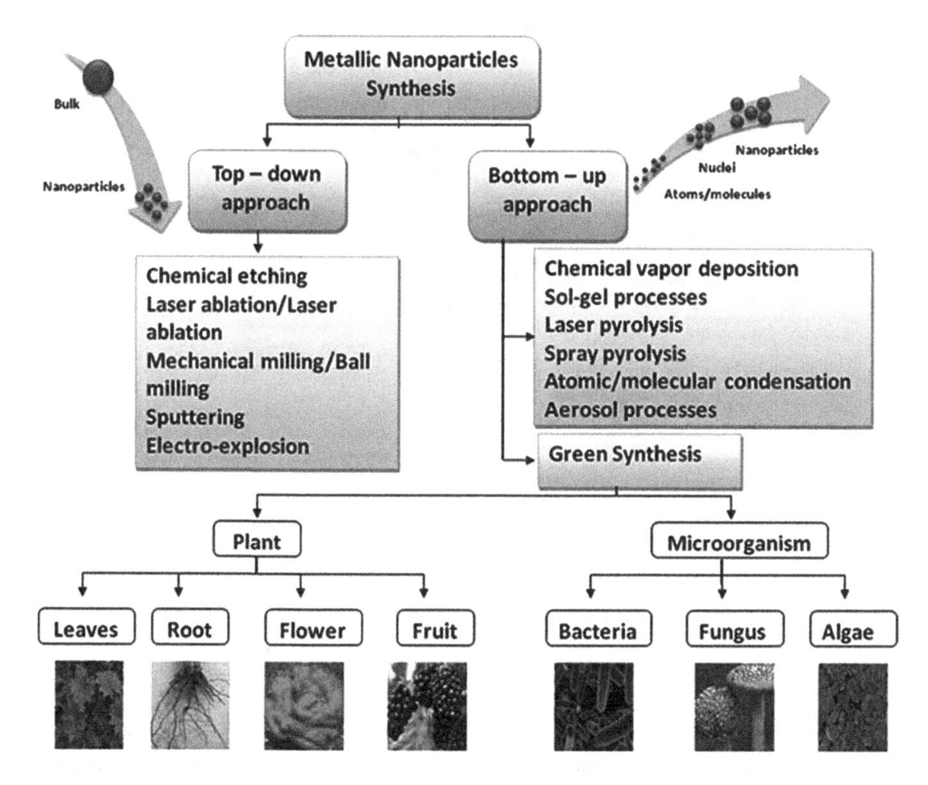

FIGURE 13.2 The two approaches to the synthesis of nanomaterials, and the different methods to be performed. Reprinted with permission from Singh 2018). (For interpretation of the references to color in this figure legend, the reader is referred to the web version of this article.)

Nanoparticles from green synthesis are used in different fields such as biology and pharmacology. The synthesis of nanometals depends on two routes of synthesis: bioreduction, in which organic compounds are used to reduce the metal ions, and biosorption, which includes the introduction of metal ions into the biological cell and their subsequent reduction using biological processes (Kumar and Rajeshkumar, 2018).

The most important point in green synthesis is that it can be used in any country, either developed or developing. It is not restricted to particular plants or microorganisms. Therefore, any plant or microorganism can be used if exhibits the metabolic substance and mechanism (Saravanan et al., 2022). Another approach to the green synthesis of nanoparticles involves the use of biowaste. Biowaste can be classified according to the original source into four categories: animal waste, forestry and agroindustry wastes, food processing waste, and municipal wastes (Ashrafi et al., 2022; Nasrollahzadeh et al., 2020; Xu et al., 2019; Baran and Nasrollahzadeh, 2019; Motahharifar et al., 2020).

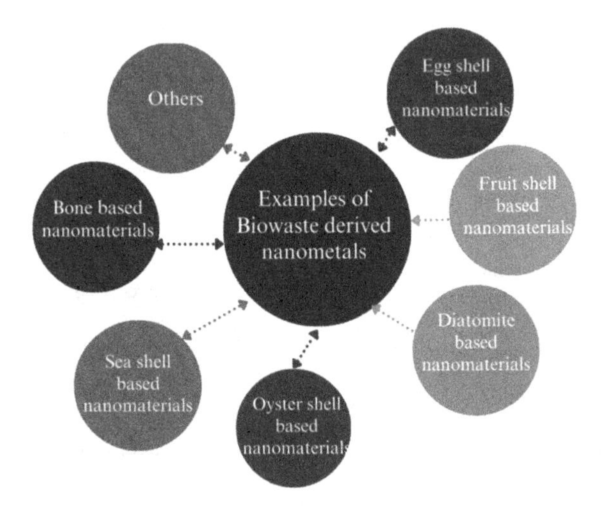

FIGURE 13.3 Different examples of biowastes.

13.5 TYPES OF NANOMATERIALS THAT CAN BE USED FOR THE TREATMENT OF CONTAMINATED WATER

The incorporation of nanotechnology is practically ubiquitous in research and technology. Finding a solution to numerous environmental issues, particularly water contamination, is also aided by nanotechnology. Nanomaterials are superior to traditional materials in several ways, including higher surface area, both polar and non-polar properties, controlled and size-tunable scaling, and simpler biodegradation, making them the perfect alternative for water and environmental treatment (Baby et al., 2022).

13.5.1 NANOMATERIALS FROM ORGANIC WASTES

The term "biodegradable waste", or "biowaste", actually refers to organic waste, which includes home, sewage, food, animal, and agricultural wastes. To manage and control the problems caused by biowastes, it is necessary to address their sources of origin and associated health risks. Bio-based wastes can be broadly categorized into four main groups as shown in Figure 13.4: animal waste, wastes from the forestry and agricultural industries, wastes from the food industry, and municipal wastes (Ashrafi et al., 2022).

The use of non-traditional additives for remediating toxic metal-contaminated water is becoming more popular recently. Strong adsorption capacity, high metal selectivity, recycling ability, eco-friendliness, and accessibility are all good attributes in sorbent materials. It has taken a lot of studies to meet these requirements for wastewater treatment. Two ecologically sustainable materials that can be used to capture toxic metals in water are among the low-cost sorbents. The first is a waste product termed "sugar beet factory lime" that is created during the purification of sugar beet juice. The second is a non-combustible by-product of burnt clay from industrial facilities, which is made from clay-containing soils and is rich in clay.

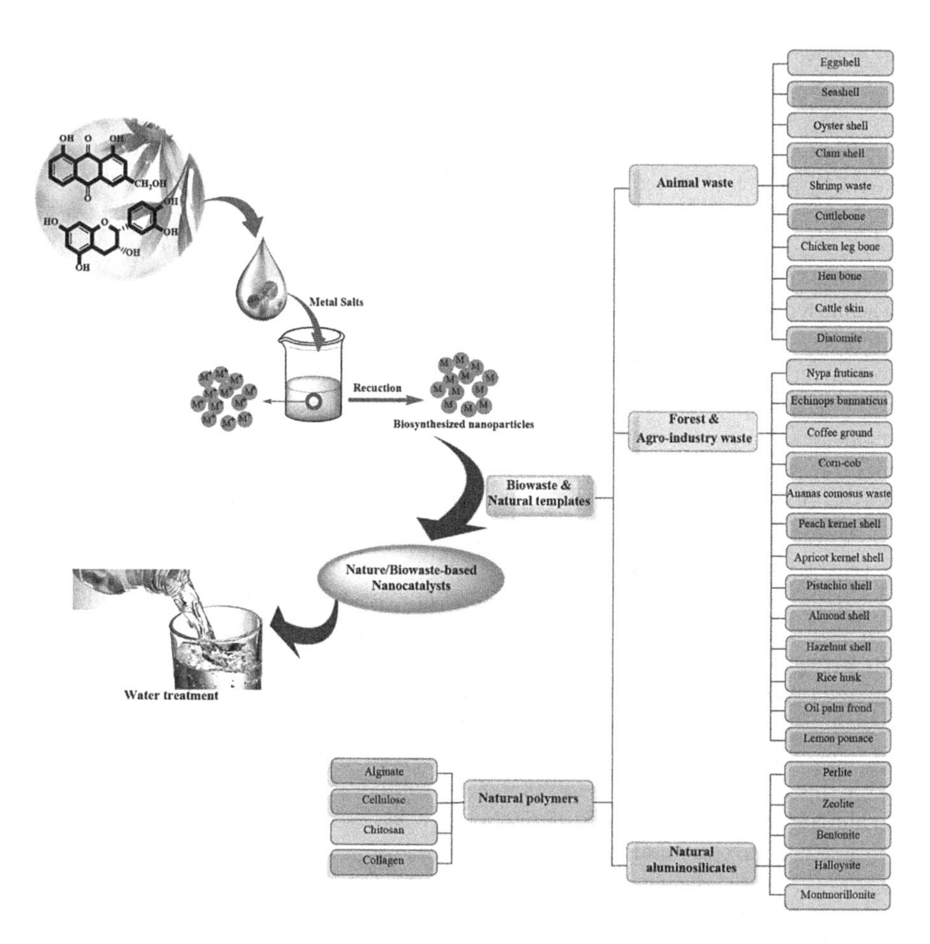

FIGURE 13.4 Categorization of biologically produced synthetic biowaste and organic nanoparticles for water treatment systems (Ashrafi et al., 2022). Adapted with permission from Elsevier B.V.

These raw materials are available as solid waste and can be freely removed and used beyond the red mud brick and sugar beetroot industries. On the other hand, their accumulation may harm the environment. Nano-based sugar beetroot residuals increased the retention of both cadmium (Cd) and copper (Cu) in the equilibrium solution more quickly than nano-based brick factory residuals did. Nano-based brick factory residues were unable to attain sorption equilibrium even 24 hours after addition. Nano-based sugar beetroot residuals outperformed nano-based brick factory residuals in terms of obtaining practically maximal sorption of both metals even at the most acidic recorded pH values (99.3% retention of Cu and 91.3% retention of Cd at pH 3) (Lashen et al., 2022).

Another study investigated the capacity of titanium-doped activated carbon–botanical cellulose nanocomposite in water purification. By physical adsorption and photocatalytic degradation, the nanomaterials obtained have been employed to remove crystal violet and methyl violet (organic contaminants) from wastewater (Maqbool et al., 2022).

Using several bio-shells, such as eggshells, oyster, and clam shells, as support for Mn_3O_4 quantum dots, a novel and straightforward approach for creating 0D/2D $Ca(OH)_2/Mn_3O_4$ (Mn-Ca) composites was created. The Mn-Ca nano catalysts generated effectively removed Co(II), Pb(II), Cd(II), Cu(II), Eu(III), and Ag(I) from wastewater. A safe procedure using *Sapindus mukorossi* extract, with the help of microwaves, was also suggested for the fabrication of Ag nanoparticles supported on eggshells (ES). The biosynthesized Ag@ES NC was used to study the degradation of the organic dyes Methyle Orange, EBT (eriochrome black T), Methylene Blue, and Rhodamine B as well as the adsorption of Cr(VI) from an aqueous solution (Ashrafi et al., 2022).

Organic waste, such as fruit and vegetable residues, may contain a variety of biomolecules that can function as reducing and stabilizing agents such as the production of silver nanoparticles from *Solanum tuberosum* (potato) peel which is used in the purification of water from bromophenol blue dye (Ureña-Castillo et al., 2022).

A modified cellulose carbon aerogel was produced via wet ball milling, 2,2,6,6-tetramethyl-1-piperidinyloxy (TEMPO)-mediated oxidation, and pyrolysis. Reports state that the treatment environment effectively transformed the cellulose fibers into planar or wrinkled structures. These graphite-like structures gave the aerogel a large specific surface area and maximum adsorption capacity for the organic dyes alizarin reds and methylene blue. It was found that the processes by which dyes adhered to the aerogel created were pore-filling, hydrophobic partition, *pi* interactions of electron donors–acceptors, and hydrogen-bonding. Electrostatic attraction strengthened the adsorption of the cationic dye methylene blue, in contrast to anionic alizarin reds, which, due to its high salt content, showed reduced electrostatic repulsion (Muhammad et al., 2023).

13.5.2 Nanomaterials from Inorganic Wastes

13.5.2.1 Nanomaterials from Mine Tailings and Acid Mine Drainage (AMD)

All mining operations generate a substantial amount of waste, such as a stack of shattered rocks or low-grade ores that are not commercially viable for metallurgical extraction. Close to the mining industry is also the slurry-based refuse of the related mineral processing industries. Many useful elements can be retrieved from the waste, in addition to the environmental issues that might result from a large pile of unusable materials. For example, magnetite nanoparticles (NPs) generated from iron ore industrial effluents have diverse purposes in sectors ranging from wastewater treatment to technological ones. Another environmental issue related to mining is acid mine drainage (AMD), which is brought on by bacterial activity in the mining region. AMD produces an acidic environment that can dissolve heavy metals from the minerals that form rocks. Heavy metals like Pb and Zn are hazardous and can harm the ecosystem in a variety of ways. The bioaccumulative nature of this pollution has an impact on both human and animal health. Hence, it may be advantageous to use mining waste for the synthesis of NPs. Magnetite nanoparticles are created

TABLE 13.2
Benefits and Drawbacks of Using Biological Waste as a Starting Point for the Creation of Metallic Nanoparticles (Ureña-Castillo et al., 2022)

Benefits	Drawbacks
It has no cost because it is a waste product.	It is vital to categorize the waste according to its source and composition.
Accessible to obtain and available in huge quantities from industries that produce them.	To guarantee reproducibility, several standardized factors must be considered in the technique.
Nanoparticles are produced using environmentally friendly, sustainable methods without the use or production of harmful materials.	To optimize the process, it could be essential to keep in mind the seasons in which the material is produced, depending on the type of waste.
More biocompatible nanoparticles are produced, similar to previous green synthesis techniques.	According to their source of origin, several variations of the same waste must be assessed.
Increased awareness of organic waste recycling.	

through processes like the chemical or ultrasonic coprecipitation of minerals and the acid leaching of mine tailings (Brar et al., 2022).

One method for utilizing the waste from iron mining is to recover iron ions and use them to make multifunctional magnetic materials. They consist of combinations of organic and inorganic elements whose complementary properties differ from those of the original materials. A hybrid magnetic material was produced using commercial iron as the inorganic phase and water with a high concentration of natural organic matter (NOM) as the organic precursor. Here, it is suggested to use the iron collected from mine tailings to make the magnetic component of the ferrite structure and to serve as the hybrid's inorganic phase. Magnetic hybrid materials using ferrite as the inorganic phase have been utilized to remove a variety of organic contaminants. Nitrophenols, which are common in effluent from the paper, agricultural, and pharmaceutical industries, are among these pollutants (Cruz et al., 2021).

13.5.2.2 Nanomaterials from Industrial Residues

Batteries, plastics, and tires are just a few examples of industrial wastes that have been used as starting feedstock for nanomaterials (NMs). NM manufacture from diverse industrial wastes could therefore be an environmentally friendly recycling technique (Samaddar et al., 2018).

A hydrothermal technique was used to successfully create hydroxysodalite nanoparticles from aluminum waste with various crystallite sizes. As a silicon source, several millimoles of sodium metasilicate pentahydrate were used. Composites made of hydroxysodalite and chitosan were also created. The apparent breakdown of the hydroxy sodalite's crystalline structure served as proof that the hydroxysodalite/chitosan composites had formed. The nanoparticles created and their chitosan composites might be used to successfully remove the ions Ni(II) and Pb(II) from contaminated water. Moreover, they can be reused repeatedly without losing their ability to adsorb Ni(II) or Pb(II) ions (Abdelrahman and Hegazey, 2019).

The widespread use of synthetic polymers is responsible for one of the largest waste material categories in the world. Non-biodegradable plastic bags, cutlery, and water bottles continue to make up a sizable portion of solid waste. Using waste polyethylene terephthalate, multi-walled carbon nanotubes (MWCNTs) and nano-channeled ultrafine nanotubes were produced using a novel catalyst- and solvent-free arc discharge method. They produced carbon nanotubes (CNTs) in a variety of topologies utilizing arc discharge techniques, including solid carbon spheres, non-branched/branched ultra-fine nano-channeled carbon tubes, and multi-walled carbon nanotubes (Samaddar et al., 2018).

13.6 CHARACTERIZATION OF NANOMATERIALS

NMs are routinely characterized using both structural and morphological characteristics. When it comes to structural characterization, researchers mainly analyze nanomaterials using Fourier transform infrared spectroscopy (FT-IR). X-ray spectroscopy, X-ray fluorescence (XRF), and X-ray diffraction (XRD) techniques, while scanning electron microscopy (SEM), transmission electron microscopy (TEM), and energy dispersive spectroscopy (EDS) techniques can be used to examine the morphology of nanomaterials (Samaddar et al., 2018).

13.6.1 Structural Characterization

The vibrational modes of precursors and products are characterized using FT-IR. Using an FT-IR spectrogram, it is also possible to identify contaminants in a finished product. Li et al. (2016) converted leftover ferrous sulfate into a nano-α-Fe_2O_3 crimson pigment powder (Li et al., 2016; Samaddar et al., 2018). Peaks at 1117.06 cm^{-1} and 1051.03 cm^{-1}, both of which can be credited to SO_4^{2-} being stretched by υ3, were visible in the final product's spectrogram (as shown in Figure 13.5A). Stretching of SO_4^{2-} in the υ4 direction resulted in another peak at 602.84 cm^{-1}. This suggests that some sulfate impurities were still present in the final α-Fe_2O_3 nanopowder. Polystyrene nanoparticles were made from waste polystyrene and analyzed by FT-IR (Mangalara and Varughese, 2016). The aromatic C-H stretching vibrations corresponded to the peaks they discovered at 3014 cm^{-1} and 2922 cm^{-1}. Aliphatic C-H stretching caused the peak at 2839 cm^{-1}, whereas aromatic C=C bond stretching caused those between 1615 and 1580 cm^{-1} and 1510–1450 cm^{-1} (Figure 13.5B). The researchers decided that no chemical modifications happened during the recycling process based on this FT-IR spectrum analysis (Samaddar et al., 2018).

X-ray diffraction (XRD) techniques were used to examine the crystalline nature of generated nano samples made from various waste sources. For instance, according to Fang et al. (2011), the three main Fe metal peaks in the XRD pattern were located at 44.66, 65.16, and 82.36. Figure 13.5C displays the zerovalent metallic nanoparticle peaks for Fe-110, Fe-200, and Fe-211, respectively. (Samaddar et al., 2018; Fang et al., 2011).

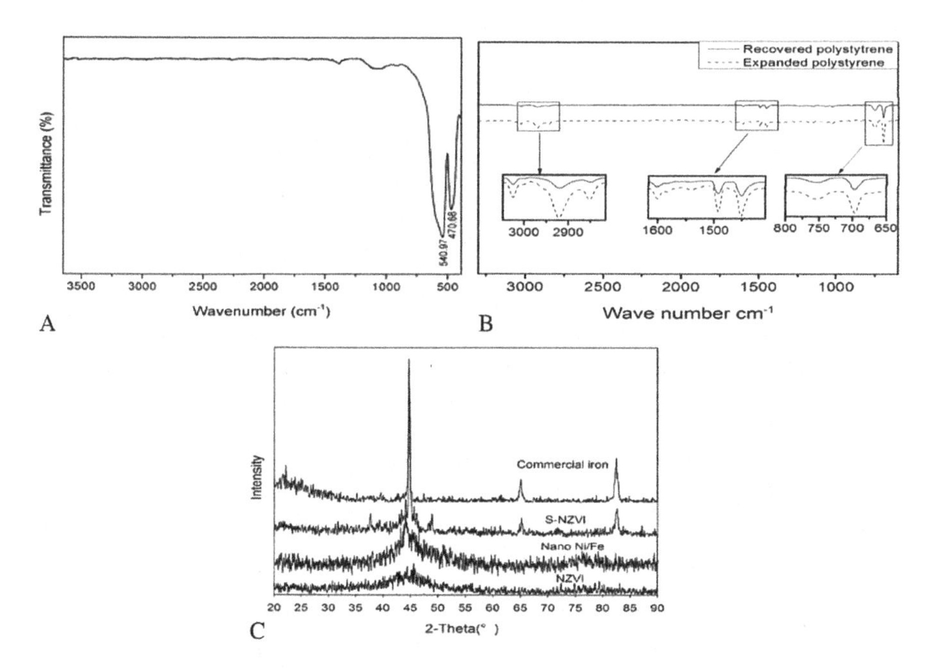

FIGURE 13.5 Some examples of the structural study of various nanomaterials (Samaddar et al., 2018). A: FT-IR spectrogram of α-Fe2O3 NPs, B: FT-IR spectrogram of recovered and expanded polystyrene, and C: XRD data of Fe NPs and commercial Fe powder produced. Adapted with permission from Elsevier B.V.

Another study used X-ray diffractometry to examine the crystallinity of a nanocomposite of titanium-doped activated carbon and botanical cellulose. In particular, Figure 13.6A shows the XRD spectra of 10% Ti-AC-Cel-NC, 5% Ti-AC-Cel-NC, and Ti-ACNs. Two sizable diffraction peaks can be identified in the spectra of Ti-ACNs at $2\theta = 23.7°$ and $2\theta = 25.06°$, as well as a smaller peak at $2\theta = 44.34°$. These peaks are all related to changed graphite and activated carbon structures that mimic graphene. The width of the peaks indicates that carbon is reduced into narrower stacks of carbon sheets, as is typically observed during the reduction and exfoliation process of graphite into graphene (Maqbool et al., 2022).

13.6.2 Morphological Characterization

It is commonly acknowledged that particle size is a significant NM feature. Therefore, rigorous examination of the size and shape of the nanomaterials produced is required to discover new applications for them. SEM can be used to determine how internal nanoparticle dispersion is distributed. This method allows for the visualization of carbon nanotube forms and nanoporous surfaces of activated carbon. Using SEM, it was determined that the nano-channeled ultrafine carbon tubes were "Y-shaped". SEM was also used to examine nanostructured activated carbon. Chen et al. (2016) described a sandwich structure of porous tubes between two layers of thin flakes in

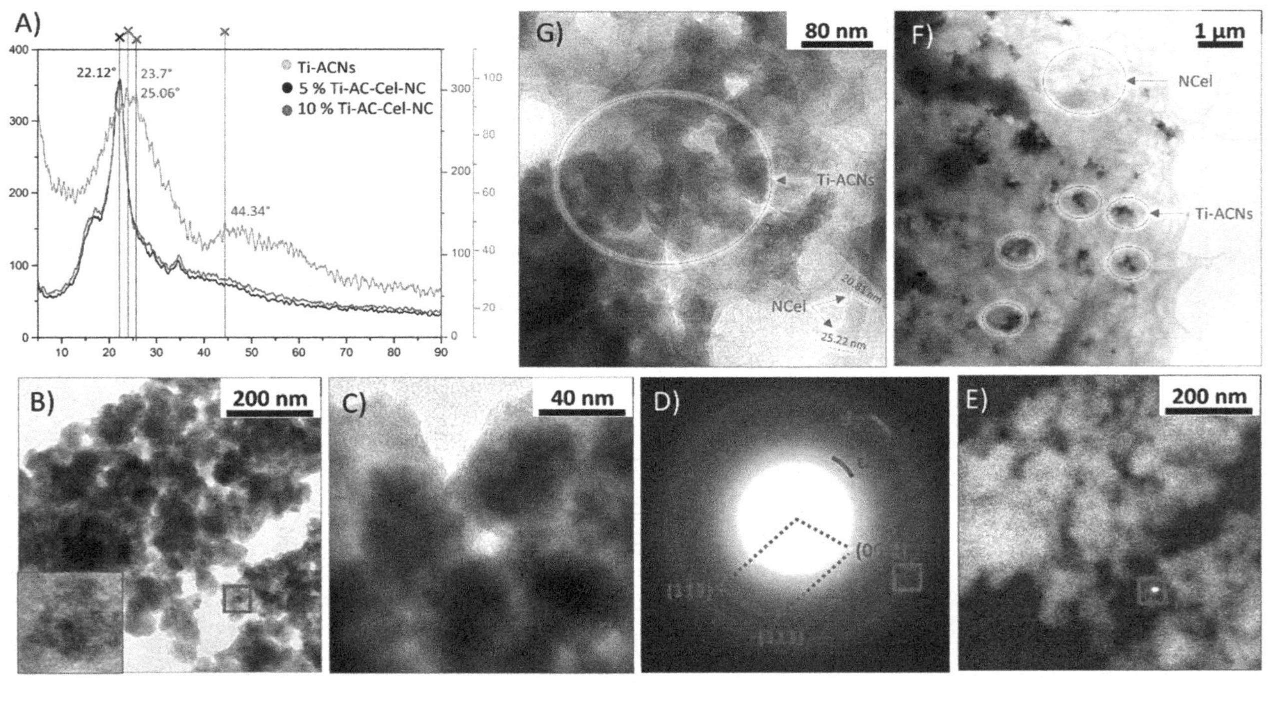

FIGURE 13.6 Experiment on the morphology and crystallography of prepared material. A: XRD pattern of Ti-ACNs showing distinctive features of mutant graphite/graphene-like carbon material, including large peaks at 23.7° and 25.05°. In samples of 5% and 10% Ti-AC-Cel-NC, the characteristic peak (at 22.12°) of polymorphic nanocrystalline cellulose-I has been observed. B: Freshly synthesized Ti-ACNs are captured in a bright field TEM image, which displays the sample's granular structure. C: An expanded TEM image of as-prepared Ti-ACNs, where the contrast shows that a carbon cloud surrounds the titanium nanostructures. D: SAED pattern of the zone shown in (B): diffuse rings (d1 and d2) are linked to modified graphene-like topologies, and diffraction spots (red cell) show the presence of a hexagonal Ti-phase in [1–10] zone axis orientation. In the dark field image, the Ti grain location and size are displayed. E: Ti was made using the diffraction spot included in (D). As can be seen in the bright field TEM images in (F) and (G), 10% Ti-ACNs were successfully and consistently absorbed into the NCel matrix (Maqbool et al., 2022). Adapted with

porous carbon nanosheets generated from elm samara plant waste (Samaddar et al., 2018; Chen et al., 2016).

The precise distribution pattern of particles in manufactured NMs can be determined by energy-dispersive X-ray spectroscopy (EDS) analysis during SEM imaging. The EDS spectrograph of nano Ni/Fe particles demonstrated that Ni was uniformly dispersed throughout the surface of Fe. Nevertheless, other than Fe, they found no other metals. After some consideration, they concluded that the other metal loadings, such as Ni and Zn, were too small to be observed on the nanoparticle surfaces (Samaddar et al., 2018).

As illustrated in Figure 13.6B, the morphological structure of Ti-ACNs takes the shape of granular particles. It has an uneven morphology because of its granular structure. At higher magnification, it is possible to detect that the granules, whose diameters typically range from 20 nm to 100 nm (Figure 13.6C), have a core-shell structure in which a substance with a darker contrast is surrounded by a less dense matrix. The carbon cloud that surrounds the titanium nanostructures in this image provides evidence in favor of that theory. In Figure 13.6D, an image of the sample fragment seen in the selected area electron diffraction (SAED) pattern of Figure 13.3B is shown. It is possible to see two diffusely intense diffraction rings: a weakly crystallized activated carbon structure that resembles graphite or graphene can be associated with the interplanar distances $d_1 = (0.207 \pm 0.003)$ nm, and $d_2 = (0.118 \pm 0.003)$ nm that are associated with the radii of these rings. Moreover, a few evenly spaced diffraction dots are evident. The necessary interplanar lengths and angles between these diffraction spots have been related to the existence of a hexagonal Ti phase in [1–10] zone axis orientation. The dark field image of Figure 13.6E was made using the contained spot of Figure 13.6C to identify the crystalline phase that produced the specific diffraction spot. The crystallite has a red square encircling it, even when zoomed in on in the comparable inset of the corresponding bright field image in Figure 13.6B. The average size is 15 nm, and the contrast is dark. Even when correctly ordered, alpha-Ti should allow for the observation of diffraction spots in the [1–10] zone axis orientation, but this is rarely the case in SAED studies conducted in other regions of the sample. In Figure 13.6F, a bright field picture taken by a TEM is displayed beside a 10% Ti-AC-Cel-NC sample. The Ti-ACNs may be easily seen inside the nano cellulose matrix because of their darker contrast. They usually congregate into particles between 0.1 and 1 m in size, which are then evenly distributed over nano cellulose fibers to maximize the amount of active surface area available for surface-based applications. By magnifying the picture, it can be shown in Figure 13.6G that nanocellulose, which is created from the residual biomass from leaves, provides an excellent network of homogeneous nanofibers that can accommodate metallic and carbonaceous materials. finding cellulose at the nanoscale from a different plant source (Maqbool et al., 2022).

13.7 APPLICATIONS OF NANOMATERIALS IN WASTEWATER TREATMENT

Many innovative nanomaterial adsorbents have been created to improve the efficacy and adsorption capabilities of effluent pollutant removal. Nanomaterials can be used as an adsorbent material to solve a variety of environmental problems since they possess several distinctive structural and morphological characteristics. The use of nanoparticles originating from a range of sources, such as bacteria, plants, animal cells, and carbon, can be used to treat wastewater and remove heavy metal pollutants from water effluents (Saravanan et al., 2022).

13.7.1 Heavy Metal Removal

Heavy metal ions have recently been removed using cheap, widely accessible biodegradable wastes, such as coffee grounds, coconut husks, tea leaves, sunflower stalks, cotton, and wool. A thorough understanding of the functional groups involved in the binding is crucial to examine the method by which the metal attaches to the biomass. The majority of the time, they are present in the cellulose-rich cell walls of plants. The effectiveness of the plant waste as a stratum is strongly influenced by the degree of polymerization, the available specific surface area, and structural characteristics like crystallinity. Depending on the sample and measurement mode, the relative content of the crystalline and amorphous fractions varies. Mesoporous silica is an effective adsorbent for gas phase pollutants due to its large surface area and varied pore diameter. Surface modifications, like the presence of a hydroxyl group, facilitate adsorption and speed up procedures like wetting. One common way to alter and enhance the performance of these adsorbents is by grafting the functional groups onto the pore walls. It is feasible to create brand-new adsorbents and catalysts because of this surface modification.

The inclusion of amine groups on silica surfaces increases the efficacy of carbon dioxide and hydrogen sulfide removal by up to 80%. Moreover, superparamagnetic iron oxide (Fe_3O_4) surface-functionalizing 2,3-dimercaptosuccinic acid (DMSA) nanoparticles can be used as highly efficient adsorbents to remove harmful compounds including the heavy metals mercury, lead, and cadmium, among others, which bind to the DMSA ligands, whereas arsenic attaches to the iron oxide lattices. Figure 13.7 depicts this process. A 1.2 T magnet can be used to remove them from the solution. They are highly functional and have a surface area of nearly 120 m^2/g with 1.85 mmol thiols/g (Kolluru et al., 2021). A study analyzed the ability of the endophytic fungus *Aspergillus niger* and the seed husk of *Jatropha curcas* to bioadsorb nickel and manganese. Results discovered that *J. curcas* seed husk and *A. niger* dry biomass could adsorb most Mn at 1.1 and 3.3 mg/L, respectively, and Ni at 4.56 mg/L and 6.84 mg/L, respectively. For Mn and Ni, respectively, pH 6 and 8 and 4–8 and 6 were found to have the maximum biosorption values for *J. curcas* seed husk (Abdel-Wareth et al., 2023).

13.7.2 Radioactive Pollutant Removal

Radioactive waste is a by-product of the nuclear energy industry and the use of various radioactive materials in numerous commercial sectors. Because of its great potential harm to the environment and all living things, this waste is governed by strict rules. Due to their particular physical and chemical properties, such as the nano-size

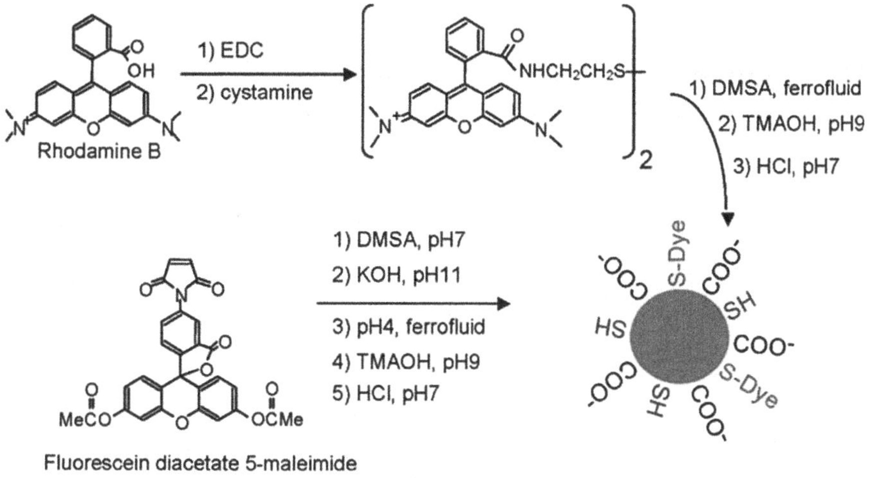

FIGURE 13.7 Diagram of the ligand absorption process using a fluorescent dye and DMSA to modify the surface of MNPs (Kolluru et al., 2021). Adapted with permission from Kolluru et al. (2021), published by Elsevier B.V.

effect, strong reactivity and selectivity, and high specific surface area, nanomaterials have become the new cleaning agents for radioactive wastewater. Natural materials having a high cellulose content, such as *Citrus limetta* (sweet lemon) peel, pineapple peel, sugarcane bagasse, and coconut coir, can be milled to form adsorbents that may be used as carbonaceous nanomaterials (CNs) to purify radioactively contaminated water. More investigation into the transformation of other agricultural wastes should be done in the context of this problem. For instance, banana peel has been used to treat radioactively contaminated water to remove Th and U. According to this study, it was discovered that banana peel that had been ground to a size of 12 nm could effectively bind to amine and carboxylic groups to remove radioactive elements from both simulated and real mine fluids. Furthermore, for the removal of radioactive UO_2^{2+} ions from water, ultrafine cellulose nanofibrile (CNFs) might be a useful adsorbent. With the chemical treatment of wood pulp with TEMPO, sodium bromide (NaBr), and sodium hypochlorite (NaClO), as well as mechanical disintegration, ultrafine CNFs were created. Due to the interaction between the positively charged UO_2^{2+} and the negatively charged carboxylate groups on the CNFs, a gel was created as soon as UO_2^{2+} was added to the CNF suspensions. These ultrafine CNFs have a maximum adsorption capacity for UO_2^{2+} of 167 mg/g (Zhang and Liu, 2020).

13.7.3 Pharmaceutical Pollutant Removal

Recently, there has been a lot of interest in the use of sustainable, eco-friendly nanomaterials for eliminating pharmaceutical contaminants. This is since depletion, accumulation, and resource degradation are significant environmental concerns. Whether they are utilized alone or in composites with other agricultural waste

TABLE 13.3
Current Uses of Nanoparticles and Their Nanocomposites for Removing Pharmaceutical Contaminants from Aqueous Solutions

Nanoparticles	Source	Removal %	Adsorbent dose (mg/L)	Contact time	Temperature	Reference
Cu-NPs	*Tilia* leaves	91.4	5–20	0–60 min	298 K	Khan et al. (2022b); Husein et al. (2019)
PAC/GAC	Vegetable	90	10	Up to 10 h	298 K	Khan et al. (2022b); Delgado et al. (2019)
AC, CX, CNT	Wood	–	–	Up to 8 h	303 K	Khan et al. (2022b); Carabineiro et al. (2012)

extracts, such as *Datura inoxia* leaf extract, green nanomaterials have gained a lot of popularity in the most promising medicinal adsorption applications. There has been substantial research on several cutting-edge methods for extracting drugs from water or wastewater. Fenton oxidation, photolysis, UV degradation, reverse osmosis, and adsorption are some of these processes (Khan et al., 2022b).

Because they are the pharmaceutical drugs that are used the most frequently, there is an increase in the number of antibiotics that are discarded into the environment. When ingested, these substances offer substantial health dangers to both people and other living creatures. Using nickel-ferrite-treated activated carbon manufactured from wood fibers for the sorption of metronidazole and levofloxacin from pharmaceutical wastewater, researchers developed MWCNTs (Ni–Fe). the surface area differential between Ni-Fe enhanced and unaltered activated carbon. Ultrafiltration, nanofiltration, membrane processes, reverse osmosis, ion exchange, and other unique element features for separation processes have all been studied in graphene-based materials (Saravanan et al., 2022; Kariim et al., 2020).

13.7.4 Chemical Removal

Textile, leather, cosmetic, culinary, and paper printing sectors all use synthetic dyes extensively. Molecules of color called reactive dyes are used to color cellulose fibers. Large amounts of effluent with intense color are produced as a result of these dyes. Most of the time, physical or chemical methods are used to treat wastewaters from the textile dying process. Examples of these methods include electrochemical coagulation, electrochemical destruction, irradiation, precipitation, ozonation, and the Katox method, which uses active carbon and a mixture of specific gases, such as air, to treat the wastewater (Khalil et al., 2016). The majority of dyes are resistant to microbial attack, light, and temperature because of their high stability. This industrial wastewater is hazardous and is distinguished by traits such as high suspended solids content, high chemical oxygen demand (COD), and high biological oxygen demand (BOD). These techniques are quite costly and have functional issues. Thus, it is necessary to create useful biological procedures for processing dye waste that can be applied to a variety of pollutants (Naguib and Hanafy, 2023; Moreira et al., 2014). Moreover, enzymes, such as microbial laccases, have been used in the degradation of dyes (Abd El-Ghany et al., Ali et al., 2015). Because of their ultrafine size, high aspect ratio, and interaction-dominating properties, nano-sized materials can be used to detoxify hazardous organic and inorganic compounds from the environment. The physical and chemical properties of iron nanoparticles are distinct due to their small size and the high density of edge surface sites (Khan et al., 2022a). In a related study, sixteen organic dyes were subjected to rapid degradation by TiO_2/WO_3-coated magnetic nanoparticles. Complete decolorization was seen for ten of the dyes when exposed to sunlight. Copper-based nanoparticles are another form of photocatalyst that effectively degrade organic dyes. They reported 98% effectiveness while decomposing Reactive Black dye using copper oxide nanorods. Moreover, these nanoparticles have comparable degradation rates while simultaneously removing Congo red, methylene blue, and methyl orange. Other photocatalytic dye degradation methods include those for rhodamine B using $Cu/Cu(OH)_2$ nanoparticles, methyl orange using CdS/CuS nanoparticles, methylene blue using reduced graphene oxide doped with copper nanoparticles, methyl red, methyl orange, and phenyl red using biologically prepared copper nanoparticles, and methylene blue using graphene oxide-doped $CuO\text{-}Cu_2O$ (Sivaraman et al., 2022).

The environment is frequently exposed to compounds known as polycyclic aromatic hydrocarbons (PAHs). The unfortunate thing about them is that they cause cancer in both people and aquatic animals. In addition to their carcinogenic activity, they also exhibit environmental persistence and are resistant to decomposition because of their hydrophobic tendency (Fouad et al., 2023). Due to their ability to reduce the toxic effects caused by the use of chemical reagents in the traditional synthesis processes of adsorbents, the use of green nano adsorbents, produced by means of natural agents (plant extracts, fruits, and microorganisms), for the removal of organic pollutants, like PAHs, from the environment, has been the central objective of some studies. Rice straw (RS) and sugar cane bagasse (SB), which are low-cost bioadsorbents, were used as residual biomass to successfully adsorb PAHs from oil industry wastewater, while another study conducted a green synthesis of a magnetic activated carbon utilizing green tea leaves (MNPs-GTAC), with the goal of employing it to remove benzo(a)anthracene (BaA), benzo(b)fluoranthene (BbF), chrysene (CHR), and benzo(a)pyrene (BaP) from wastewater (Queiroz et al., 2022).

13.8 ROLE OF BIONANOSENSORS IN THE MONITORING OF POLLUTANTS

Because pollution detection technology is more accessible and less expensive, process control, ecosystem monitoring, and environmental decision-making are all increased as part of this technology. The ability of humans to support sustainable human health and the environment is improved by quick and precise sensors that can find pollutants at the molecular level (Beni, 2022).

According to the International Union of Pure and Applied Chemistry, a biosensor is a self-contained integrated receptor transducer that can reliably deliver quantitative or semi-quantitative data via a biological recognition component. The second category of analytical tools is the biosensor, which has a wide range of uses. By immobilizing a biologically sensitive recognition factor in a physicochemical transducer and connecting it to a detector, a biosensor may detect the presence, concentration, and dynamics of one or more species in a sample, as shown in Figure 13.8. The specificity and selectivity of the biosensor are determined by the biological element's affinity or catalytic properties. The signal is sent to an appropriate reader, such as an optical or electrical reader, depending on the relationship between the target analyte and a biological analytical feature (Adam and Gopinath, 2022).

Nanosensors can be used in a range of label-free detection applications and can deliver a strong signal-to-noise ratio even at extremely low substrate concentrations. It's difficult to define the term "nanosensor," but most definitions point to a sensing system with at least one dimension smaller than 100 nm that gathers data at the nanoscale and transforms it into information that can be used for analysis. Nanosensors have recently piqued the interest of many researchers because they can be used as tools for other nanoproducts, such as semiconductor chips or nanoscale machinery. Nanosensors have demonstrated advantages over conventional mechanical, electrical, and optical sensors (Adam and Gopinath, 2022).

Microbiological contaminants must be detected in water bodies to protect public health. Also, the contaminated species must be identified and isolated before the proper management measures may be put in place. Traditional microbial growth markers are inefficient and incorrect, missing a variety of potentially harmful

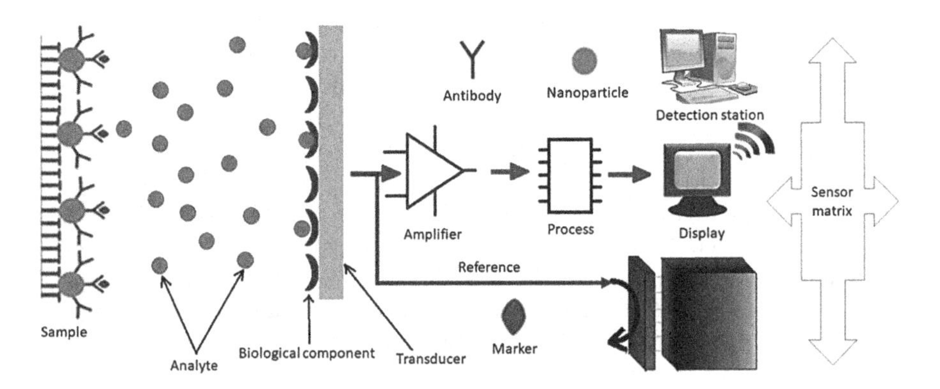

FIGURE 13.8 Principle of operation of nano biosensor-based nanosensors; using signal transduction, which emits a signal; after measuring the interaction, biosensors are operated. The fundamental idea, on the other hand, is to bind bio analytes to bioreceptors, which changes the physiochemical signal associated with the binding (Adam & Gopinath, 2022). Adapted with permission from Elsevier Ltd.

viruses, bacteria, and protozoa that may be present. As a result, over the past ten years, a brand-new class of sensing and monitoring devices based on nanomaterials has emerged. Quantum dots have a lot of potential for sensing applications because of their easily customizable optical and photo-electrochemical properties. It is simple to identify the presence of microbial pollutants because quantum dots can be used to monitor changes in fluorescence spectra or photoelectrochemical behavior when they interact with pathogens and biomolecules in water. Gold nanoparticles are frequently employed for sensing applications due to the size- and shape-dependent surface plasmon resonance phenomenon. Gram-negative bacteria are extremely sensitive to gold nanoparticles that contain cysteine (Thanigaivel et al., 2022).

Numerous studies have tried to use nanomaterials to modify and/or functionalize an electrode before integrating it with microbial biosensors. The most prevalent nanomaterial seems to be carbon nanotubes (CNTs), which can easily switch electrodes. By increasing electrical conductivity, functionalizing cationic surfactants to stabilize certain molecules, and enhancing reaction time, they can enhance microbial biosensors. Nanotechnology-enabled sensors could replace many of the current water-quality sensors. Researchers have distinguished between the terms "nanoprobe" and "nanosensor", asserting that a nanoprobe measures an analyte and necessitates quick and reversible binding while a nanosensor measures bacteria or chemicals with high sensitivity but does not require reversibility or quantitative measurement. The biosensor has Ag and Au nanoparticles as well as specific bio-elements for detecting ions of heavy metals like Cu, Hg, Pb, and Cr (detected by Ag), as well as Cu, Hg, Pb, Pt, Pd, and Co (detected by Au). An Au-nanoparticle-rhodamine 6 G-based fluorescent nanosensor for the detection of Hg ions with a detection limit of 0.012 ppb has been reported. Additionally, a nano silica-based biosensor with a detection limit of 50 ppb specifically responded to Fe and Cu ions with a recognizable color shift (Laad and Ghule, 2022).

13.9 CHALLENGES AND PROSPECTS

Given that the rate at which waste is being produced is increasing and endangering the earth's finite resources, turning waste materials into useful assets, and controlling pollutants in such a low-cost, environmentally friendly, and adaptable manner shed light on the potential utility of this testing technique. The use of bio-safe, waste- and nature-based nanomaterials and catalysts for the remediation and mitigation of organic and inorganic pollutants is remarkably encouraging, but some challenging issues relating to toxicity and biosafety issues and their mechanistic aspects should be thoroughly and systematically investigated. Nanomaterials generated from waste are expected to have a promising future:

- The industrial production of biowaste- and nature-derived nanomaterials can be achieved at a relatively low cost.
- Extensive analyses of toxicity and biocompatibility testing can be carried out by biosynthesized nanomaterials.

- The stability, dispersibility, catalytic activity, and adsorption capacity of ecologically friendly nanomaterials produced from biowaste and other natural sources are high.
- Mild, water-soluble, and eco-friendly reductants can be used for the production of eco-friendly NPs and the degradation/reduction of toxic pollutants.
- Isolation/purification systems, ideal circumstances, and effective factors can be improved based on the size, morphology, and performance of biomaterials.

REFERENCES

Abdel-Wareth, M. T. A., Abdel-Rahman, T. M., Abdel-Ghany, M. N. & Hamed, K. A. 2023. Consortium effect of Jatropha curcas seed husk and its endophyte Aspergillus niger on biosorption of manganese and nickel from wastewater. *International Journal of Environmental Studies*, 80, 1617–1636.

Abdelrahman, E. A. & Hegazey, R. 2019. Utilization of waste aluminum cans in the fabrication of hydroxysodalite nanoparticles and their chitosan biopolymer composites for the removal of Ni (II) and Pb (II) ions from aqueous solutions: Kinetic, equilibrium, and reusability studies. *Microchemical Journal*, 145, 18–25.

Adam, T. & Gopinath, S. C. 2022. Nanosensors: Recent perspectives on attainments and future promise of downstream applications. *Process Biochemistry*, 117, 153–173.

Alhassani, H. A., Rauf, M. A. & Ashraf, S. S. 2007. Efficient microbial degradation of Toluidine Blue dye by Brevibacillus sp. *Dyes and Pigments*, 75, 395–400.

Ali, M., Ouf, S., Khalil, N. & Abd El-Ghany, M. 2015. Biosynthesis of laccase by Aspergillus flavus NG85 Isolated from Saint Catherine protectorate. *Egyptian Journal of Botany*, 55, 127–147.

Anjana, K., Hinduja, M., Sujitha, K. & Dharani, G. 2020. Review on plastic wastes in marine environment–Biodegradation and biotechnological solutions. *Marine Pollution Bulletin*, 150, 110733.

Ashrafi, G., Nasrollahzadeh, M., Jaleh, B., Sajjadi, M. & Ghafuri, H. 2022. Biowaste-and nature-derived (nano) materials: Biosynthesis, stability and environmental applications. *Advances in Colloid and Interface Science*, 301, 102599.

Awan, T. I., Bashir, A. & Tehseen, A. 2020. *Chemistry of nanomaterials: Fundamentals and applications*. Elsevier.

Baby, R., Hussein, M. Z., Abdullah, A. H. & Zainal, Z. 2022. Nanomaterials for the treatment of heavy metal contaminated water. *Polymers*, 14, 583.

Baran, T. & Nasrollahzadeh, M. 2019. Facile synthesis of palladium nanoparticles immobilized on magnetic biodegradable microcapsules used as effective and recyclable catalyst in Suzuki-Miyaura reaction and p-nitrophenol reduction. *Carbohydrate Polymers*, 222, 115029.

Beni, A. A. 2022. Nanomaterial application for environmental. *Results in Engineering*, 15, 100467.

Bhatia, S. K., Mehariya, S., Bhatia, R. K., Kumar, M., Pugazhendhi, A., Awasthi, M. K., Atabani, A., Kumar, G., Kim, W. & Seo, S.-O. 2021. Wastewater based microalgal biorefinery for bioenergy production: Progress and challenges. *Science of the Total Environment*, 751, 141599.

Brar, K. K., Magdouli, S., Othmani, A., Ghanei, J., Narisetty, V., Sindhu, R., Binod, P., Pugazhendhi, A., Awasthi, M. K. & Pandey, A. 2022. Green route for recycling of low-cost waste resources for the biosynthesis of nanoparticles (NPs) and nanomaterials (NMs)-A review. *Environmental Research*, 207, 112202.

Brodin, T., Fick, J., Jonsson, M. & Klaminder, J. 2013. Dilute concentrations of a psychiatric drug alter behavior of fish from natural populations. *Science*, 339, 814–815.

Cao, D.-Q., Wang, X., Wang, Q.-H., Fang, X.-M., Jin, J.-Y., Hao, X.-D., Iritani, E. & Katagiri, N. 2020. Removal of heavy metal ions by ultrafiltration with recovery of extracellular polymer substances from excess sludge. *Journal of Membrane Science*, 606, 118103.

Carabineiro, S., Thavorn-Amornsri, T., Pereira, M., Serp, P. & Figueiredo, J. 2012. Comparison between activated carbon, carbon xerogel and carbon nanotubes for the adsorption of the antibiotic ciprofloxacin. *Catalysis Today*, 186, 29–34.

Chen, C., Yu, D., Zhao, G., Du, B., Tang, W., Sun, L., Sun, Y., Besenbacher, F. & Yu, M. 2016. Three-dimensional scaffolding framework of porous carbon nanosheets derived from plant wastes for high-performance supercapacitors. *Nano Energy*, 27, 377–389.

Chen, J., Zhang, W., Wan, Z., Li, S., Huang, T. & Fei, Y. 2019. Oil spills from global tankers: Status review and future governance. *Journal of Cleaner Production*, 227, 20–32.

Chiu, S.-Y., Kao, C.-Y., Chen, T.-Y., Chang, Y.-B., Kuo, C.-M. & Lin, C.-S. 2015. Cultivation of microalgal Chlorella for biomass and lipid production using wastewater as nutrient resource. *Bioresource Technology*, 184, 179–189.

Corso, C. R. & Maganha de Almeida, A. C. 2009. Bioremediation of dyes in textile effluents by Aspergillus oryzae. *Microbial Ecology*, 57, 384–390.

Cruz, D. R., Silva, I. A., Oliveira, R. V., Buzinaro, M. A., Costa, B. F., Cunha, G. C. & Romão, L. P. 2021. Recycling of mining waste in the synthesis of magnetic nanomaterials for removal of nitrophenol and polycyclic aromatic hydrocarbons. *Chemical Physics Letters*, 771, 138482.

Daneshvar, N., Ayazloo, M., Khataee, A. & Pourhassan, M. 2007. Biological decolorization of dye solution containing Malachite Green by microalgae Cosmarium sp. *Bioresource Technology*, 98, 1176–1182.

Delgado, N., Capparelli, A., Navarro, A. & Marino, D. 2019. Pharmaceutical emerging pollutants removal from water using powdered activated carbon: Study of kinetics and adsorption equilibrium. *Journal of Environmental Management*, 236, 301–308.

Ensafi, A. A. & Karimi-Maleh, H. 2010. Ferrocenedicarboxylic acid modified multiwall carbon nanotubes paste electrode for voltammetric determination of sulfite. *International Journal of Electrochemical Science*, 5, 392–406.

Fang, Z., Qiu, X., Chen, J. & Qiu, X. 2011. Degradation of the polybrominated diphenyl ethers by nanoscale zero-valent metallic particles prepared from steel pickling waste liquor. *Desalination*, 267, 34–41.

Fouad, F. A., Youssef, D. G., Shahat, F. M. & Abd El-Ghany, M. N. 2023. Role of microorganisms in biodegradation of pollutants. In Ali, G.A.M. and Makhlouf, A.S.H. (eds), *Handbook of Biodegradable Materials* (pp. 221–260). Cham: Springer. https://doi.org/10.1007/978-3-030-83783-9_17-1

Gan, P. P. & Li, S. F. Y. 2012. Potential of plant as a biological factory to synthesize gold and silver nanoparticles and their applications. *Reviews in Environmental Science and Bio/Technology*, 11, 169–206.

Horký, P., Grabic, R., Grabicová, K., Brooks, B. W., Douda, K., Slavík, O., Hubená, P., Sancho Santos, E. M. & Randák, T. 2021. Methamphetamine pollution elicits addiction in wild fish. *Journal of Experimental Biology*, 224, jeb242145.

Huang, R., Yang, J., Cao, Y., Dionysiou, D. D. & Wang, C. 2022. Peroxymonosulfate catalytic degradation of persistent organic pollutants by engineered catalyst of self-doped iron/carbon nanocomposite derived from waste toner powder. *Separation and Purification Technology*, 291, 120963.

Husein, D. Z., Hassanien, R. & Al-Hakkani, M. F. 2019. Green-synthesized copper nano-adsorbent for the removal of pharmaceutical pollutants from real wastewater samples. *Heliyon*, 5, e02339.

Jacob, J. M., Karthik, C., Saratale, R. G., Kumar, S. S., Prabakar, D., Kadirvelu, K. & Pugazhendhi, A. 2018. Biological approaches to tackle heavy metal pollution: A survey of literature. *Journal of Environmental Management*, 217, 56–70.

Jefferson, B., Laine, A., Parsons, S., Stephenson, T. & Judd, S. 2000. Technologies for domestic wastewater recycling. *Urban Water*, 1, 285–292.

Joseph, L., Jun, B.-M., Flora, J. R., Park, C. M. & Yoon, Y. 2019. Removal of heavy metals from water sources in the developing world using low-cost materials: A review. *Chemosphere*, 229, 142–159.

Kariim, I., Abdulkareem, A. & Abubakre, O. 2020. Development and characterization of MWCNTs from activated carbon as adsorbent for metronidazole and levofloxacin sorption from pharmaceutical wastewater: Kinetics, isotherms and thermodynamic studies. *Scientific African*, 7, e00242.

Khalil, N., Ali, M., Ouf, S. & Abd El-Ghany, M. 2016. Characterization of Aspergillus flavus NG 85 laccase and its dye decolorization efficiency. *Research Journal of Pharmaceutical, Biological and Chemical Sciences*, 7, 829–817.

Khan, A., Roy, A., Bhasin, S., Emran, T. B., Khusro, A., Eftekhari, A., Moradi, O., Rokni, H. & Karimi, F. 2022a. Nanomaterials: An alternative source for biodegradation of toxic dyes. *Food and Chemical Toxicology*, 164, 112996.

Khan, A. H., Khan, N. A., Zubair, M., Shaida, M. A., Manzar, M. S., Abutaleb, A., Naushad, M. & Iqbal, J. 2022b. Sustainable green nanoadsorbents for remediation of pharmaceuticals from water and wastewater: A critical review. *Environmental Research*, 204, 112243.

Khan, S. T. & Malik, A. 2019. Engineered nanomaterials for water decontamination and purification: From lab to products. *Journal of Hazardous Materials*, 363, 295–308.

Kidd, K. A., Blanchfield, P. J., Mills, K. H., Palace, V. P., Evans, R. E., Lazorchak, J. M. & Flick, R. W. 2007. Collapse of a fish population after exposure to a synthetic estrogen. *Proceedings of the National Academy of Sciences*, 104, 8897–8901.

Kolluru, S. S., Agarwal, S., Sireesha, S., Sreedhar, I. & Kale, S. R. 2021. Heavy metal removal from wastewater using nanomaterials-process and engineering aspects. *Process Safety and Environmental Protection*, 150, 323–355.

Kumar, S. V. & Rajeshkumar, S. 2018. Plant-based synthesis of nanoparticles and their impact. In Tripathi, D.K., Ahmad, P., Sharma, S., Chauhan, D.K. and Dubey, N.K. (eds), *Nanomaterials in plants, algae, and microorganisms*. Academic Press, 33–57.

Laad, M. & Ghule, B. 2022. Removal of toxic contaminants from drinking water using biosensors: A systematic review. *Groundwater for Sustainable Development*, 20, 100888.

Lahrich, S., Laghrib, F., Farahi, A., Bakasse, M., Saqrane, S. & El Mhammedi, M. 2021. Review on the contamination of wastewater by COVID-19 virus: Impact and treatment. *Science of The Total Environment*, 751, 142325.

Lashen, Z. M., Shams, M. S., El-Sheshtawy, H. S., Slaný, M., Antoniadis, V., Yang, X., Sharma, G., Rinklebe, J., Shaheen, S. M. & Elmahdy, S. M. 2022. Remediation of Cd and Cu contaminated water and soil using novel nanomaterials derived from sugar beet processing-and clay brick factory-solid wastes. *Journal of Hazardous Materials*, 428, 128205.

Li, X., Wang, C., Zeng, Y., Li, P., Xie, T. & Zhang, Y. 2016. Bacteria-assisted preparation of nano α-Fe2O3 red pigment powders from waste ferrous sulfate. *Journal of Hazardous Materials*, 317, 563–569.

Liu, X.-Y. & Hong, Y. 2021. Microalgae-based wastewater treatment and recovery with biomass and value-added products: A brief review. *Current Pollution Reports*, 7, 227–245.

Mangalara, S. C. H. & Varughese, S. 2016. Green recycling approach to obtain nano-and microparticles from expanded polystyrene waste. *ACS Sustainable Chemistry & Engineering*, 4, 6095–6100.

Manikandan, G. & Saravanan, A. 2018. Modelling and analysis on the removal of methylene blue dye from aqueous solution using physically/chemically modified Ceiba pentandra seeds. *Journal of Industrial and Engineering Chemistry*, 62, 446–461.

Maqbool, Q., Barucca, G., Sabbatini, S., Parlapiano, M., Ruello, M. L. & Tittarelli, F. 2022. Transformation of industrial and organic waste into titanium doped activated carbon–cellulose nanocomposite for rapid removal of organic pollutants. *Journal of Hazardous Materials*, 423, 126958.

Maurya, R., Zhu, X., Valverde-Pérez, B., Kiran, B. R., General, T., Sharma, S., Sharma, A. K., Thomsen, M., Mohan, S. V. & Mohanty, K. 2022. Advances in microalgal research for valorization of industrial wastewater. *Bioresource Technology*, 343, 126128.

Moreira, S., Milagres, A. M. & Mussatto, S. I. 2014. Reactive dyes and textile effluent decolorization by a mediator system of salt-tolerant laccase from Peniophora cinerea. *Separation and Purification Technology*, 135, 183–189.

Motahharifar, N., Nasrollahzadeh, M., Taheri-Kafrani, A., Varma, R. S. & Shokouhimehr, M. 2020. Magnetic chitosan-copper nanocomposite: A plant assembled catalyst for the synthesis of amino-and N-sulfonyl tetrazoles in eco-friendly media. *Carbohydrate Polymers*, 232, 115819.

Muhammad, S., Yahya, E. B., Abdul Khalil, H., Marwan, M. & Albadn, Y. M. 2023. Recent advances in carbon and activated carbon nanostructured aerogels prepared from agricultural wastes for wastewater treatment applications. *Agriculture*, 13, 208.

Naguib, A. E.-G. M. & Hanafy, N.A. 2023. Laccase as Key Enzyme among: Fungi-Pollutant-Environment and sustainable development goals nexus. *Biomedical Journal of Scientific and Technical Research*, 48, 39485–39491.

Nasrollahzadeh, M., Shafiei, N., Nezafat, Z., Bidgoli, N. S. S. & Soleimani, F. 2020. Recent progresses in the application of cellulose, starch, alginate, gum, pectin, chitin and chitosan based (nano) catalysts in sustainable and selective oxidation reactions: A review. *Carbohydrate Polymers*, 241, 116353.

Queiroz, R. N., Prediger, P. & Vieira, M. G. A. 2022. Adsorption of polycyclic aromatic hydrocarbons from wastewater using graphene-based nanomaterials synthesized by conventional chemistry and green synthesis: A critical review. *Journal of Hazardous Materials*, 422, 126904.

Samaddar, P., Ok, Y. S., Kim, K.-H., Kwon, E. E. & Tsang, D. C. 2018. Synthesis of nanomaterials from various wastes and their new age applications. *Journal of Cleaner Production*, 197, 1190–1209.

Saratale, R., Saratale, G., Chang, J.-S. & Govindwar, S. 2009a. Decolorization and biodegradation of textile dye Navy blue HER by Trichosporon beigelii NCIM-3326. *Journal of Hazardous Materials*, 166, 1421–1428.

Saratale, R., Saratale, G., Kalyani, D., Chang, J.-S. & Govindwar, S. 2009b. Enhanced decolorization and biodegradation of textile azo dye Scarlet R by using developed microbial consortium-GR. *Bioresource Technology*, 100, 2493–2500.

Saravanan, A., Kumar, P. S., Hemavathy, R., Jeevanantham, S., Jawahar, M. J., Neshaanthini, J. & Saravanan, R. 2022. A review on synthesis methods and recent applications of nanomaterial in wastewater treatment: Challenges and future perspectives. *Chemosphere*, 307, 135713.

Saravanan, A., Kumar, P. S., Jeevanantham, S., Karishma, S., Tajsabreen, B., Yaashikaa, P. & Reshma, B. 2021. Effective water/wastewater treatment methodologies for toxic pollutants removal: Processes and applications towards sustainable development. *Chemosphere*, 280, 130595.

Saravanan, A., Kumar, P. S., Vardhan, K. H., Jeevanantham, S., Karishma, S. B., Yaashikaa, P. R. & Vellaichamy, P. 2020. A review on systematic approach for microbial enhanced oil recovery technologies: Opportunities and challenges. *Journal of Cleaner Production*, 258, 120777.

Selvaraj, V., Karthika, T. S., Mansiya, C. & Alagar, M. 2021. An over review on recently developed techniques, mechanisms and intermediate involved in the advanced azo dye degradation for industrial applications. *Journal of Molecular Structure*, 1224, 129195.

Singh, J., Dutta, T., Kim, K.-H., Rawat, M., Samddar, P. & Kumar, P. 2018. 'Green'synthesis of metals and their oxide nanoparticles: Applications for environmental remediation. *Journal of Nanobiotechnology*, 16, 1–24.

Sivaraman, C., Vijayalakshmi, S., Leonard, E., Sagadevan, S. & Jambulingam, R. 2022. Current developments in the effective removal of environmental pollutants through photocatalytic degradation using nanomaterials. *Catalysts*, 12, 544.

Thanigaivel, S., Priya, A., Gnanasekaran, L., Hoang, T. K., Rajendran, S. & Soto-Moscoso, M. 2022. Sustainable applicability and environmental impact of wastewater treatment by emerging nanobiotechnological approach: Future strategy for efficient removal of contaminants and water purification. *Sustainable Energy Technologies and Assessments*, 53, 102484.

Udaiyappan, A. F. M., Hasan, H. A., Takriff, M. S. & Abdullah, S. R. S. 2017. A review of the potentials, challenges and current status of microalgae biomass applications in industrial wastewater treatment. *Journal of Water Process Engineering*, 20, 8–21.

Ullmann, A., Brauner, N., Vazana, S., Katz, Z., Goikhman, R., Seemann, B., Marom, H. & Gozin, M. 2013. New biodegradable organic-soluble chelating agents for simultaneous removal of heavy metals and organic pollutants from contaminated media. *Journal of Hazardous Materials*, 260, 676–688.

Ureña-Castillo, B., Morones-Ramírez, J. R., Rivera-De la Rosa, J., Alcalá-Rodríguez, M. M., Cerdán Pasarán, A. Q., Díaz-Barriga Castro, E. & Escárcega-González, C. E. 2022. Organic waste as reducing and capping agents for synthesis of silver nanoparticles with various applications. *ChemistrySelect*, 7, e202201023.

Vickers, N. J. 2017. Animal communication: When i'm calling you, will you answer too? *Current Biology*, 27, R713–R715.

Wellington, E. M., Boxall, A. B., Cross, P., Feil, E. J., Gaze, W. H., Hawkey, P. M., Johnson-Rollings, A. S., Jones, D. L., Lee, N. M. & Otten, W. 2013. The role of the natural environment in the emergence of antibiotic resistance in Gram-negative bacteria. *The Lancet Infectious Diseases*, 13, 155–165.

Wen, X., Chen, F., Lin, Y., Zhu, H., Yuan, F., Kuang, D., Jia, Z. & Yuan, Z. 2020. Microbial indicators and their use for monitoring drinking water quality—A review. *Sustainability*, 12, 2249.

Wilkinson, J. L., Boxall, A. B., Kolpin, D. W., Leung, K. M., Lai, R. W., Galbán-Malagón, C., Adell, A. D., Mondon, J., Metian, M. & Marchant, R. A. 2022. Pharmaceutical pollution of the world's rivers. *Proceedings of the National Academy of Sciences*, 119, e2113947119.

Xu, C., Nasrollahzadeh, M., Sajjadi, M., Maham, M., Luque, R. & Puente-Santiago, A. R. 2019. Benign-by-design nature-inspired nanosystems in biofuels production and catalytic applications. *Renewable and Sustainable Energy Reviews*, 112, 195–252.

Xu, Z., Wang, H., Cheng, P., Chang, T., Chen, P., Zhou, C. & Ruan, R. 2020. Development of integrated culture systems and harvesting methods for improved algal biomass productivity and wastewater resource recovery: A review. *Science of the Total Environment*, 746, 141039.

Yin, K., Wang, Q., Lv, M. & Chen, L. 2019. Microorganism remediation strategies towards heavy metals. *Chemical Engineering Journal*, 360, 1553–1563.

You, X., Yang, L., Zhou, X. & Zhang, Y. 2022. Sustainability and carbon neutrality trends for microalgae-based wastewater treatment: A review. *Environmental Research*, 209, 112860.

Zhang, X. & Liu, Y. 2020. Nanomaterials for radioactive wastewater decontamination. *Environmental Science: Nano*, 7, 1008–1040.

14 Natural Adsorbents for Effective Treatment of Dye Wastewater from Textile Industries

A Comprehensive Review

Chirag Yogendra Chaware
Department of Civil Engineering, COEP Technological University Pune, Maharashtra, India

Moni Udhaorao Khobragade
Department of Civil Engineering, COEP Technological University Pune, Maharashtra, India

Ashish Kumar Nayak
Department of Civil Engineering,
Prasad V. Potluri Siddhartha Institute of Technology,
Vijayawada, Andhra Pradesh, India

14.1 INTRODUCTION

Global concern surrounds the harmful effects of dyes used in the textile industry on water bodies (Kant, 2012). India is one of the world's largest textile producers, and the country is grappling with the negative consequences that dyes used in the textile sector have on waterbodies. Water pollution has increased as a result of the textile industry's rapid expansion, particularly in major textile production centers like Tirupur, Ahmedabad, and Surat. Rivers, lakes, and groundwater sources in these areas have been severely impacted by the release of dye- and chemical-laden textile wastewater that has not been treated (Dave, 2020). Without proper treatment, textile wastewater containing these substances can result in serious water pollution. Heavy metals, hazardous chemicals, and artificial compounds that are harmful to amphibians are typically found in textile wastewater (Al-Tohamy et al., 2022). The natural balance and functioning of ecosystems are further disrupted by textile dyes altering water characteristics, such as color, pH, and temperature (Bhatia et al., 2017). Textile

 DOI: 10.1201/9781003432869-14

dye spills or inappropriate disposal can also contaminate groundwater, lowering the quality of drinking water. Sustainable methods, effective wastewater treatment procedures, and the use of ecologically friendly dyes and chemicals are critical in the textile sector for mitigating these negative consequences. Furthermore, government laws and stricter environmental standards have considerably reduced the industry's impact on waterbodies (Madhav et al., 2020).

Wastewater treatment for dye removal is an important step in decreasing the detrimental consequences of textile industry effluent on aquatic bodies. A range of technologies and procedures are used to efficiently remediate textile wastewater (Azanaw et al. 2022). To remove suspended particulates and color, wastewater is cleaned using physical treatment processes such as sedimentation, filtration, and flotation (Dinkar Gosavi and Sharma, 2013). In chemical treatment, coagulants and flocculants are added to dye particles to encourage their aggregation and precipitation (Azbar et al., 2004). In activated sludge systems, microorganisms are used in biological treatment methods to break down organic compounds, like dyes, into harmless by-products. To break down dyes and other pollutants, advanced oxidation processes (AOPs) use powerful oxidizing agents (Manavi et al., 2017). Dye removal at the molecular level is accomplished through membrane filtration methods like ultrafiltration and reverse osmosis (Sunil et al., 2021). Utilizing activated carbon or other adsorbents, adsorption techniques effectively adhere dyes to their surfaces to be removed. The wastewater's particular characteristics and desired treatment objectives guide the choice of treatment methods. For complete dye removal, integrated strategies often combine multiple treatment processes (Yagub et al., 2014). The sustainable treatment of dyes in wastewater entails recycling and desorption of adsorbent materials. The goal of desorption research is to free the dyes that have been adsorbed from the natural adsorbents' surfaces so that they can be regenerated and used again (Aragaw and Bogale, 2021). The objective of this review paper is to highlight the effective use of various natural adsorbents, primarily waste material adsorbents, for effective and sustainable remediation of dye pollutants by applying a circular economy approach.

14.2 TYPES OF DYES USED IN THE TEXTILE INDUSTRY

The use of natural dyes for textile coloring has a long and rich history in many cultures around the world. The incredible range of materials used in natural dyeing is one of its most remarkable aspects. Dyes can be made from a wide variety of sources, including plants, animals, and minerals, and several cultures have developed their own special methods for extracting and applying these dyes to fabrics. The Indus Valley Civilization (2500 BCE) in India was the first known civilization to utilize natural colors derived from plants such as madder, indigo, and turmeric. Egyptians utilized natural dyes derived from diverse plants, animals, and minerals as early as 2500 BCE. By combining copper, silica, and calcium, they produced an abundance of the now-famous "Egyptian Blue" color. Indigo-dyed fabric dating to 5000 BCE reveals that the Chinese have been using natural dyes since the Neolithic period (Mussak and Bechtold, 2009; Gürses et al., 2016). Synthetic dyes were first invented in the mid-nineteenth century. William Henry Perkin, discovered Mauveine, the first synthetic dye, in 1856. This unintentional discovery allowed the textile industry to

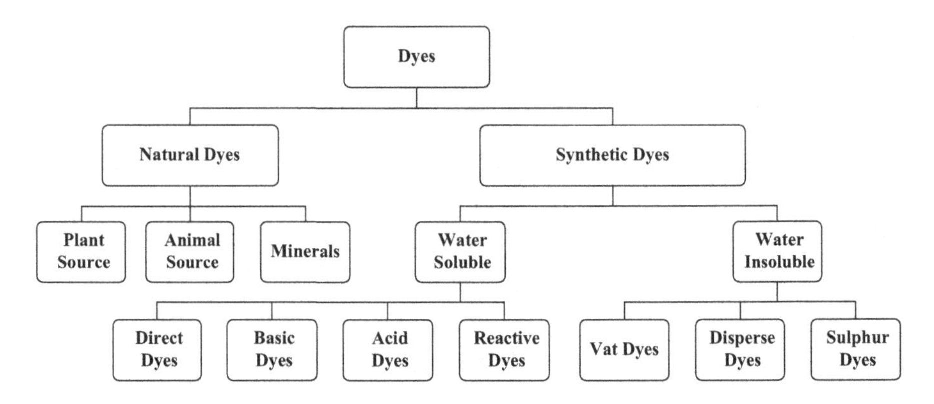

FIGURE 14.1 Dye classification.

manufacture a wide range of vibrant and long-lasting colors. After that, Verguin and Griess developed Fuchsine and diazo compounds in 1859, paving the path for modern synthetic dyes. Martius created Bismarck brown, the first azo dye, in 1863. The further development of synthetic dyes was facilitated by the development of synthetic fibers like nylon, polyester, and polyacrylonitrile, which were developed in 1930–1950 (Hunger, 2003).

Chemists discovered new chemicals and dyeing methods as synthetic dyes replaced natural dyes, creating a variety of dyes with various qualities. To manage and classify the expanding number of textile dyes, categorization was developed. Classification of dyes based on their chemical structure, color, and dyeing qualities was developed to make them easier to comprehend and use. The early twentieth-century Color Index system assigns each dye a unique number depending on its chemical structure and other features like color fastness, date invented, and application (Clark, 2011a, 2011b).

Dyes are divided into two types: natural dyes and synthetic dyes. Natural colours are derived from insects, minerals, and plant matter. Indigo is a well-known natural dye made by fermenting leaves of the *Indigofera* plant, while red and brown hues are made from Lac (from the scale insect *Laccifer lacca*) and iron oxide powder, respectively. Synthetic dyes, on the other hand, are obtained from petrochemical feedstock and are created by combining diverse organic chemical components. Perkin developed the first synthetic dye from coal tar. Direct dyes, azoic dyes, acid dyes, disperse dyes, vat dyes, reactive dyes, basic dyes, moderate dyes, sulfur dyes, and direct dyes are all examples of synthetic dyes. Figure 14.1 shows a schematic diagram of the dye classification.

14.3 AVAILABLE TREATMENT TECHNOLOGIES

The various methods for removing dyes from wastewater can be classified into three basic categories: physical, chemical, and biological. Physical processes for dye removal include adsorption, filtration, and membrane technologies. These procedures are typically straightforward and effective and do not need a significant amount of input from either energy or chemicals. On the other hand, they might not be efficient for the removal of all kinds of dyes, and certain kinds of dyes might call for further treatment operations (Nigam et al., 2000). To break down or remove dyes from wastewater, chemical methods entail the employment of chemical processes as the primary mechanism of action. Oxidation, reduction, precipitation, and coagulation are the four types of chemical processes that are utilized most frequently. These techniques are capable of effectively removing a wide variety of dyes, but they may call for the use of harmful chemicals, require a significant amount of input energy, and produce poisonous by-products (Raghu and Ahmed Basha, 2007). Microorganisms such as bacteria, fungi, and algae are used in biological processes to break down or remove dyes from wastewater. These processes are referred to as "biological techniques". These techniques are not only economical but also kind to the environment, as they do not generate any potentially harmful waste products. However, they may require longer treatment times, and their efficacy may be contingent on a number of variables, including pH, temperature, and the type of microorganisms that are utilized in the process (Bhatia et al., 2017). The choice of method is determined by several criteria, including the type of dye, concentration, pH, temperature, and the level of treatment that is wanted. Each of these methods offers a few benefits as well as several drawbacks (Bustos-Terrones et al., 2021). Table 14.1 contains a description of the benefits and drawbacks associated with the various treatment techniques used for dye removal.

TABLE 14.1
Advantages and Disadvantages of Dye Wastewater Treatment Methods

Treatment Methods	Advantages	Disadvantages	References
Adsorption	Effective for a diverse array of dyes, inexpensive, easy to operate	Requires frequent replacement of adsorbent material, not effective for all types of dyes	Garg et al. (2004)
Coagulation/flocculation	Effective for large-scale treatment, can remove a variety of pollutants in addition to dyes	Generates large amounts of sludge, can be expensive for small-scale treatment	Sadri Moghaddam et al. (2010)
Biological treatment	Sustainable, low energy consumption, can be combined with other treatment processes	Requires long treatment times, may not be effective for all types of dyes	Rai et al. (2005)
Membrane filtration	High removal efficiency, low chemical consumption	Can be expensive for large-scale treatment, requires frequent cleaning and maintenance	Rashidi et al. (2015)
Advanced oxidation processes	Effective for hard-to-treat dyes, can be used in combination with other treatment processes	High energy consumption, can produce toxic by-products	Chan et al. (2011)

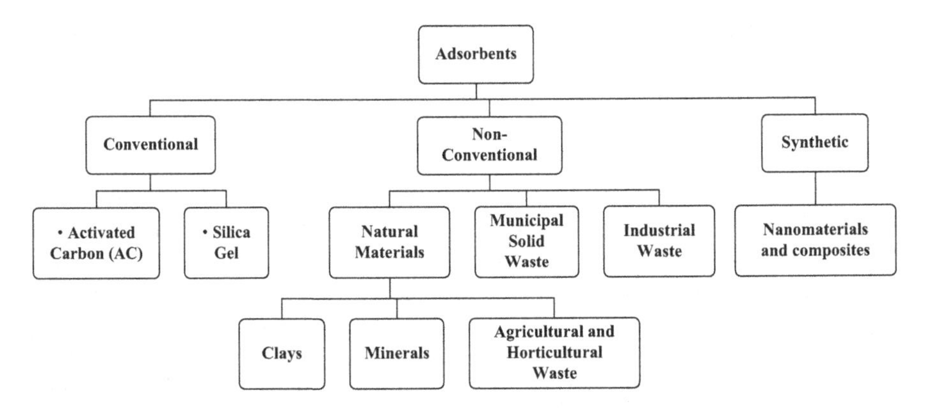

FIGURE 14.2 Types of adsorbents used for dye removal from wastewater.

14.4 ADSORPTION PROCESS AND TYPES OF ADSORBENTS

Adsorption is an effective approach for extracting dyes from wastewater. This approach uses an adsorbent substance that attracts and binds the dye molecules to its surface, allowing them to be extracted from the water. In the dye removal process, adsorbent materials such as activated carbon, zeolites, silica gel, chitosan, and clay minerals can be used. The type of adsorbent material employed is governed by parameters such as the dye content in the wastewater and the operating conditions of the adsorption process (Husien et al., 2022). Adsorption for dye removal can be done in batch or continuous mode. A predetermined amount of adsorbent material is added to the wastewater in batch mode, and the mixture is agitated for a predetermined period until an equilibrium is attained. The wastewater is continuously fed through a bed of adsorbent material, and the dye molecules are removed when they come into contact with the adsorbent surface (Adegoke and Bello, 2015). The surface area of the adsorbent material, the pH of the solution, the temperature, and the contact duration between the adsorbent and the wastewater all have an impact on the efficacy of the dye removal adsorption process. The factors evaluated are then optimized to achieve optimal dye removal efficiency (Zhang et al., 2021). The adsorption technique for dye removal is an efficient and cost-effective way of wastewater treatment. It is a straightforward procedure that may be quickly scaled up for industrial uses. However, spent adsorbent material must be disposed of appropriately to avoid environmental pollution (Rial and Ferreira, 2022). The various types of adsorbents used in dye wastewater treatment can be classified based on their application and are shown in Figure 14.2.

14.5 UTILIZING LOW-COST MATERIALS FOR AN ADSORPTION SYSTEM – THE ROLE OF THE CIRCULAR ECONOMY IN REPLACING CONVENTIONAL ACTIVATED CARBON

Adsorption is a method that is often used to get rid of pollutants present in water. Considering its high surface area, activated carbon (AC) is one of the most common adsorbent materials used for adsorption. On the other hand, commercial AC can be expensive and isn't always readily available. This is where the idea of a "circular economy", which aims to cut down on waste and make the most of resources, can be applied. The ideas behind the circular economy encourage the use of low-cost materials or AC derived from low-cost materials, instead of commercial activated carbon, for adsorption. Pyrolysis can be used to turn various materials into activated carbon that has the same properties as regular activated carbon which can be applied for effective and economical treatment of dye wastewater (Ettish et al., 2021). It can be low-cost materials (treated/untreated) like rice husks, coconut waste, or sawdust, as well as things like fly ash and used coffee grounds from factories. Studies have shown that heavy metals, dyes, and pharmaceuticals can be taken out of the air and water by activated carbon made from agricultural and industrial waste (Neolaka et al., 2023). Using waste materials for adsorption is a low-cost alternative to using traditional activated carbon. Circular economy ideas are put into action when waste materials are reused. This reduces the amount of trash that goes to landfills or is burned. Also, making activated carbon from waste materials can leave less of a carbon footprint than making activated carbon in the traditional way (Kim et al., 2019).

Erichrome-Black-T (EBT) is an azo dye used in the textile industry to dye fabrics made of nylon, silk, and wool. Biological processes and traditional aeration treatments fail to eliminate EBT from the wastewater. Household tea waste (brewed in only water), chemically activated with sulfuric acid, was used as a low-cost adsorbent to eliminate EBT and found to be 95% efficient in remediation at optimal conditions of pH 2 and an initial dye concentration of 100 mg/L, at 25 ± 5°C for 60 min. The acid treatment of tea waste resulted in an increase in the surface area and pore size and showed that tea waste could be utilized as a low-cost resource to eliminate EBT dye (Bansal et al., 2020). Methylene blue and methyl green dyes were spontaneously adsorbed onto timber sawdust (alkali-treated) due to the presence of new functional groups available after chemical activation and an increase in the number of sorption sites on the adsorbent surface. Rapid and uniform adsorption of dye pollutants was noted within 10 min of contact time at different pH ranges. Also, desorption studies carried out with 1M NaCl indicate that dyes can readily detach from the adsorbent surface without significantly reducing the adsorption capacity for further cycles, making timber sawdust suitable for adsorption of cationic dyes as a sustainable adsorbent and increasing the overall efficiency of the adsorption process (Djilali et al., 2016). Etim et al. (2016) studied the potential of raw coconut coir dust for the effective remediation of methylene blue dye from wastewater. The adsorption capacity of the selected adsorbent was reported as 29.50 mg/g and a removal efficiency of above 99% was recorded for the optimal conditions at pH 6 and an initial dye concentration of 50 mg/L, suggesting the possible industrial application of coconut coir dust as an effective and low-cost alternative to treat dye wastewater. Another low-cost alternative for the removal of malachite green dye was reported by Sartape et al. (2017), utilizing wood apple shell-derived adsorbent for the dye remediation from aqueous solutions, which is an economic and effective alternative, achieving effective removal of malachite green dye above 98% at optimal conditions of pH 7–9 and a contact time of 3.3 h.

14.6 AGRICULTURAL AND HORTICULTURAL RESIDUES AS NATURAL ADSORBENTS FOR EFFECTIVE TREATMENT OF DYE WASTEWATER

Residues from agriculture and horticulture, including plant stalks, leaves, and peel, are abundant sources of cellulose, lignin (Karić et al., 2022), and other organic compounds that have the potential to be utilized as adsorbent material with or without surface or chemical modifications. These materials have a large surface area and contain functional groups that have the potential to interact with dye molecules through a variety of different mechanisms (Zulfajri et al., 2021). Some of these mechanisms include hydrogen bonding, electrostatic interactions, and Van der Waals forces (Bhat et al., 2023). Malhotra et al. (2022) developed a biosorbent by utilizing waste palm leaves to remove malachite green (MG, a cationic dye) from aqueous solutions, achieving a maximum removal of 95% at a 60-min contact time with the initial adsorbate concentration ranging up to 1 ppm. Waste palm leaves were suggested as an efficient natural adsorbent for MG dye removal; Alam et al. (2022) used statistical optimization of adsorption removal efficiency for examining the removal of methylene blue dye using banana leaf ash, coconut coal, rice husk ash, and neem seeds as adsorbent, of which banana leaf ash was found to be the most efficient in the removal of MB dye, with 93.75% being the highest removal efficiency achieved for a 3-h contact period and 23.9 mg/100 mL dose of adsorbent. Birch tree waste heteroatom-doped with melamine clay-modified adsorbent was developed with improved adsorptive properties to eliminate Acid Red-18 dye from aqueous solutions. The adsorbent developed showed an increase in surface area after doping. The maximum adsorption capacity increased after doping by 22%, with a non-dope capacity of 444.5 mg/g and a doped adsorbent capacity of 545.2 mg/g (dos Reis et al., 2023). Rice is the staple food grown and consumed around the world, and rice husk contributes to the solid waste generated after harvesting of the crop. Haider et al. (2022) developed an efficient method to extract silica from burnt rice husk which could then be used as an adsorbent for the removal of dyes like methylene blue, thus creating a valuable product from agricultural waste and reintroducing the material to the circular economy. Adsorption of MB was rapid and a maximum removal efficiency of 107 mg/g was observed as pH 8.

14.7 NATURAL ADSORBENTS FROM MINERALS, CLAYS, AND BIOPOLYMERS FOR EFFECTIVE TREATMENT OF DYE WASTEWATER

Natural adsorbents have been proven to have a high capacity for adsorption and to be inexpensive, making them useful at removing dyes from wastewater. Some natural adsorbents, like biopolymers, can be utilized successfully to treat dye wastewater. The ability of minerals to adsorb dye molecules, such as zeolites, bentonite, and kaolin, has been well investigated. Due to their distinct chemical and physical characteristics, these minerals have a high surface area, high cation exchange capacity, and high adsorption capacity. Clays are natural adsorbents that can be utilized for the treatment of colored wastewater (Rahman et al., 2013). For instance, it has been discovered that the clay mineral montmorillonite has a high capacity for adsorbing cationic dyes (Jawad and Abdulhameed, 2020). The capacity of other clays to remove dyes from wastewater has also been investigated, including smectite and chlorite (Yadav and Sahu, 2023). Natural adsorbents that can be employed for dye wastewater treatment include chitosan, cellulose, and lignin. It has been discovered that the biopolymer chitosan, which is derived from chitin, has a high capacity for adsorbing anionic dyes (Lima et al., 2019). The ability of cellulose, a biopolymer present in plant cell walls, to remove dyes from wastewater has also been investigated. Due to its high adsorption capacity and chemical durability, lignin, a complex biopolymer found in plant cell walls, has been proven to be efficient at removing dyes from wastewater (Subash et al. 2023). Dolomite waste (alkali-activated) was utilized for the efficient removal of crystal violet and methylene blue dye. Alkali-activated dolomite waste adsorbent, prepared by simply mixing with sodium silicate and distilled water at room temperature, has shown promising results in the removal of dyes due to the availability of functional groups after the activation process. The maximum adsorption capacity reported for MB was 442.1 mg/g and 466 mg/g for crystal violet (CV) dye (Elshimy et al., 2023). Natural clays were also used for removal of crystal violet dye by Boulahbal et al. (2022) and found to be environmentally friendly alternatives to treat dye waste as well as being cost-efficient. Single and binary mixtures of Brilliant Violet 16 and Reactive Red 195 were treated using a mixture of Bentonite charred dolomite and Persian bentonite as adsorbents. The removal efficiencies of both adsorbents exceeded 50%, while desorption studies carried out with 1 M sodium hydroxide as the eluent also showed the high recyclability of the adsorbent, which is important for large-scale implementation and sustainability of the process (Khalilzadeh Shirazi et al., 2020).

14.8 USE OF MUNICIPAL SOLID WASTE AND INDUSTRIAL WASTE AS NATURAL ADSORBENTS FOR EFFECTIVE TREATMENT OF DYE WASTEWATER

An interesting possibility for the removal of dye is the use of activated carbon made from sewage sludge (Wang et al., 2021), which is a type of municipal solid waste generated after treatment of sewage. Because it has a large pore volume and a large surface area, it can adsorb a significant number of pollutants present in wastewater (Zhou et al. 2022). In addition to this, it is generated regularly and can be acquired with little effort and expense, which makes it a more practical choice than commercial adsorbents. Several different types of industrial waste, such as fly ash, rice husk ash, tannery waste, and municipal solid waste, have been investigated as possible adsorbents for the removal of different pollutants (Hossain et al. 2023). It has been demonstrated that fly ash, which is a by-product of the combustion of coal, possesses excellent adsorption properties for dye removal. It is essential, however, to check that the fly ash being utilized for the adsorption process does not include any heavy metals or other types of contaminants. Plastic waste makes up a large share of the overall municipal solid waste generated, resulting in various environmental hazards and disposal concerns. Polyurethane plastic waste, when upcycled and utilized as a low-cost adsorbent, can be used to treat malachite green dye from aqueous solutions. Polyurethane plastic waste, when carbonized and activated, has shown a high surface area of around 1034 m^2/g which proved effective in eliminating the malachite green dye with an adsorption capacity of 1428 mg/g (Li et al., 2021). This indicates the applicability of the circular economy concept to utilize a waste material as a resource and reintroduce it to the circular economy cycle (Waikar and Sadgir, 2023). An adsorbent developed from a mixture of municipal solid waste like paper waste, yard waste, and corn stover can be effective in the remediation of Indigo Carmine dye. Also, a mathematical model developed by using experimental data of dye removal, using artificial neural networks, shows the role of different adsorption-affecting parameters (Ahmad et al., 2021). Metal industry solid waste like Jarosite, which contains various metal oxides, metals, and composites, can be used as an adsorbent to treat dye wastewater, solving the solid waste disposal problem and addressing wastewater treatment at the same time (Kushwaha and Agarwal, 2023).

14.9 SUSTAINABLE MANAGEMENT OF SPENT ADSORBENTS – RECOVERY AND REGENERATION FOR A CIRCULAR ECONOMY

The spent adsorbent is a solid waste produced by the adsorption treatment; to reduce this generated waste, its reuse is vital. When evaluating the effectiveness of an adsorbent for use in industrial application, one of the most significant factors to consider is the reusability performance of the adsorbent material (Brahma and Saikia, 2022). Adjusting the pH, varying the temperature, utilizing organic solvents, and introducing competitive adsorbents are some of the desorption methods that can be used. To dissociate the dyes from the adsorbent surfaces, enabling their release, pH adjustment involves altering the pH of the solution. By encouraging the diffusion of dyes from the adsorbents, temperature variation boosts desorption efficiency (Coltre et al., 2020). Natural solvents with a higher affinity for the dyes can elute the adsorbed dyes, working with their desorption. The dyes in the natural adsorbents are displaced by the introduction of competitive adsorbents, allowing for their recovery. The recovered dyes can be given additional treatment or disposed of appropriately, while the regenerated natural adsorbents can be used again for subsequent dye adsorption cycles. By extending the lifespan and performance of natural adsorbents, reducing the requirement for additional adsorbents, and minimizing the generation of waste, effective desorption techniques contribute to the sustainability and cost-effectiveness of natural adsorbents (Patel, 2021). The adsorbent's desorption study reflects the reusability of the adsorbent. Tests for desorption may disclose the sort of adsorption that has taken place, as well as whether or not the depleted material can be reused (Al-Zawahreh et al., 2022). It is essential to carry out research on adsorption and desorption in order to improve the economic viability of the adsorption process and gain a knowledge of all aspects of adsorbate desorption from adsorbent (Yu et al., 2012).

TABLE 14.2
List of Natural Adsorbents (Activated/Non-activated) Used for Treatment of Various Dye Pollutants and Their Performance

Sr. No.	Natural Adsorbents	Adsorbate	Adsorption Parameters	Adsorption Capacity	References
1	Dragon fruit peel activated carbon	Methylene blue	Temperature (T): 30°C, initial concentration (IC): 25–300 mg/L	233 mg/g	Ahmad et al. (2021)
2	Rice husk	Reactive blue-4	T: 60°C, I: 500mg/L, pH 2	151.3 mg/g	Hong and Wang (2017)
3	Lemon peel	Eosin dye	IC: 5mg/L, Adsorbent dose: 0.5 g, pH: 2	8.240 mg/g	Bukhari et al. (2022)
4	Wheat straw (phytic acid modified)	Methylene blue	pH: 10, T: 25°C	205.43g/g	You et al. (2016)
5	Jackfruit leaf powder	Methyl orange	pH: 2, IC: 10 mg/L, Adsorbent dose: 1 g/L	32.89 mg/g	Dutta et al. (2022)
6	Wood apple shell	Methylene blue (MB), Crystal violet (CV)	pH: 10, Adsorbent dose: 1.0 mg/L	MB: 95.2 mg/g; CV: 129.87 mg/g	Jain and Jayaram (2010)
7	Walnut shell activated carbon	Methylene blue	Adsorbent dose: 1 g/L, IC: 400 mg/L, Contact time: 3 h, agitation speed: 150 rpm.	307.4 mg/g	Vakili et al. (2023)
8	*Lawsonia inermis* seed powder (chemically treated)	Brilliant green	T: 50°C, Contact time: 180 min, Agitation speed:50 rpm	34.96 mg/g	Ahmad & Ansari (2020)
9	Banana stem (activated carbon, AC)	Methylene blue	pH: 7, IC: 50 mg/L, Contact time: 90 min.	101.01 mg/g	Misran et al. (2022)
10	Date seeds	Methyl violet	IC: 20–80 mg/L, Contact time: 60–120 min	59.5 mg/g	Ali et al. (2022)

(Continued)

TABLE 14.2 (CONTINUED)
List of Natural Adsorbents (Activated/Non-activated) Used for Treatment of Various Dye Pollutants and Their Performance

Sr. No.	Natural Adsorbents	Adsorbate	Adsorption Parameters	Adsorption Capacity	References
11	*Moringa* seed waste	Disperse red-60, Congo red	IC: 100 mg/L, pH: 7, T: 25°C	Disperse Red: 60–170.7 mg/g, Congo Red: 196.8 mg/g	Soliman et al. (2019)
12	Carpet waste	Methylene blue, methyl orange	Adsorbent dose: 10 g/L, IC: 20 mg/L	MB: 2 mg/g, MO: 0.99 mg/g	Janbooranapinij et al. (2021)
13	Rice husk-based pristine biochar	Erichrome black T	pH: 2, Adsorbent dose: 1.5 g/L, Contact time: 2 h	–	Sudan et al. (2023)
14	Oil palm waste-based AC	Methylene blue, Acid orange-10	pH:2–3, IC: 50 mg/L, Adsorbent dose: 5 g/L	MB: 24 mg/g, AO: 10–18.76 mg/g	Baloo et al. (2021)
15	Waste char (from municipal solid waste)	Congo red, crystal violet	-	CR: 49.7 mg/g, CV: 356 mg/g	Jung et al. (2019)
16	Acid functionalized Ceramic	Malachite green	pH: 4.5, T: 27°C, IC: 3 mg/L	0.0401 mg/g	Yadav and Sahu (2023)
17	Mesoporous clay ceramic pellets	Methylene blue	Adsorbent dose: 5.0 g/L, Contact time: 1500 min	–	Sharma et al. (2021)

TABLE 14.3
Efficiencies of Various Adsorbents after Desorption Cycles

Sr. No.	Adsorbate	Adsorbent	Eluents Used	Desorption Cycle Nos.	Adsorption Efficiency (%) after Complete Cycles	Reference
1	Indigo carmine	Amorphous zirconium phosphate–graphitic carbon nitride composite	Ethanol and deionized water	5	95	Bakry et al. (2022)
2	Rhodamine B	Poly (acrylic acid-co-aniline) grafted itaconic acid hydrogel	0.05M NaOH and 0.05M HCl	5	87.9–82.2	Thakur et al. (2022)
3	Methyl orange	Biochar	0.1M NaOH	3	81–51	Aichour et al. (2022)
4	Eosin dye	Lemon peel	0.1M HCL and 0.1M NaOH	5	98–85	Bukhari et al. (2022)
5	Methylene blue (MB) and Safranine O (SO)	*Saccharum munjal* carbon nanotube composites	Acetone, hot water, water, 1N HCl	10	99	Yadav et al. (2021)
6	Methyl orange	Polyaniline-multiwalled carbon nanotubes	1M HCL and others	6	90	Pete et al. (2021)
7	Methyl blue and Congo red	Hydrophobic silica aerogels	Heating in muffle	10	95.5–89.5	Chen et al. (2021)
8	Reactive red, 24% Congo red	Waste polystyrene foam	Ultrasonic- assisted 0.25 mol/L NaOH	6	90	Liu et al. (2021)
9	Orange-II	Surfactant modified magnetic expanded graphite composites	Acid base and ethanol	6	47	Tian et al. (2021)

Not only does a desired adsorbent need to have a high adsorption capacity, but it also must be stable throughout the course of a prolonged period. This is because long-term stability is essential for reducing the costs of manufacture, with implications for the circular economy.

14.10 CONCLUSION AND FUTURE RECOMMENDATIONS

To summarize, the treatment of dye effluent from textile industry wastewater necessitates comprehensive and long-term solutions to reduce its environmental impact. This comprehensive review investigated several available treatment strategies and highlighted the potential of natural adsorbents for efficient dye removal from wastewater. The use of low-cost materials as natural adsorbents, such as agricultural and horticultural wastes, minerals, clays, biopolymers, municipal solid waste, and industrial waste, has yielded promising results in dye removal from wastewater. These materials not only provide cost-effective alternatives, but also contribute to the circular economy principles by substituting traditional activated carbon and utilizing waste resources. Furthermore, the study underlines the significance of long-term adsorbent management through recovery and regeneration processes. The circular economy concept can be applied to the adsorption system by employing these measures, lowering waste generation, and boosting resource efficiency. In conclusion, by embracing the potential of natural adsorbents and adopting a circular economy strategy, the textile sector may significantly reduce the environmental impact of dye wastewater and move toward more sustainable and responsible wastewater treatment processes.

Based on the findings of this review, recommendations can be made for future research and practical implementation:

1. Continued research should focus on finding and testing new natural adsorbents generated from a variety of sources, such as agricultural waste, plant-based materials, microbial biomass, and other sustainable alternatives. The adsorption capacity, selectivity, and cost-effectiveness of these materials should be investigated.
2. Adsorption should be investigated in conjunction with other treatment technologies such as biological treatment, advanced oxidation processes (AOPs), or membrane filtration to develop integrated hybrid systems that can improve overall treatment efficiency and ensure complete removal of dyes and other pollutants.
3. Pilot-scale and industrial-scale studies should be done to determine the viability and scalability of employing natural adsorbents in large-scale textile wastewater treatment plants. Cost analysis, commercialization, and life-cycle assessments are critical for assessing these technologies' economic viability and environmental sustainability.

CONFLICT OF INTEREST

The authors declare no conflict of interest.

REFERENCES

Adegoke, K.A. & Bello, O.S., 2015. Dye sequestration using agricultural wastes as adsorbents. *Water Resources and Industry*, 12, pp.8–24. http://dx.doi.org/10.1016/j.wri.2015.09.002.

Ahmad, M.A., Ahmad, M.A., Eusoff, M.A., Adegoke, K.A. & Bello, O.S., 2021. Sequestration of methylene blue dye from aqueous solution using microwave assisted dragon fruit peel as adsorbent. *Environmental Technology and Innovation*, 24, p.101917. https://doi.org/10.1016/j.eti.2021.101917.

Ahmad, R. & Ansari, K., 2020. Chemically treated Lawsonia inermis seeds powder (CTLISP): An eco-friendly adsorbent for the removal of brilliant green dye from aqueous solution. *Groundwater for Sustainable Development*, 11(May), p.100417. https://doi.org/10.1016/j.gsd.2020.100417.

Aichour, A., Zaghouane-Boudiaf, H. & Djafer Khodja, H., 2022. Highly removal of anionic dye from aqueous medium using a promising biochar derived from date palm petioles: Characterization, adsorption properties and reuse studies. *Arabian Journal of Chemistry*, 15(1), p.103542. https://doi.org/10.1016/j.arabjc.2021.103542.

Al-Tohamy, R. et al., 2022. A critical review on the treatment of dye-containing wastewater: Ecotoxicological and health concerns of textile dyes and possible remediation approaches for environmental safety. *Ecotoxicology and Environmental Safety*, 231, p.113160.

Al-Zawahreh, K. et al., 2022. Competitive removal of textile dyes from solution by pine bark-compost in batch and fixed bed column experiments. *Environmental Technology and Innovation*, 27, p.102421. https://doi.org/10.1016/j.eti.2022.102421.

Alam, M.Z., Bari, M.N. & Kawsari, S., 2022. Statistical optimization of Methylene Blue dye removal from a synthetic textile wastewater using indigenous adsorbents. *Environmental and Sustainability Indicators*, 14(September 2021), p.100176. https://doi.org/10.1016/j.indic.2022.100176.

Ali, N.S. et al., 2022. Adsorption of methyl violet dye onto a prepared bio-adsorbent from date seeds: Isotherm, kinetics, and thermodynamic studies. *Heliyon*, 8(8), p.e10276. https://doi.org/10.1016/j.heliyon.2022.e10276.

Aragaw, T.A. & Bogale, F.M., 2021. Biomass-based adsorbents for removal of dyes from wastewater: A review. *Frontiers in Environmental Science*, 9. https://doi.org/10.3389/fenvs.2021.764958.

Azanaw, A. et al., 2022. Textile effluent treatment methods and eco-friendly resolution of textile wastewater. *Case Studies in Chemical and Environmental Engineering*, 6 (June), p.100230.

Azbar, N., Yonar, T. & Kestioglu, K., 2004. Comparison of various advanced oxidation processes and chemical treatment methods for COD and color removal from a polyester and acetate fiber dyeing effluent. *Chemosphere*, 55(1), pp.35–43.

Bakry, A.M. et al., 2022. Facile synthesis of amorphous zirconium phosphate graphitic carbon nitride composite and its high performance for photocatalytic degradation of indigo carmine dye in water. *Journal of Materials Research and Technology*, 20, pp.1456–1469. https://doi.org/10.1016/j.jmrt.2022.07.161.

Baloo, L. et al., 2021. Adsorptive removal of methylene blue and acid orange 10 dyes from aqueous solutions using oil palm wastes-derived activated carbons. *Alexandria Engineering Journal*, 60(6), pp.5611–5629. https://doi.org/10.1016/j.aej.2021.04.044.

Bansal, M., Patnala, P.K. & Dugmore, T., 2020. Adsorption of Eriochrome Black-T (EBT) using tea waste as a low cost adsorbent by batch studies: A green approach for dye effluent treatments. *Current Research in Green and Sustainable Chemistry*, 3(October), p.100036. https://doi.org/10.1016/j.crgsc.2020.100036.

Bhat, S. et al., 2023. Abundant cilantro derived high surface area activated carbon (AC) for superior adsorption performances of cationic/anionic dyes and supercapacitor application. *Chemical Engineering Journal*, 459(January), p.141577. https://doi.org/10.1016/j.cej.2023.141577.

Bhatia, D. et al., 2017. Biological methods for textile dye removal from wastewater: A review. *Critical Reviews in Environmental Science and Technology*, 47(19), pp.1836–1876.

Boulahbal, M. et al., 2022. Removal of the industrial azo dye crystal violet using a natural clay: Characterization, kinetic modeling, and RSM optimization. *Chemosphere*, 306(July), p.135516. https://doi.org/10.1016/j.chemosphere.2022.135516.

Brahma, D. & Saikia, H., 2022. Synthesis of ZrO2/MgAl-LDH composites and evaluation of its isotherm, kinetics and thermodynamic properties in the adsorption of congo red dye. *Chemical Thermodynamics and Thermal Analysis*, 7(January), p.100067. https://doi.org/10.1016/j.ctta.2022.100067.

Bukhari, A. et al., 2022. Removal of Eosin dye from simulated media onto lemon peel-based low cost biosorbent. *Arabian Journal of Chemistry*, 15(7), p.103873. https://doi.org/10.1016/j.arabjc.2022.103873.

Bustos-Terrones, Y.A. et al., 2021. Removal of BB9 textile dye by biological, physical, chemical, and electrochemical treatments. *Journal of the Taiwan Institute of Chemical Engineers*, 121, pp.29–37.

Chan, S.H.S. et al., 2011. Recent developments of metal oxide semiconductors as photocatalysts in advanced oxidation processes (AOPs) for treatment of dye waste-water. *Journal of Chemical Technology and Biotechnology*, 86(9), pp.1130–1158.

Chen, K. et al., 2021. Hydroxyl modification of silica aerogel: An effective adsorbent for cationic and anionic dyes. *Colloids and Surfaces A: Physicochemical and Engineering Aspects*, 616(December 2020), p.126331. https://doi.org/10.1016/j.colsurfa.2021.126331.

Clark, M. 2011a. *Handbook of Textile and Industrial Dyeing Volume 1: Principles, Processes and Types of Dyes*. Woodhead Publishing Limited. https://doi.org/10.1533/9780857093974.1.1.

Clark, M. 2011b. *Handbook of Textile and Industrial Dyeing Volume 2: Applications of Dyes*. Woodhead Publishing Limited.

Coltre, D.S. de C. et al., 2020. Study of dye desorption mechanism of bone char utilizing different regenerating agents. *SN Applied Sciences*, 2(12), pp.1–14.

Dave, R.H., 2020. Current trends of textile dyes on the environment: A review. *International Journal for Innovative Research in Multidisciplinary Field*, 15, pp.194–200.

Dinkar Gosavi, V. & Sharma, S., 2013. A general review on various treatment methods for textile wastewater. *Journal of Environmental Science*, 3(1), pp.29–39.

Djilali, Y. et al., 2016. Alkaline treatment of timber sawdust: A straightforward route toward effective low-cost adsorbent for the enhanced removal of basic dyes from aqueous solutions. *Journal of Saudi Chemical Society*, 20, pp.S241–S249.

dos Reis, G.S. et al., 2023. Preparation of highly porous nitrogen-doped biochar derived from birch tree wastes with superior dye removal performance. *Colloids and Surfaces A: Physicochemical and Engineering Aspects*, 669(February), p.131493. https://doi.org/10.1016/j.colsurfa.2023.131493.

Dutta, S.K. et al., 2022. Removal of toxic methyl orange by a cost-free and eco-friendly adsorbent: Mechanism, phytotoxicity, thermodynamics, and kinetics. *South African Journal of Chemical Engineering*, 40(December 2021), pp.195–208. https://doi.org/10.1016/j.sajce.2022.03.006.

Elshimy, A.S. et al., 2023. Utilization of alkali-activated dolomite waste toward the fabrication of an effective adsorbent: Experimental study and statistical physics formalism for the removal of methylene blue and crystal violet. *Journal of Physics and Chemistry of Solids*, 180(February), p.111442.

Etim, U.J., Umoren, S.A. & Eduok, U.M., 2016. Coconut coir dust as a low cost adsorbent for the removal of cationic dye from aqueous solution. *Journal of Saudi Chemical Society*, 20, pp.S67–S76. http://dx.doi.org/10.1016/j.jscs.2012.09.014.

Ettish, M.N. et al., 2021. Preparation and characterization of new adsorbent from Cinnamon waste by physical activation for removal of Chlorpyrifos. *Environmental Challenges*, 5(May), p.100208. https://doi.org/10.1016/j.envc.2021.100208.

Garg, V.K. et al., 2004. Basic dye (methylene blue) removal from simulated wastewater by adsorption using Indian Rosewood sawdust: A timber industry waste. *Dyes and Pigments*, 63(3), pp.243–250.

Gürses, A., Açıkyıldız, M., Güneş, K. & Gürses, M.S., 2016. Classification of dye and pigments. *Springer Briefs in Molecular Science Dyes and Pigments*. Cham: Springer. https://doi.org/10.1007/978-3-319-33892-7_3.

Haider, J.B. et al., 2022. Efficient extraction of silica from openly burned rice husk ash as adsorbent for dye removal. *Journal of Cleaner Production*, 380(P2), p.135121. https://doi.org/10.1016/j.jclepro.2022.135121.

Hong, G.B. & Wang, Y.K., 2017. Synthesis of low-cost adsorbent from rice bran for the removal of reactive dye based on the response surface methodology. *Applied Surface Science*, 423, pp.800–809. http://dx.doi.org/10.1016/j.apsusc.2017.06.264.

Hossain, M.A. et al., 2023. A novel bio-adsorbent development from tannery solid waste derived biodegradable keratin for the removal of hazardous chromium: A cleaner and circular economy approach. *Journal of Cleaner Production*, 413(May), p.137471. https://doi.org/10.1016/j.jclepro.2023.137471.

Hunger, K. 2003. *Industrial Dyes: Chemistry, Properties, Applications*. https://doi.org/10.1002/3527602011.

Husien, S. et al., 2022. Review of activated carbon adsorbent material for textile dyes removal: Preparation, and modelling. *Current Research in Green and Sustainable Chemistry*, 5(June), p.100325. https://doi.org/10.1016/j.crgsc.2022.100325.

Jain, S. & Jayaram, R.V., 2010. Removal of basic dyes from aqueous solution by low-cost adsorbent: Wood apple shell (Feronia acidissima). *Desalination*, 250(3), pp.921–927. http://dx.doi.org/10.1016/j.desal.2009.04.005.

Janbooranapinij, K. et al., 2021. Conversion of industrial carpet waste into adsorbent materials for organic dye removal from water. *Cleaner Engineering and Technology*, 4, p.100150. https://doi.org/10.1016/j.clet.2021.100150.

Jawad, A.H. & Abdulhameed, A.S., 2020. Mesoporous Iraqi red kaolin clay as an efficient adsorbent for methylene blue dye: Adsorption kinetic, isotherm and mechanism study. *Surfaces and Interfaces*, 18(December 2019), p.100422. https://doi.org/10.1016/j.surfin.2019.100422.

Jung, H. et al., 2019. Characterization and adsorption performance evaluation of waste char by-product from industrial gasification of solid refuse fuel from municipal solid waste. *Waste Management*, 91, pp.33–41. https://doi.org/10.1016/j.wasman.2019.04.053.

Kant, R., 2012. Textile dyeing industry an environmental hazard. *Natural Science*, 4(1), pp.22–26.

Karić, N. et al., 2022. Bio-waste valorisation: Agricultural wastes as biosorbents for removal of (in)organic pollutants in wastewater treatment. *Chemical Engineering Journal Advances*, 9, p.100239.

Khalilzadeh Shirazi, E. et al., 2020. Removal of textile dyes from single and binary component systems by Persian bentonite and a mixed adsorbent of bentonite/charred dolomite. *Colloids and Surfaces A: Physicochemical and Engineering Aspects*, 598(February), p.124807. https://doi.org/10.1016/j.colsurfa.2020.124807.

Kim, M.H. et al., 2019. Analysis of environmental impact of activated carbon production from wood waste. *Environmental Engineering Research*, 24(1), pp.117–126.

Kushwaha, P. & Agarwal, M., 2023. Adsorption of cationic dye by using metal industry solid waste as an adsorbent. *Materials Today: Proceedings*, 78, pp.198–203. https://doi.org/10.1016/j.matpr.2023.02.203.

Li, Z. et al., 2021. Removal of malachite green dye from aqueous solution by adsorbents derived from polyurethane plastic waste. *Journal of Environmental Chemical Engineering*, 9(1), p.104704. https://doi.org/10.1016/j.jece.2020.104704.

Lima, V.V.C. et al., 2019. Synthesis and characterization of biopolymers functionalized with APTES (3-aminopropyltriethoxysilane) for the adsorption of sunset yellow dye. *Journal of Environmental Chemical Engineering*, 7(5), p.103410. https://doi.org/10.1016/j.jece.2019.103410.

Liu, M. et al., 2021. Waste polystyrene foam – Chitosan composite materials as high-efficient scavenger for the anionic dyes. *Colloids and Surfaces A: Physicochemical and Engineering Aspects*, 627(May), p.127155. https://doi.org/10.1016/j.colsurfa.2021.127155.

Madhav, S. et al., 2020. Water pollutants: Sources and impact on the environment and human health. *Advanced Functional Materials and Sensors*, pp.43–62. https://doi.org/10.1007/978-981-15-0671-0_4.

Malhotra, A. et al., 2022. Study of adsorbent characteristics of palm leaves powder as a bio sorbent for removal of malachite green (MG) dye. *Materials Today: Proceedings*, 67, pp.900–904. https://doi.org/10.1016/j.matpr.2022.07.347.

Manavi, N., Kazemi, A.S. & Bonakdarpour, B., 2017. The development of aerobic granules from conventional activated sludge under anaerobic-aerobic cycles and their adaptation for treatment of dyeing wastewater. *Chemical Engineering Journal*, 312, pp.375–384.

Misran, E. et al., 2022. Banana stem based activated carbon as a low-cost adsorbent for methylene blue removal: Isotherm, kinetics, and reusability. *Alexandria Engineering Journal*, 61(3), pp.1946–1955. https://doi.org/10.1016/j.aej.2021.07.022.

Mussak, R.A.M. & Bechtold, T., 2009. *Natural Colorants in Textile Dyeing*. https://doi.org/10.1002/9780470744970.ch18.

Neolaka, Y.A.B. et al., 2023. Potential of activated carbon from various sources as a low-cost adsorbent to remove heavy metals and synthetic dyes. *Results in Chemistry*, 5(December 2022), p.100711. https://doi.org/10.1016/j.rechem.2022.100711.

Nigam, P. et al., 2000. Physical removal of textile dyes from effluents and solid-state fermentation of dye-adsorbed agricultural residues. *Bioresource Technology*, 72(3), pp.219–226.

Patel, H., 2021. Review on solvent desorption study from exhausted adsorbent. *Journal of Saudi Chemical Society*, 25(8), p.101302. https://doi.org/10.1016/j.jscs.2021.101302.

Pete, S., Kattil, R.A. & Thomas, L., 2021. Polyaniline-multiwalled carbon nanotubes (PANI-MWCNTs) composite revisited: An efficient and reusable material for methyl orange dye removal. *Diamond and Related Materials*, 117(May), p.108455. https://doi.org/10.1016/j.diamond.2021.108455.

Raghu, S. & Ahmed Basha, C., 2007. Chemical or electrochemical techniques, followed by ion exchange, for recycle of textile dye wastewater. *Journal of Hazardous Materials*, 149(2), pp.324–330.

Rahman, A., Urabe, T. & Kishimoto, N., 2013. Color removal of reactive procion dyes by clay adsorbents. *Procedia Environmental Sciences*, 17, pp.270–278. http://dx.doi.org/10.1016/j.proenv.2013.02.038.

Rai, H.S. et al., 2005. Removal of dyes from the effluent of textile and dyestuff manufacturing industry: A review of emerging techniques with reference to biological treatment. *Critical Reviews in Environmental Science and Technology*, 35(3), pp.219–238.

Rashidi, H.R. et al., 2015. Synthetic reactive dye wastewater treatment by using nano-membrane filtration. *Desalination and Water Treatment*, 55(1), pp.86–95.

Rial, J.B. & Ferreira, M.L., 2022. Potential applications of spent adsorbents and catalysts: Re-valorization of waste. *Science of the Total Environment*, 823, p.153370.

Sadri Moghaddam, S., Alavi Moghaddam, M.R. & Arami, M., 2010. Coagulation/flocculation process for dye removal using sludge from water treatment plant: Optimization through response surface methodology. *Journal of Hazardous Materials*, 175(1–3), pp.651–657.

Sartape, A.S. et al., 2017. Removal of malachite green dye from aqueous solution with adsorption technique using Limonia acidissima (wood apple) shell as low cost adsorbent. *Arabian Journal of Chemistry*, 10, pp.S3229–S3238.

Sharma, P., Olufemi, A.F. & Qanungo, K., 2021. Development of green geo-adsorbent pellets from low fire clay for possible use in methylene blue removal in aquaculture. *Materials Today: Proceedings*, 49, pp.1556–1565. https://doi.org/10.1016/j.matpr.2021.07.343.

Soliman, N.K. et al., 2019. Effective utilization of Moringa seeds waste as a new green environmental adsorbent for removal of industrial toxic dyes. *Journal of Materials Research and Technology*, 8(2), pp.1798–1808. https://doi.org/10.1016/j.jmrt.2018.12.010.

Subash, A. et al., 2023. Biopolymer – A sustainable and efficacious material system for effluent removal. *Journal of Hazardous Materials*, 443(PA), p.130168. https://doi.org/10.1016/j.jhazmat.2022.130168.

Sudan, S., Khajuria, A. & Kaushal, J., 2023. Adsorption potential of pristine biochar synthesized from rice husk waste for the removal of Eriochrome black azo dye. *Materials Today: Proceedings*. https://doi.org/10.1016/j.matpr.2023.01.258.

Sunil, K. et al., 2021. Prolific approach for the removal of dyes by an effective interaction with polymer matrix using ultrafiltration membrane. *Journal of Environmental Chemical Engineering*, 9(6), p.106328.

Thakur, S. et al., 2022. Highly efficient poly(acrylic acid-co-aniline) grafted itaconic acid hydrogel: Application in water retention and adsorption of rhodamine B dye for a sustainable environment. *Chemosphere*, 303(P1), p.134917. https://doi.org/10.1016/j.chemosphere.2022.134917.

Tian, Y., Ma, H. & Xing, B., 2021. Preparation of surfactant modified magnetic expanded graphite composites and its adsorption properties for ionic dyes. *Applied Surface Science*, 537(September 2020), p.147995. https://doi.org/10.1016/j.apsusc.2020.147995.

Vakili, A. et al., 2023. The impact of activation temperature and time on the characteristics and performance of agricultural waste-based activated carbons for removing dye and residual COD from wastewater. *Journal of Cleaner Production*, 382(May 2022), p.134899. https://doi.org/10.1016/j.jclepro.2022.134899.

Waikar, M. & Sadgir, P., 2023. Sustainable development in circular economy: A review. *Lecture Notes in Civil Engineering*, 260, pp.629–639. https://doi.org/10.1007/978-981-19-2145-2_48.

Wang, X. et al., 2021. Co-pyrolysis of sewage sludge and organic fractions of municipal solid waste: Synergistic effects on biochar properties and the environmental risk of heavy metals. *Journal of Hazardous Materials*, 412(January), p.125200. https://doi.org/10.1016/j.jhazmat.2021.125200.

Yadav, A. et al., 2021. Adsorptive studies on the removal of dyes from single and binary systems using Saccharum munja plant-based novel functionalized CNT composites. *Environmental Technology and Innovation*, 24, p.102015. https://doi.org/10.1016/j.eti.2021.102015.

Yadav, J. & Sahu, O., 2023. Total Environment Research Themes Dye removal of cationic dye from aqueous solution through acid functionalized ceramic. *Total Environment Research Themes*, 6(January), p.100038. https://doi.org/10.1016/j.totert.2023.100038.

Yagub, M.T. et al., 2014. Dye and its removal from aqueous solution by adsorption: A review. *Advances in Colloid and Interface Science*, 209, pp.172–184. http://dx.doi.org/10.1016/j.cis.2014.04.002.

You, H. et al., 2016. Selective removal of cationic dye from aqueous solution by low-cost adsorbent using phytic acid modified wheat straw. *Colloids and Surfaces A: Physicochemical and Engineering Aspects*, 509, pp.91–98. http://dx.doi.org/10.1016/j.colsurfa.2016.08.085.

Yu, J.X. et al., 2012. Desorption and photodegradation of methylene blue from modified sugarcane bagasse surface by acid TiO2 hydrosol. *Applied Surface Science*, 258(8), pp.4085–4090. http://dx.doi.org/10.1016/j.apsusc.2011.12.106.

Zhang, T. et al., 2021. Removal of heavy metals and dyes by clay-based adsorbents: From natural clays to 1D and 2D nano-composites. *Chemical Engineering Journal*, 420(P2), p.127574. https://doi.org/10.1016/j.cej.2020.127574.

Zhou, C. et al., 2022. Pyrolysis of typical solid wastes in a continuously operated microwave-assisted auger pyrolyser: Char characterization, analysis and energy balance. *Journal of Cleaner Production*, 373(May), p.133818. https://doi.org/10.1016/j.jclepro.2022.133818.

Zulfajri, M., Kao, Y.T. & Huang, G.G., 2021. Retrieve of residual waste of carbon dots derived from straw mushroom as a hydrochar for the removal of organic dyes from aqueous solutions. *Sustainable Chemistry and Pharmacy*, 22(June), p.100469. https://doi.org/10.1016/j.scp.2021.100469.

15 Circular Economy on Biogas Production from Wastewater

Laura Helena dos Santos
Laboratório de Microbiologia e Bioprocessos,
Universidade Federal da Fronteira Sul,
UFFS – Campus Erechim, Erechim, RS – Brasil

Caroline Berto
Laboratório de Microbiologia e Bioprocessos,
Universidade Federal da Fronteira Sul,
UFFS – Campus Erechim, Erechim, RS – Brasil

Gabriel Henrique Klein
Laboratório de Microbiologia e Bioprocessos,
Universidade Federal da Fronteira Sul,
UFFS – Campus Erechim, Erechim, RS – Brasil

Larissa Capeletti Romani
Laboratório de Microbiologia e Bioprocessos,
Universidade Federal da Fronteira Sul,
UFFS – Campus Erechim, Erechim, RS – Brasil

Guilherme Cabral Wancura
Laboratório de Microbiologia e Bioprocessos,
Universidade Federal da Fronteira Sul,
UFFS – Campus Erechim, Erechim, RS – Brasil

Helen Treichel
Laboratório de Microbiologia e Bioprocessos,
Universidade Federal da Fronteira Sul,
UFFS – Campus Erechim, Erechim, RS – Brasil

DOI: 10.1201/9781003432869-15

15.1 INTRODUCTION

Wastewater has several origins and can be classified as domestic, industrial, or agricultural. Composed mainly of human waste, domestic wastewater is complex in nature and characterized by high concentrations of organic matter and nutrients, such as phosphorus and nitrogen (Musa et al., 2018). In particular, domestic wastewater contains considerable pollutant loads in its composition. In general, Karunanithi et al. (2015) report that the composition of municipal wastewater can vary from 32 to 202 mg/L of nitrogen, 5 to 106 mg/L of phosphorus for a COD ranging between 300 and 4,294 mg/L, where 75% of the phosphorus load can reach watercourses without proper treatment, causing eutrophication. Consequently, wastewater can transmit infectious agents that cause diarrheal diseases such as dysentery, cholera, and typhoid, resulting in thousands of deaths per year (Farkas et al., 2020).

In developing countries, around 90% of domestic wastewater is discharged without any treatment into waterbodies or coastal areas, causing severe ecological problems, mainly associated with the deterioration of water quality and damage to biodiversity (Werkneh and Gebru, 2023). As the leading cause of these alarming data, disorganized urbanization and industrialization are considered among the main pivots, generating impacts on society and the environment. As a consequence of this anthropic development, guidelines on sustainable development began to be debated globally. In a more targeted way, it is of fundamental importance to recognize the problems that wastewater can cause in the population and the economy due to its environmental impacts and management costs (Agarwal et al., 2022).

In this sense, the United Nations (UN) created in 2015 the Sustainable Development Goals (SDGs), aiming at a future with a greater degree of sustainability, including access to safe drinking water for all. However, according to data from the UN itself, efforts to reach this goal are still delayed in most countries. On this basis, several studies have focused on developing technologies for better wastewater management to open new paths for a circular economy involved in new business models based on the reuse/recovery of valuable by-products (Obaideen et al., 2022).

The circular economy refers to a concept that guarantees a longer-lasting permanence of raw materials and other products in the network of resources. This also allows waste to be treated as secondary raw materials, thus being reused. The primary example in the context of wastewater is the treatment plants for these effluents, as they can reuse water and recover energy (Karla et al., 2022; Sanchez et al., 2022; Vaz et al., 2023). Wastewater treatment plants play an essential role in the availability of water and the supply of clean energy. This applies to two SDGs focused on water and energy: SDG 6: "clean water and sanitation", and SDG 7: "clean and affordable energy". Concerning SDG 6, it is clear that the main contribution is associated with increasing the availability and quality of water. As for SDG 7, contributions are focused on improving energy availability, mainly by biogas production (Obaideen et al., 2022).

Renewable energy sources have been sought as a viable alternative to fossil fuels since the concern with the climate crisis, and ways to reduce environmental impacts are being increasingly studied. Given this, the National Organization of the United Nations (UN) aims to significantly increase the production of renewable energies in the global energy matrix and achieve universal and equitable access to safe drinking water for all. These are some Sustainable Development Goals (SDGs).

Wastewater from industrial, agricultural, and municipal sectors generates soil and groundwater contamination, compromising the environment and human health. Because of this, it is necessary to seek techniques and methods that effectively use the effluents generated. The manufacture of disposable products from fossil fuels and natural resources leads to excessive emission of greenhouse gases (GHGs); this is one of the characteristics of the current economic model of the linear economy. In contrast, the circular economy is a method of the financial system that seeks to deal with the intensive consumption of resources and emissions of polluting gases; one of its objectives is to reduce the generation of waste and the use of materials. The need to manage water resources more circularly in countries with a need for water supply is given to recovering drinking water and its energy matrix.

Biogas is an alternative for improving air quality, as it promotes the retention of CO_2, and wastewater that has contaminants is a threat to public health and the environment. Effluent treatment and associated biogas production are linked, as they reinforce the circular economy; wastewater can also be used as a raw material (François et al., 2023). To maximize the storage and use of CO_2 for conversion into products with high added value, wastewater treatment technologies using physical, chemical, or biological approaches are becoming increasingly sought after, such as the use of bioprocesses involving bioreactors and bioelectrochemical systems for the removal of active sludge contaminants.

The main objective of these wastewater treatments is the efficient removal of contaminants; however, one of the challenges to using large-scale carbon sequestration technologies is the process costs and energy demand. Because of this, using microalgae becomes an exciting approach for using carbon dioxide and wastewater treatment since microalgae have a flexible metabolism. They use CO_2 and sunlight for photosynthesis, and to generate organic carbon as a source of carbon and energy for cell growth.

Biogas is a renewable energy source produced from organic matter; microorganisms degrade this material under anaerobic conditions and it can be used to produce electricity or fuel. Biogas consists of methane (about 50 to 70%) and carbon dioxide (about 30 to 50%). In addition to producing clean energy, biogas plants offer other advantages, such as managing manure on farms, using domestic sewage, allowing recycling of nutrients, and reducing methane emissions. These advantages make biogas production a technology that can benefit the environment.

Between 590 million and 800 million tons of methane come into the atmosphere each year by natural biodegradation of organic matter. Biogas generation systems use these biochemical processes to generate clean energy sources. During the combustion of biogas, methane is transformed into water; this is one of the advantages of using biogas, as it reduces negative impacts and makes waste processing activity profitable. Recent advances in biogas production focus on improving solid waste and sewage treatment processes, anaerobic digestion, heat and energy production, and purification and extraction of chemicals.

On a large scale, biogas production is a technology used in developed countries because, for the proper production of biogas, challenges with materials and construction models, efficiency, and performance need to be overcome. In Europe, biogas production has increased significantly in terms of the number of plants and capacity; countries such as Germany, Italy, Spain, France, and the UK produce most of the biogas on the continent. In Germany, the most significant amount of biogas is made from agricultural waste, while in

Spain, France, Italy, and the UK, the most critical capacity to produce biogas comes from landfills (Ahlberg-Eliasson et al., 2017).

As much as biogas production in Europe is highlighted, other regions, such as Asia and Africa, have been growing in biogas capacity. In Europe, most of it is released by anaerobic digestion, while, in the USA, it comes from landfill sites. In addition to the production of clean and sustainable energy, the nutrients retained in the digestate from the biogas manufacturing process make the residue potentially an organic fertilizer, attracting interest from the agricultural sector (Ahlberg-Eliasson et al., 2017).

Throughout this chapter, recent studies on the use of biogas from wastewater treatment and effluent treatment will be discussed, as its performance in the face of the circular economy, as well as the new technologies used to reduce the cost of biogas production.

15.2 CIRCULAR ECONOMY AND WASTEWATER TREATMENT

In the current context, effluent treatment terminals have the potential to become authentic sources of wealth, in the form of energy, fuels, nutrients, and clean water, and, consequently, cooperate for sustainable development and the construction of a circular economy. However, for this process to become real, it must develop advanced techniques for treating wastewater and the resultant sludge. Furthermore, to avoid any unfortunate results, it is necessary to evaluate both the effectiveness and the sustainability of these methodologies (Ratola et al., 2012).

The main wastewater treatment techniques used in this area are microfiltration, ultrafiltration, membrane bioreactor, reverse osmosis, granular activated carbon, and ultraviolet disinfection (Tarpani et al., 2023). Still, as adequate proof of the beneficial role of sustainability in general processes and sewage treatment, the potable reuse of water is practiced to alleviate the lack of water in arid and intensely populated areas such as Namibia, South Africa, and Big Spring, TX, USA.

In addition to rescuing and recycling wastewater, it is also necessary to recover resources from the sludge – a material with a high concentration of microorganisms, organic solids, and minerals and with considerable potential for reuse. In addition, it is currently known that some techniques can be used for this purpose, including incineration, gasification, oxidation with humid air, and pyrolysis of sludge as a working method (Tarpani et al., 2023).

Based on three pillars, namely production, consumption, and disposal, this model takes advantage of an organizational pattern characterized by the massive extraction of natural resources, production of goods, and disposal of waste, building a form of consumption with a beginning, middle, and end – without expectation of return or recycling. Historically consolidated, the linear pattern of economic production and consumption harms nature and accelerates the shortage of natural resources.

In the research by Deng et al. (2023), regardless of whether the genesis of the waters are residential, industrial, or runoff, their formation is induced by their origin. In addition, domestic and industrial effluents can be associated with rainwater in municipal treatment terminals, mixing and dissolving components previously agglutinated in their sources. Thus, as wastewater is increasingly recognized as a resource due to the existence of a rich range of nutrients and compounds with reusable potential, restoration possibilities need to be pointed out along with management possibilities. However, it is worth noting that the formation of effluents can raise awareness of both resource recovery goals and treatment efficiency. Therefore, a complete understanding of what integrates wastewater and where its components come from corroborates its practical use and treatment (Deng et al., 2023). Below, we can better understand the circular economy concept, illustrated in the diagram.

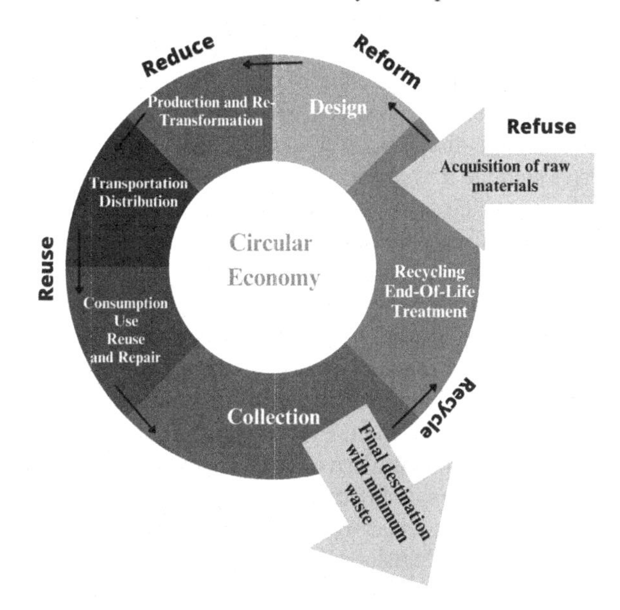

Source: author

15.2.1 Types of Waste Found in Wastewater Treatment Plants

In the sanitary context, it is known that wastewater management sites receive different types of sediments with diverse genesis: urban, industrial, and agricultural, as well as wet and/or dry atmospheric deposition. Thus, to better explore this phenomenon, a study promoted by Ratola et al. (2012) detected in wastewater notable classes of contaminants, such as persistent organic pollutants (POPs), polycyclic aromatic hydrocarbons (PAHs), and pesticides, in addition to compounds usually used in pharmaceutical and cosmetic products, the PPCPs (pharmaceuticals and personal care products).

Throughout history, wastewater treatment centers (ETEs) were, at first, designed to remove and minimize usual pollution criteria such as BOD-5, COD, and total suspended solids from the effluent stream, so that, at the end, the process does not create a new source of contamination for the body of water. However, when witnessing high levels of harmful organic pollutants, it becomes evident, therefore, the need to include other treatment and control actions in the treatment plants (Ratola et al., 2012). In this context, in recent decades, the massive-scale manufacturing of different pharmaceutical compounds and their increasing consumption is notorious. Consequently, their presence in the aquatic environment has drawn the attention of scientific communities.

Therefore, there is a need to exercise cleaner production and correct disposal of these substances to minimize harmful environmental consequences. In the study by Hong et al. (2019), about 52 drugs were found in the effluents and effluents of two sewage management sites (ETEs) and their receiving waters. Furthermore, these components were quantified with an analytical approach using online solid-phase-extraction coupled with liquid chromatography–tandem mass spectrometry.

Thus, because micropollutants are emerging as a new problem for the environment and consequently for the scientific community, the review by Luo et al. (2014) provides an overview of the possible entry of these compounds into the aquatic environment, covering sewage, surface water, groundwater, and drinking water. Thus, it was highlighted that the discharge of treated effluents from wastewater treatment plants (WWTPs) is suitable for introducing contaminants into surface waters.

In addition, the existence of advanced management methods such as activated carbon adsorption, oxidative processes, and nanofiltration that effect a greater efficiency of removal of such micropollutants is remarkable. However, regardless of the applied technique, removing these components depends on their physicochemical properties and treatment conditions. Thus, the analysis of eliminating pollutants from effluents must contain a range of criteria that rationalize sources and end uses. Also, understanding and modeling the fate of these substances in surface waters is essential to predicting their results in the receiving environment (Luo et al., 2014).

The presence of micropollutants derived from chemicals dumped in urban wastewater and the use of coagulants, disinfectants, and neutralizers in treatment processes leads to the generation of chemical waste. These residues require adequate treatment required by the specific legislation of each location, which generally consists of incineration, stabilization, neutralization, and recycling. In addition, management methodologies such as transportation and proper identification are also critical to avoid accidents that endanger public health and the environment (Achon et al., 2013).

Chemical waste, gases, and sludge are the leading waste produced in effluent treatment stations and have been the subject of several recent studies to value them economically (Heberle, 2013). Methane, carbon dioxide, and hydrogen sulfide are the primary examples of gases released in WWTPs. There is the possibility of producing biogas through methane, but there is still a need for more research in this regard, in addition to the fact that the process depends on the amount of this gas released. Some currently applied treatment methods are capture and burning, deodorization, biofiltration, or absorption (Achon et al., 2013). Obtaining biogas through sludge from wastewater treatment plants is discussed in more detail Section 15.4 of this chapter.

The treated water from the WWTP must also be directed appropriately according to its mode of treatment to avoid further pollution. The water quality parameters classify treated water and limit its final use; these parameters were created to prevent inappropriate uses that could generate environmental impacts. Among the various purposes used to reuse effluents, such as replenishing aquifers or returning water to lakes and oceans, irrigating crops, industrial activities, and/or cleaning and washing, there is the possibility of producing renewable energy and biofertilizers. However, the quality parameters to be followed and the final use of effluents will depend on the needs of each region (Bagheri et al., 2023).

There are several reuse purposes for treated water, also representing a solution for regions where this resource is scarce in potable form. Despite this, the treatments used are expensive and often produce harmful waste that returns to the environment, making the whole process non-viable in the long term (Baresel et al., 2020).

The proposal made by Baresel et al. (2020) was to create a circular synergy of water, that is, to ensure its reuse from advanced treatment for industrial processes, agriculture, or to return it to groundwater, since the region where the study was carried out had problems with water shortages. The authors produced a full-scale project of advanced water treatment, put into operation in 2019 in Simrishamn, Sweden. The treatment plant removes micropollutants, such as pharmaceutical substances, endocrine disruptors, and antibiotics. Analyses performed by the authors revealed that the rate of removal of these pollutants was almost 100%.

Tarpani et al. (2018) evaluated the Granular Activated Carbon (GAC), Nanofiltration (NF), Solar Photo-Fenton (SPF), and Ozonation methods, considered the most promising for removing Pharmaceuticals and Personal Care Products (PPCPs) in terms of their environmental viability. Individually, the procedures did not show great potential for producing harmful elements. However, when evaluating the set of operations, the authors conclude that advanced wastewater treatment for agricultural purposes is not feasible since the water has a high potential for ecotoxicity derived from the treatments themselves, which is even more significant than the harmfulness found in PPCPs.

15.2.2 The Economic Viability of Wastewater Treatment

According to the report "The United Nations World Water Development Report" by the UN, published in 2017, treating wastewater costs, on a world average, about US$2 per cubic meter. In developed countries, the price is generally a little higher, reaching values such as US$4–6, while, for less developed countries, such as India, the values are around US$0.1 per cubic meter. Other factors that influence these values, in addition to each country's economy and regional aspect, are the type of technology used in the treatment and the physical and chemical characteristics of the effluent.

Sequencing unit operations, used to increase efficiency in removing pollutants and reduce waste produced, using solar energy through photovoltaic panels, and producing biogas as a source of energy for the treatment plant itself, are some of the current proposals to reduce energy costs, for example, and make wastewater treatment a cheaper and consequently more viable goal worldwide (Souza et al., 2021).

Souza et al. (2021) analyzed the operating and maintenance costs of a sewage treatment plant located in Brazil for three years. Expenses per treated cubic meter reached values close to US$0.1, so the main expense factor was energy consumption, representing up to around 50% of total expenses. Operating with an anaerobic reactor and activated sludge, the station has an average flow of 406 L/s, and serves a population of around 202,000 inhabitants.

Looking for alternatives to reduce energy costs in WWTPs, Laramee et al. (2018) analyzed the costs and benefits of biogas generation by a community anaerobic digester installed in a peri-urban region of Zambia, in southern Africa, to treat wastewater from community homes. Installing the biodigester provided the community with a reduction in energy costs, fossil fuel usage, and greenhouse gas emissions. It improved the population's health conditions, such as improving air quality. As for the economic benefit, the high costs of installing the biodigester soon became viable because of the consequent energy savings, raising a new possibility for treating urban wastewater and improving environmental, economic, and public health conditions.

In the study by Baş et al. (2022), 456 urban ETEs in Turkey were selected to participate in an estimate of the possibility of producing biogas based on the location of each one and the treatment process employed. Using the HOMER software developed by the National Renewable Energy Laboratory (NREL), 25 of the 456 ETEs were selected to undergo another assessment regarding the feasibility of installing photovoltaic panels for energy generation. Implementing photovoltaic panels and a biogas production system was feasible for 40% of the secondary and tertiary ETEs, proving to be effective methods for stations that carry out treatment on a large scale. According to the authors, for low-flow ETEs, it is cheaper to use the network's energy than to spend on optimizations since they produce only a small amount of sludge (biomass).

Although access to basic sanitation and drinking water has advanced significantly worldwide, there is still a great need for work in seeking alternatives for treating urban wastewater and other activities. According to the UN, data from 2018 shows that only 39% of global wastewater is treated. This value represents a significant concern with the percentage of water that is not treated and returns to the environment consisting of contaminants that can lead to severe environmental damage, including risks to public health. Given this, investments need to look for alternatives to control emissions from untreated effluents, guaranteeing well-being and public health in all regions of the world and more drinking water sources for places that need it.

15.3 BIOGAS PRODUCTION TECHNOLOGIES

Although the United Nations organizations have promoted the goals of sustainable development, the production of effluents and the generation of sewage sludge is still a challenge. The recovery of nutrients and the generation of clean energy through anaerobic digestion is a fundamental point of the advancement of society in the coming years. Public health, management techniques, and sustainability are becoming increasingly recurrent. Because of this, the need to produce clean energy, such as biogas, has become fundamental, and methods of methane extraction in effluents are becoming increasingly researched (Bagheri et al., 2023).

Due to the high concentration of carbohydrates and proteins, the production of microalgae has become widely exploited in the energy sector; microalgae are organisms that quickly adapt to different climatic conditions and changes in pH. These characteristics facilitate cultivation methods. Some processes convert microalgal biomass into biofuels, such as thermochemical conversion, fermentation, and anaerobic digestion. During biomass hydrolysis, volatile fatty acids are generated, which increase biomethane production (Thakur et al., 2022). Anaerobic digestion is used to convert microalgal biomass into biogas. During the thermochemical conversion and fermentation, dehydration and energy input occur, in the process of obtaining biogas; the algae already offer all the necessary nutrients without requiring dehydration and energy input. Among the most studied macroalgae in the production of biogas are the green algae Chlorophyta and the brown algae, namely the genera *Laminaria*, *Fucus*, and *Saccharina*.

In the study by Vassalle et al. (2020), an up flow anaerobic sludge blanket reactor and an anaerobic hybrid reactor were used to evaluate the efficiency of sewage treatment and biogas production; the study lasted about one year, and the mixing time of 30% (w/w). With sewage sludge, methane production increased by about 25% in microalgal anaerobic digestion. In the study by Beltran et al. (2015), when mixing 25% (w/w) with residual activated sludge, methane production increased by about 22% compared with sludge-only monodigestion. In the same study, an energy assessment was performed, and it showed a positive energy balance.

Some studies indicate that anaerobic co-digestion of sewage sludge decreases the process efficiency. In the study by Olsson et al. (2014), conducted under thermophilic conditions, when adding 40% (w/w) and using microalgae as co-substrate, there was a decrease of about 26% in methane production when compared with sewage sludge monodigestion; the authors attributed this factor to the high content of nutrients present in the microalgae, which result in the release of ammonia at high temperatures.

Sewage sludge is a residue from the treatment of water and effluents, and this process aims at economic development and the reduction of risks to the environment and health; sewage sludge contains several components that can be used for energy production. Wastewater has a high amount of organic compounds necessary for biogas production. Due to the high cost of water treatment, strategies aimed at the circular economy in production are sought; given this scenario, the production of biogas from the sludge resulting from water treatment plants (ETAs) is a fundamental strategy due to the high amount of energy used in ETAs (Bagheri et al., 2023).

The study by Christensen et al. (2022) used sludge collected from a biological tank of an ETA to produce a hydrolyzate with a liquid cocktail containing volatile fatty acids that can be used in biogas production. The total amount of ammonia increased by about 120% in the hydrolyzate. It was assumed that 70% of the biochemical oxygen demand (BOD) that was produced in the hydrolyzate was used for denitrification, so a mass balance could be used to obtain the amount of methane that could be removed if the hydrolyzate were added to the tank or a biodigester. In this case, about 330 Nm^3 of methane could be produced from the primary sludge hydrolyzate.

The production of biogas is made from the anaerobic decomposition of sewage sludge; the yield of the heat capacity of the biogas varies according to the nutrient composition of the wastewater and other characteristics such as temperature, pH, salinity, and the mode of operation of the reactor employed. The methane yield must be monitored to evaluate the biogas performance; it varies according to the anaerobic membrane bioreactor (AnMBR) system used (Elmoutez et al., 2023). The pilot-scale study by Seco et al. (2018) using an AnMBR obtained 66% methane in biogas composition, while the study by Quek et al. (2017) on a laboratory scale for wastewater treatment reported production of about 66% methane, which could be recovered to produce energy.

Another substrate that can be useful in biogas production is using agricultural waste, such as animal waste. Using agricultural waste, anaerobic treatment is effective, as it is responsible for the degradation of organic matter and is efficient in terms of energy (Coluna, 2016). The incorrect disposal of agricultural waste is a socio-economic problem, as it may lead to environmental and health issues for the population.

The study by Mutungwazi et al. (2023) evaluated various animal wastes for biogas production; the study aimed to discover new strategies for improving the biogas production process under different conditions. The following biomasses were considered for biogas production: cow, horse, chicken, and pig manure. Cow manure had about 56.6% of methane in its biogas composition and 2,146.9 NmL of biogas production. In contrast, horse, chicken, and pig manure had 35.9%, 41.0%, and 2% methane in their biogas composition and 1870.2 NmL, 2122.9 NmL, and 2122.9 NmL in total biogas production. One of the critical points for the production of biogas from animal manure is that the biomass has high levels of nitrogen, which results in ammonia that causes the inhibition of microbial processes in systems using anaerobic treatment.

The ability to generate energy from biogas is dirctly linked to its methane production capacity, so one of the challenges of biogas production using animal waste as substrates is the production of ammonia; due to this, increasing the inhibition of ammonia production and increasing the methane production rate, biogas generation and its calorific value become more efficient. Menzel et al. (2020) identified that manure undergoing anaerobic digestion has inhibition by ammonia, and, due to its amount of fibers, it also suffers from biodegradability. In contrast, the study by Molaey et al. (2019) proposes to potentiate acetate-oxidizing syntrophic (methanogenic) bacteria by using oligo-element supplementation, as these bacteria are resistant to ammonia inhibition.

The production of biogas from waste is still a challenge in today's society due to its high cost; currently, it provides a methodology for using waste from water treatment plants, effluents, industrial and agricultural. Effluents that undergo extraction technologies are becoming increasingly attractive to researchers because they have a high amount of organic matter and nutrients. The implementation of advanced technologies for methane recovery requires scientific development and economic benefits. Biogas production is shown to be promising to boost and encourage the circular economy (Bagheri et al., 2023).

15.4 PRODUCTION OF BIOGAS FROM WASTEWATER TREATMENT

The production of biogas is given through a biochemical fermentation process that consists of four phases: hydrolysis, acidogenesis, acetogenesis, and methanogenesis. In each of the phases, a corresponding class of microorganism is involved. The hydrolysis of organic matter dissolves insoluble organic polymers; in the acidogenesis phase, bacteria produce carbon dioxide, ammonia, and hydrogen. In methanogenesis, bacteria convert the products into carbon dioxide and methane (Ghosh et al., 2020). In acetogenesis bacteria produce acetate either by the reduction of organic acids.

Methanogenesis is a crucial phase for methane production, which can affect pH values at low concentrations. In methane conversion, some characteristics of wastewater influence the process, such as pH and temperature, while the ratio of carbon, nitrogen, and phosphorus (C:N:P) are also an important factor. The essential nutrients in the production of biogas must be in the correct C:N:P proportions; some studies report that the adequate balance is 100:3:1, and in this proportion, the yield of methane gas is higher.

Assuming the need to treat wastewater all over the planet, for reasons of public health and conservation of the environment, there is a strong need to optimize wastewater treatment processes, aimed at reducing operating costs and the waste generated in these processes. The production of biogas from sludge from ETEs is a relatively new alternative, which is gaining more and more attention from researchers looking for techniques to consolidate the circular economy (Tarpani et al., 2023).

These optimizations are necessary for the processes allied with biogas production from ETE effluents, seeking economic and environmental viability. To consolidate this possibility, changing how the industry sees the effluents generated in the treatment processes is essential. Much more than waste to be dumped in landfills or at the bottom of the sea, the effluents generated are a potential source of resources to produce energy, fertilizers, nutrients, and other products with high added value (Bagheri et al., 2023).

Sludge is one of the residues resulting from the wastewater treatment process, where, to remove organic matter, there are insuffIcent aerobic bacteria, leading to the precipitation of organic matter, thus forming sludge (Heberle, 2013). For some time, this residue has been the subject of research that seeks sustainable purposes for it, creating new applications and benefits of its use so that it is of added value and thus making it a bioproduct, following the concept of the circular economy (Bertanza et al., 2016).

According to Bagheri et al. (2023), possible existing sludge management methods involve the disposal of sludge in landfills, dumping it in the sea, soil application, or using it as a raw material for resource extraction and energy recovery by anaerobic digestion. Disposal is a method of increasingly decreasing interest since this material is a source of multiple possibilities for industrial benefits.

The anaerobic digestion process is responsible for the production of biogas, where the biomass is decomposed by bacteria generating gases such as methane and carbon dioxide, elements that are promising sources for energy production. More studies are needed, regarding optimizing these processes, aiming at a more significant production of biomass per unit area, thus favoring the obtaining of these products on a large scale and consequently making the process more economically viable.

Starting from the context of the scarcity of reuse techniques for effluents generated in the treatment of sludge, Tarpani et al. (2023) evaluated several treatment operations to identify the most sustainable one to be used, addressing the possibility of a lifecycle in the treatment, considering environmental, economic and social impacts. Operations that generate marketable products become viable, whereas methods such as incineration generate a lot of handling and maintenance costs and end up being unusable.

Sludge from wastewater is rich in nutrients such as nitrogen and phosphorus, essential for microalga cultivation. Combining microalgal biomass production with sludge residues is a technique that has been evaluated since the 1990s. Although it is already used in several places worldwide, it is still considered a new method to be analyzed regarding its economic and environmental viability (Arashiro et al., 2018).

Arashiro et al. (2022) evaluated the lifecycle assessment (LCA) of using microalgal systems for industrial and urban wastewater treatment plants, using conventional means for valuing bioproducts. These systems were identified as promising to boost a circular economy within the recovery of microalga-based products. The microalgae production capacity of food industry wastewater is better as it contain more nutrient and less contaminatas than the other industrial wastewater.

Implementing biogas production processes using microalgal biomass, combined with wastewater treatment, perfectly follows the synergy of a circular economy. Some studies analyze residues from sludge digestion as a possible source of natural fertilizer production, such as the study by Ezemagu et al. (2021), who composted the digestate with sawdust to obtain fertilizer. This possibility of production adds a lot to the concept of circular economy, favoring the closure of a cycle of nutrients (Arashiro et al., 2022).

Biogas production technology from microalgal biomass still faces several challenges to be consolidated in industrial processes. Among them are the need for greater efficiency in the conversion to nutrients of interest or the removal of contaminants via anaerobic digestion, as well as the investigation of the best strains of microalgae for these purposes since there are hundreds of species of these microorganisms, each with its own metabolic particularities (Luo et al., 2020).

Given this, there is a clear need for more research on this topic, which seeks new management methods and combinations of techniques that make the process increasingly favorable to industrial application on a large scale, guaranteeing the treatment of wastewater and valorization. of the waste generated, fitting it into the concepts of the circular economy (Hong et al., 2019).

15.5 CONCLUSION

Some technologies make the possibility of effluent treatment plants becoming sources of wealth a reality, thus producing energy, fuels, nutrients, and clean water, and contributing to sustainable development and the circular economy. The reuse of effluents is an effective practice to mitigate water scarcity in arid and densely populated areas, with examples being practiced in Namibia, South Africa, and Big Spring, TX, USA. In addition to wastewater reuse, resource recovery from sludge is an important highlight that includes incineration, gasification, wet air oxidation, and pyrolysis techniques.

Thus, the importance of treating and managing wastewater, including domestic, industrial, and agricultural wastewater, which have high levels of pollutants, is highlighted. The lack of proper treatment of these waters can cause environmental problems, with disorganized urbanization and industrialization being identified as the leading causes. Therefore, the circular economy is presented as a solution for managing effluents, seeking to prolong the use of resources in the production chain. Treatment plants are cited as technologies that can help manage this type of waste and produce clean energy, contributing to the UN Sustainable Development Goals related to water and energy.

The need to adopt a circular economy approach, promoting the recycling and reuse of wastewater resources, is indisputable. It is observed that the influence of the origin of the wastewater in its composition and the importance of fully understanding this composition to develop an effective treatment expands the horizons of the use of the effluent generated in its treatment. The generation of effluents and the consequent production of sewage sludge, resulting in the contamination of water reserves due to their inadequate management, is an obstacle in the search for a sustainable future. The guidelines established by the circular economy for these wastes for the generation of clean energy, such as the production of biogas, is an exciting approach for the disposal of these contaminants and an essential step toward achieving the goals defined by the UN for the 2030 Agenda.

The production of biogas by anaerobic decomposition of sewage sludge is an exciting way to generate this sustainable energy source, mainly due to the high load of nutrients in the effluent. In this context, recent research also proposes the treatment of agricultural residues for producing this biofuel, a well-recognized residue associated with environmental and public health problems. Although interesting results have been reported when applying similar anaerobic treatments to sewage sludge, more detailed approaches related to the economic viability aspects of these processes are still needed. It is undeniable that the proposal to produce biogas using wastewater with high pollution power is closely associated with the concept of the circular economy; however, the processing of this type of waste has considerable costs, mainly liabilities necessary for the installation of the essential structure in treatment plants. In any case, the environmental and technological appeal of this theme is evident, where the improvement of the process tends to make the use of this biofuel feasible to simultaneously alleviate the dependence of the energy matrix of different nations on fossil fuels, as well as to propose a way of managing domestic wastewater.

REFERENCES

Achon, C. L., Achon, C. L., Barroso, M. M. and Cordeiro, J.S. (2013) *Residues of Water Treatment Plants and ISO 24512: Challenge of the Brazilian Sanitation*. Fapesp, pp. 115–122. doi: 10.1590/S1413-41522013000200003.

Agarwal, S., Darbar, S. and Saha, S. (2022) *Challenges in Management of Domestic Wastewater for Sustainable Development*. Elsevier, pp. 531–552. doi: 10.1016/B978-0-323-91838-1.00019-1.

Arashiro, L. T. et al. (2018) 'Life cycle assessment of high rate algal ponds for wastewater treatment and resource recovery', *Science of The Total Environment*, 622–623, pp. 1118–1130. doi: 10.1016/j.scitotenv.2017.12.051.

Arashiro, L. T. et al. (2022) 'Life cycle assessment of microalgae systems for wastewater treatment and bioproducts recovery: Natural pigments, biofertilizer and biogas', *Science of The Total Environment*, 847, p. 157615. doi: 10.1016/j.scitotenv.2022.157615.

Bagheri, M. et al. (2023) 'Fifty years of sewage sludge management research: Mapping researchers' motivations and concerns', *Journal of Environmental Management*, 325, p. 116412. doi: 10.1016/j.jenvman.2022.116412.

Beltrán, C. et al. (2016) 'Batch anaerobic co-digestion of waste activated sludge and microalgae (Chlorella sorokiniana) at mesophilic temperature', *Journal of Environmental Science and Health, Part A*, 51(10), pp. 847–850. doi: 10.1080/10934529.2016.1181456.

Baresel, C. et al. (2020) *The Municipal Wastewater Treatment Plant of the Future - A Water Reuse Facility*. IVL Swedish Environmental Research Institute Ltd, pp. 24–51. https://urn.kb.se/resolve?urn=urn%3Anbn%3Ase%3Aivl%3Adiva-43.

Bertanza, G., Baroni, P. and Canato, M. (2016) 'Ranking sewage sludge management strategies by means of decision support systems: A case study', *Resources, Conservation and Recycling*, 110, pp. 1–15. doi: 10.1016/j.resconrec.2016.03.011.

Christensen, M. L. et al. (2022) 'Pilot-scale hydrolysis of primary sludge for production of easily degradable carbon to treat biological wastewater or produce biogas', *Science of The Total Environment*, 846, p. 157532. doi: 10.1016/j.scitotenv.2022.157532.

Coluna, N. M. E. (2016) *Análise do potencial energético dos resíduos provenientes da cadeia agroindustrial da proteína animal no Estado de São Paulo.* Universidade de São Paulo. doi: 10.11606/D.106.2016.tde-07062016-084519.

Deng, S. et al. (2023) 'Biological nutrient recovery from wastewater for circular economy', in *Current Developments in Biotechnology and Bioengineering.* Elsevier, pp. 355–412. doi: 10.1016/B978-0-323-99920-5.00010-X.

Elmoutez, S. et al. (2023) 'Design and operational aspects of anaerobic membrane bioreactor for efficient wastewater treatment and biogas production', *Environmental Challenges*, 10, p. 100671. doi: 10.1016/j.envc.2022.100671.

Ezemagu, I. G. et al. (2021) 'Biofertilizer production via composting of digestate obtained from anaerobic digestion of post biocoagulation sludge blended with saw dust: Physiochemical characterization and kinetic study', *Environmental Challenges*, 5, p. 100288. doi: 10.1016/j.envc.2021.100288.

Farkas, K. et al. (2020) 'Viral indicators for tracking domestic wastewater contamination in the aquatic environment', *Water Research*, 181, p. 115926. doi: 10.1016/j.watres.2020.115926.

François, M. et al. (2023) 'Advancement of biochar-aided with iron chloride for contaminants removal from wastewater and biogas production: A review', *Science of The Total Environment*, 874, p. 162437. doi: 10.1016/j.scitotenv.2023.162437.

Ghosh, P. et al. (2020) 'Biogas production from waste: Technical overview, progress, and challenges', in *Bioreactors.* Elsevier, pp. 89–104. doi: 10.1016/B978-0-12-821264-6.00007-3.

Heberle, A. N. A. (2013) 'Biogas generated from sludge from the treatment of effluents supplemented with residual vegetable oil', *Bioreactor Laboratory, Univates*, pp. 13–57. http://hdl.handle.net/10737/368.

Hong, Y. et al. (2019) 'Mass-balance-model-based evaluation of sewage treatment plant contribution to residual pharmaceuticals in environmental waters', *Chemosphere*, 225, pp. 378–387. doi: 10.1016/j.chemosphere.2019.03.046.

Karla, M.-R. et al. (2022) 'Operational performance of corncobs/sawdust biofilters coupled to microbial fuel cells treating domestic wastewater', *Science of The Total Environment*, 809, p. 151115. doi: 10.1016/j.scitotenv.2021.151115.

Karunanithi, R. et al. (2015) 'Phosphorus recovery and reuse from waste streams', *Advances in Agronomy*, pp. 173–250. doi: 10.1016/bs.agron.2014.12.005.

Laramee, J., Tilmans, S. and Davis, J. (2018) 'Costs and benefits of biogas recovery from communal anaerobic digesters treating domestic wastewater: Evidence from peri-urban Zambia', *Journal of Environmental Management*, 210, pp. 23–35. doi: 10.1016/j.jenvman.2017.12.064.

Luo, Y. et al. (2014) 'A review on the occurrence of micropollutants in the aquatic environment and their fate and removal during wastewater treatment', *Science of The Total Environment*, 473–474, pp. 619–641. doi: 10.1016/j.scitotenv.2013.12.065.

Luo, Y., Le-Clech, P. and Henderson, R. K. (2020) 'Characterisation of microalgae-based monocultures and mixed cultures for biomass production and wastewater treatment', *Algal Research*, 49, p. 101963. doi: 10.1016/j.algal.2020.101963.

Menzel, T., Neubauer, P. and Junne, S. (2020) 'Role of microbial hydrolysis in anaerobic digestion', *Energies*, 13(21), p. 5555. doi: 10.3390/en13215555.

Musa, M. et al. (2018) 'Wastewater treatment and biogas recovery using anaerobic membrane bioreactors (AnMBRs): Strategies and achievements', *Energies*, 11(7), p. 1675. doi: 10.3390/en11071675.

Mutungwazi, A., Awosusi, A. and Matambo, T. S. (2023) 'Comparative functional microbiome profiling of various animal manures during their anaerobic digestion in biogas production processes', *Biomass and Bioenergy*, 170, p. 106728. doi: 10.1016/j.biombioe.2023.106728.

Obaideen, K. et al. (2022) 'The role of wastewater treatment in achieving sustainable development goals (SDGs) and sustainability guideline', *Energy Nexus*, 7, p. 100112. doi: 10.1016/j.nexus.2022.100112.

Olsson, J. et al. (2014) 'Co-digestion of cultivated microalgae and sewage sludge from municipal waste water treatment', *Bioresource Technology*, 171, pp. 203–210. doi: 10.1016/j.biortech.2014.08.069.

Quek, P. J., Yeap, T. S. and Ng, H. Y. (2017) 'Applicability of upflow anaerobic sludge blanket and dynamic membrane-coupled process for the treatment of municipal wastewater', *Applied Microbiology and Biotechnology*, 101(16), pp. 6531–6540. doi: 10.1007/s00253-017-8358-6.

Ratola, N. et al. (2012) 'Occurrence of organic microcontaminants in the wastewater treatment process: A mini review', *Journal of Hazardous Materials*, 239–240, pp. 1–18. doi: 10.1016/j.jhazmat.2012.05.040.

Sanchez, L. et al. (2022) 'Enhanced organic degradation and biogas production of domestic wastewater at psychrophilic temperature through submerged granular anaerobic membrane bioreactor for energy-positive treatment', *Bioresource Technology*, 353, p. 127145. doi: 10.1016/j.biortech.2022.127145.

Seco, A. et al. (2018) 'Exploring the limits of anaerobic biodegradability of urban wastewater by AnMBR technology', *Environmental Science: Water Research & Technology*, 4(11), pp. 1877–1887. doi: 10.1039/C8EW00313K.

Souza, B. de M., Duarte, M. A. C. and Tinôco, J. D. (2021) 'Custos de operação e manutenção de estação de tratamento de esgotos por reator anaeróbio e lodos ativados', *Engenharia Sanitaria e Ambiental*, 26(3), pp. 505–515. doi: 10.1590/s1413-415220190228.

Tarpani, R. R. Z. and Azapagic, A. (2023) 'Life cycle sustainability assessment of advanced treatment techniques for urban wastewater reuse and sewage sludge resource recovery', *Science of The Total Environment*, 869, p. 161771. doi: 10.1016/j.scitotenv.2023.161771.

Tarpani, R. R. Z. and Azapagic, A. (2018) 'Life cycle environmental impacts of advanced wastewater treatment techniques for removal of pharmaceuticals and personal care products (PPCPs)', *Science of The Total Environment*, pp. 258–272. doi: 10.1016/j.jenvman.2018.03.047.

Thakur, N. et al. (2022) 'Efficient utilization and management of seaweed biomass for biogas production', *Materials Today Sustainability*, 18, p. 100120. doi: 10.1016/j.mtsust.2022.100120.

Vassalle, L. et al. (2020) 'Upflow anaerobic sludge blanket in microalgae-based sewage treatment: Co-digestion for improving biogas production', *Bioresource Technology*, 300, p. 122677. doi: 10.1016/j.biortech.2019.122677.

Vaz, S. A. et al. (2023) 'Recent reports on domestic wastewater treatment using microalgae cultivation: Towards a circular economy', *Environmental Technology & Innovation*, 30, p. 103107. doi: 10.1016/j.eti.2023.103107.

Werkneh, A. A. and Gebru, S. B. (2023) 'Development of ecological sanitation approaches for integrated recovery of biogas, nutrients and clean water from domestic wastewater', *Resources, Environment and Sustainability*, 11, p. 100095. doi: 10.1016/j.resenv.2022.100095.

16 Sustainable Valorization of Lignocellulose-based Green Biomass Bio-adsorbents for Toxic Pollutant Removal Using Wastewater Treatments

Shweta Gupta
University School of Chemical Technology,
Guru Gobind Singh Indraprastha University,
Dwarka Sector 16 C, New Delhi, India

Deepak Garg
University School of Chemical Technology,
Guru Gobind Singh Indraprastha University,
Dwarka Sector 16 C, New Delhi, India

Arinjay Kumar
University School of Chemical Technology,
Guru Gobind Singh Indraprastha University,
Dwarka Sector 16 C, New Delhi, India

16.1 INTRODUCTION

Wastewaters containing heavy metals and dyes generated in various industrial activities pose a threat to the environment and a matter of concern for many countries, especially developing countries. When heavy metals, such as arsenic, cadmium, chromium, copper, lead, nickel, and zinc, contained in wastewater are discharged into rivers or onto soils, etc., without proper treatment, such rivers or soil masses get polluted and thus cause damaging impacts on biota and ecology. Many heavy metals and dyes accumulate in animals, are not easily decomposed, and endanger human health through the food chain. Sources of industrial wastewaters containing

 DOI: 10.1201/9781003432869-16

pollutants include organic chemical manufacturing plants, power plants, electronic manufacturing, electroplating, steel manufacture, mining, and quarrying operations.

16.2 HEAVY METALS – THEIR HEALTH EFFECTS AND SOURCES

Heavy metals are generally metals with a density exceeding 5 g/m^3. A large number of elements fall into this category, but the elements listed in Table 16.1 are heavy metals in the context of the environment. This table includes limits on the type and concentration of heavy metals that may be present in the discharged wastewater. The Maximum Contaminant Levels (MCL) standards for those heavy metals established by USEPA (Babel and Kurniawan, 2003) are also summarized in Table 16.1

Metals are naturally present in the Earth's crust and their content in the environment can vary between different regions, resulting in a spatial variation in background concentrations. The distribution of metals in the environment depends on the nature of the metals and the influence of environmental factors (Khlifi and Hamza-Chaffai, 2010). Of the 92 normally occurring elements, approximately 30 metals and metalloids may be fatal to humans, namely B, Li, Al, Ti, V, Cr, Me, Ni, Cu, As, Se, So, Mo, Pd, Ag, Co, Cs, W, Pt, Au, Hg, and Bi. Heavy metals enter the environment through natural and artificial means. These sources include: natural weathering of the Earth's crust, mining, soil erosion, industrial emissions, urban runoff, sewage discharge, fertilizers, pest control or disease control agents used on plants, air pollution effects, and others. While some people are primarily exposed to these contaminants in the workplace, for most people, the primary route of exposure to these toxic elements is through diet (food and water). Heavy metal pollution chains almost always follow a circular order: industry, atmosphere, soil, water, food, and humans. Although the toxicity of any contaminant and its threat to human health is of course a function of concentration, it is well known that long-term exposure to relatively low levels of heavy metals and metalloids can cause adverse effects (Castro-Gonzalez and Méndez-Armenta, 2008).

TABLE 16.1
The MCL Standards for the Most Hazardous Heavy Metals (Babel and Kurniawan, 2003)

Heavy Metal	Toxicities	MCL (mg/L)
Arsenic	Skin manifestations, visceral cancers, vascular disease	0.050
Cadmium	Kidney damage, renal disorder, human carcinogen	0.01
Chromium	Headache, diarrhea, nausea, vomiting, carcinogen	0.05
Copper	Liver damage, Wilson disease, insomnia	0.25
Nickel	Dermatitis, nausea, chronic asthma, coughing, human carcinogen	0.20
Zinc	Depression, lethargy, neurological signs, increased thirst	0.80
Lead	Damage to the fetal brain, diseases of the kidneys, circulatory system, nervous system	0.006
Mercury	Rheumatoid arthritis, diseases of the kidneys, circulatory system, nervous system	0.00003

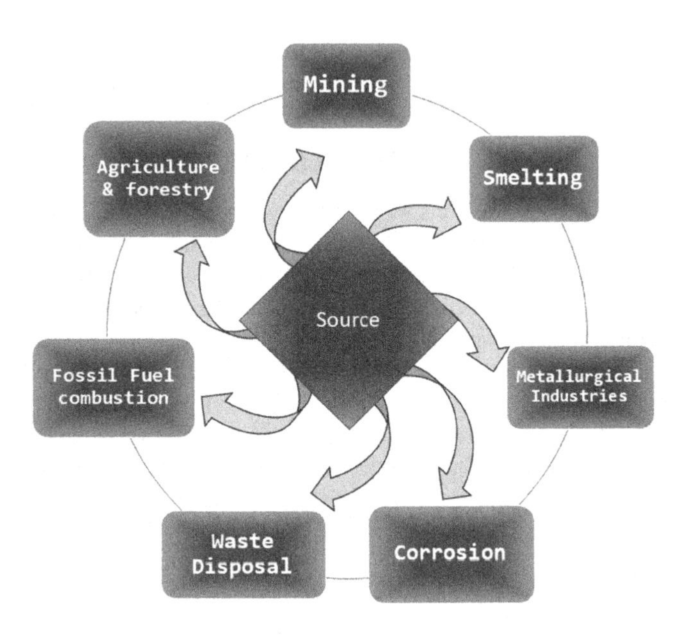

FIGURE 16.1 Sources of heavy metal pollutants.

Industrial wastewater streams containing heavy metals come from different industries, as shown in Figure 16.1. Electroplating and metal surface treatment processes produce large amounts of wastewater containing heavy metals, such as cadmium, zinc, lead, chromium, nickel, copper, vanadium, platinum, silver, and titanium from a variety of applications. These include electroplating, less deposition, conversion coatings, anodizing - cleaning, grinding, and etching processes. Printed circuit board (PCB) manufacturing is another important source of heavy metal scrap. Table 16.2 lists different types of heavy metals as shown in Figure 16.1. All of these

TABLE 16.2
Types of Heavy Metals and Their Sources

Pollutants	Major Sources
Arsenic	Pesticides, fungicides, metal smelters
Cadmium	Welding, electroplating, pesticides, fertilizers, Cd-Ni batteries, nuclear fission plant
Lead	Paint, pesticides, automobile emission, mining, burning of coal
Manganese	Welding, ferromanganese production
Mercury	Pesticides, batteries, paper industry
Zinc	Refineries, brass manufacture, metal plating, plumbing
Chromium	Mines, mineral sources
Copper	Mining, pesticide production, chemical industry, metal piping
Nickel	Cd-Ni battery industries, stainless steel, electric guitar strings, microphone capsules, plating on plumbing fixtures

processes generate large amounts of wastewater, residues, and sludges, which can be classified as hazardous wastes requiring specific waste disposal.

The heavy metals present in wastewater are persistent and non-degradable. Moreover, they are soluble in the aquatic environment and are therefore easily absorbed by living cells. By entering the food chain, the heavy metals can bioaccumulate and bio-magnify to higher levels. If heavy metals are absorbed beyond the permitted level, they can cause serious health problems.

16.3 DYE-CONTAINING EFFLUENTS

Chemical compounds which are used to impart color to surfaces are termed dyes. Dyes may be natural or synthetic. Several industries use dyestuffs, such as manufacturers of carpets, paper, textiles, food, plastics, and cosmetics, who make use of dyes to give color to their products.

Dyes are discarded in industrial wastewater and finally are discharged to the waterbody (Aktar et al., 2008). The presence of dyestuffs in the wastewater affects waterbodies in different ways.

Many dyes or their metabolites have mutagenic, carcinogenic, or teratogenic effects. Dye-containing effluent is responsible for contributing to many health issues in human beings like skin irritation, allergy, and dermatitis. If water polluted with dyestuff is used for irrigation, cultivated land will be covered with the dye which would negatively affect the plant growth. Due to growing industrialization and increasing demand of water for domestic uses, authorities are focusing their attention on this serious problem (Cengiz and Cavas, 2008).

16.4 DYE STRUCTURE AND CLASSIFICATION

There are two basic components of the structure of a dye: chromophore and auxochrome. The chromophore imparts color to the dye, while the auxochrome performs two functions; it enhances the color of the chromophore and, secondly, it makes the dye soluble in water to attach the dye to the surface of the item, as shown in Figure 16.2.

Dyes are classified depending on their chemical structure and solubility (Al-Degs et al., 2000).

On the basis of chemical structure, dyes are classified as follows:

1. Anionic dyes include acid, direct, and reactive dyes.
2. Cationic dyes include basic dyes and nonionic dyes. Disperse dyes are chromophores of nonionic or ionic dyes which are either azo groups or anthraquinones. Due to a reduction in azo linkage, toxic amines are released in the wastewater.
3. Dyes which have anthraquinone in their structure resist degradation due to fused aromatic rings. On the other hand, in cationic dyes, the chromophores are substituted aromatic groups
4. On the basis of solubility, dyes are of two types: soluble dyes include direct, reactive, mordant, acidic, and basic dyes, while insoluble dyes include azoic, sulfur, vat, and disperse dyes.

FIGURE 16.2 Classification of dyes.

16.5 BASIC CONCEPT OF BIOSORPTION AND MECHANISM FOR LIGNOCELLULOSE-BASED BIOSORPTION

Several technologies have been used to treat pollutant-containing aqueous solutions for the past few decades (Wang et al., 2009). Major disdvantages of novel methods, relative to conventional processes, include high consumption of chemicals, high initial and running costs, low flow rates, removal (%) decreases with the presence of other metals (O'Connell et al., 2008).

16.5.1 Biosorption

Biosorption has become a better alternative treatment. It is said to be a relatively low-cost, efficient, and economical method to treat heavy-metal-containing wastewater. The main advantages of the biosorption-based heavy metal/dye removal method are low capital and operating costs, with low control system requirements. Generally, pollutants are present in wastewater at low concentrations, and biosorption is suitable even when metal ions are present at concentrations as low as 1 mg/L. This makes biosorption an economical and advantageous technique for removing heavy metals from wastewater. The bio-adsorbents can be of mineral, organic, or biological origin. They can be zeolites, industrial by-products, agricultural waste, biomass, or polymeric materials. However, the high cost of the base material and the energy-intensive activation process limit its use in wastewater treatment processes. Current research activities aim to promote the search for efficient and cost-effective bio-adsorbents from natural sources with their suitability for recycling and the removal capability for heavy metals from industrial wastewater (Robinson et al. 2002; Garg et al. 2003).

16.5.1.1 Advantages of Biosorption

- Metals at low concentration can be selectively removed.
- Effluent discharge concentration meets the government regulation standards.
- System operates over a broad pH range (2–9).
- System is effective over a temperature range of 4–90°C.
- System offers low capital investment and low operation cost.
- Converts metal pollutant to metal product.
- System offers simple design and easy operation.

The mechanism of biosorption is complex, mainly ion exchange, chelation, physical biosorption, entrainment in bio-adsorbent due to concentration gradients and diffusion through the bioadsorbent cell wall as shown in Figure 16.3

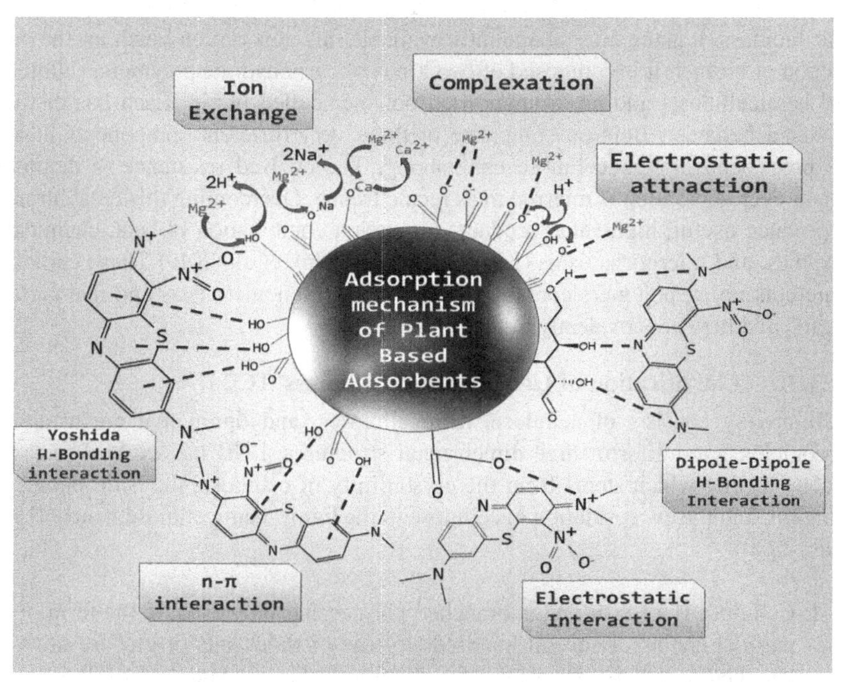

FIGURE 16.3 Plausible mechanisms of biosorption of heavy metals and dyes (adapted from Sud et al., 2008).

16.5.2 Types of Biosorption

There are two types of biosorption phenomena: physical biosorption and chemical biosorption.

16.5.2.1 Physical Biosorption (Van der Waals Biosorption)

Physical biosorption is the result of the intermolecular attraction between the solid bioadsorbent molecules and the adsorbed species. This is a reversible phenomenon. In industrial biosorption operations, this reversibility is used to recover the bioadsorbent for reuse, to recover adsorbate, or to separate the mixture (Megha et al., 2013).

16.5.2.2 Chemisorption

Chemical biosorption is the result of a chemical interaction between a solid bioadsorbent and an adsorbate. Adhesion and thermal dissolution are greater than those found in physical biosorption. This process is often irreversible. Some substances under low-temperature conditions can only be physically adsorbed. However, they exhibit chemisorption at high temperatures, and sometimes these two phenomena may occur simultaneously. Chemical biosorption is especially important in catalysis.

16.5.3 Lignocellulose Biomass and Classification for Use as a Bio-adsorbent

Lignocellulose, derived from plant dry matter (solid biomass), is called lignocellulosic biomass. It is the most abundantly available raw material on Earth for the production of biofuel. It is composed of two kinds of carbohydrate polymers, cellulose, and hemicellulose, and an aromatic-rich polymer called lignin. Each has distinct chemical behavior. Being a composite of three very different components makes the processing of lignocellulose challenging. The evolved resistance to degradation or even separation is referred to as recalcitrance. Overcoming this recalcitrance to produce useful, high-value products requires a combination of heat, chemicals, enzymes, and microorganisms (Liao et al., 2020; Yux et al., 2019) These carbohydrate-containing polymers contain different sugar monomers (six- and five-carbon sugars) and they are covalently bound to lignin.

16.5.3.1 Classification of Lignocellulosic Biomass (LCB)

LCB mainly consists of cellulose, hemicelluloses, and lignin that are arranged in complex non-uniform three-dimensional structures. LCB has evolved to resist deconstruction which stems from the crystallinity of cellulose, the hydrophobicity of lignin, and the encapsulation of cellulose in the lignin-hemicellulose matrix (Liao et al., 2020).

1. Cellulose is a linear and unbranched polysaccharide present in the form of parallel chains of several hundreds to tens of thousands of glucose units attached together through β-1,4-glycosidic linkages (Mueller and Brown, 1980).

(i) Hemicellulose is the second most abundant biopolymer with a random and amorphous structure. It is based on the types of substituent. Hemicelluloses are classified as xyloglucans, xylans, mannans, or galactans.

Lignin acts as a cement to hold cell wall components together, providing integrity to the biomass. In addition, lignin is always considered to be a hindrance in economic biomass processing. During the processing of LCB for conversion to fuels and chemicals, lignin is made freely available for its conversion into value-added products, rather than its fuel value.

The partially hydrolyzed lignin can also be used as a substitute for phenol-formaldehyde resins, polyurethane foams, adhesives, insulation materials, rubber processing, antioxidants, etc. Several approaches, such as thermal, catalytic, and biological, have been studied for many decades to break down the lignin into monomers and oligomers with further upgradation of the resulting monomers to fuels and chemicals (Beckham et al., 2016). The valorization of the lignin has been afforded great importance recently in economic evaluations of various biorefinery concepts (Ragauskas et al., 2014). Current research has identified the use of several multidisciplinary approaches to tackle challenges in lignin valorization.

16.6 MODIFICATION TECHNIQUES FOR LIGNOCELLULOSIC BIOMASS AS A BIO-ADSORBENT

There are three ways to prepare the chemical, biological, and thermochemical processes for the conversion of biomass into adsorbents. However, the thermochemical approach is often preferred because of its faster processing time, higher product, employment of the entire biomass, and energy proficiency (Mosier et al., 2005). Therefore, charcoal (CC), biochar (BC), and activated carbon are pyrogenic carbonaceous materials (PCM) often derived from biomass (Cazacu et al., 2013).

Preparation of hydro-char entails application of hydrothermal carbonization (HTC). HTC involves the use of water for the transformation of biomass at about 100–350 °C to form hydro-char (Taherzadeh et al., 2008). The traditional HTC synthesis of carbon materials is affected under harsh conditions. The improved HTC synthesis has become a more promising route since it has potential to exhibit high carbon efficiency and abundant functional groups on the product surface.

Furthermore, the production of porous carbon material in a one-pot step in an ionic liquid (IL) medium without any support is achievable under the isothermal carbonization method of (Zheng et al., 2014).

In addition, for efficient use and increased sustainability, the bio-adsorbent can be recycled after use so as to avoid waste accumulation and possible pollution

Other modifications pretreatment of lignocellulosic biomass are shown in Figure 16.4.

16.6.1 Role of Bio-adsorbents in Reducing Toxic Pollutants

The separation of certain components from the fluid onto the solid surface of the bio-adsorbent is referred to as biosorption. The transfer of molecules from the bulk

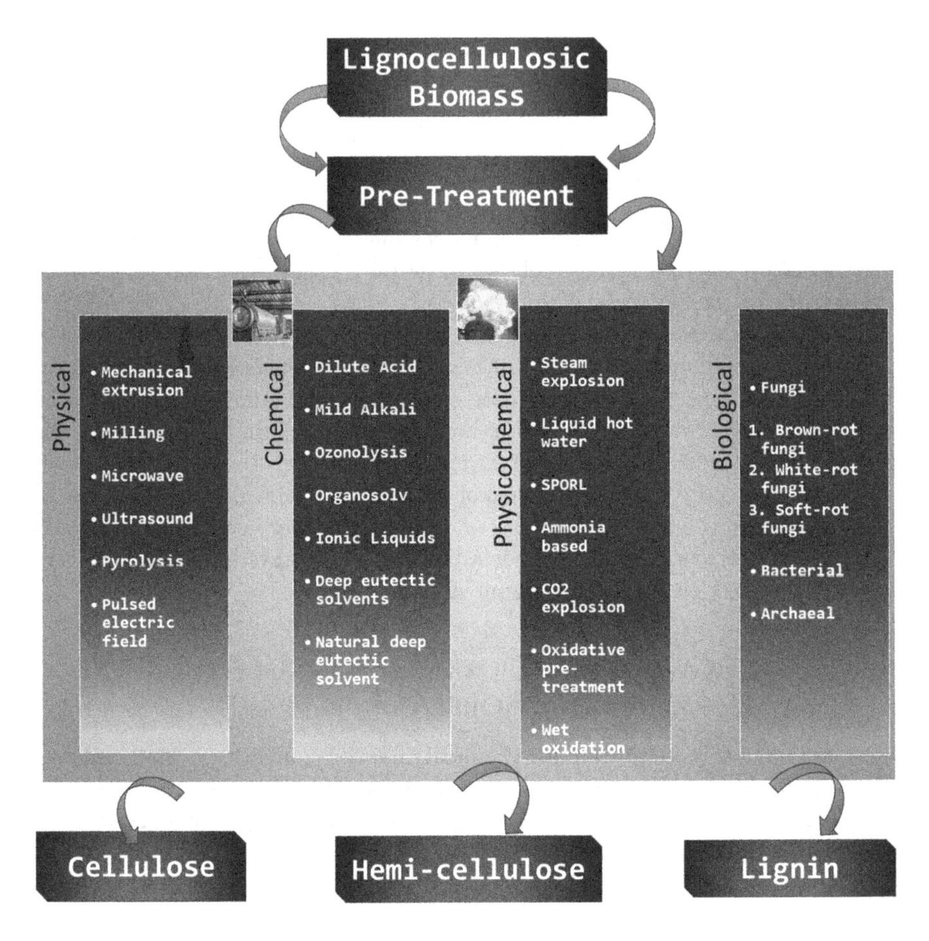

FIGURE 16.4 Pretreatments of lignocellulosic biomass.

solution to the solid surface occurs based on the concentration gradient. Here, when the solid surface is exposed to the fluid phase, molecules from the fluid phase are accumulated or concentrated on the solid surface. All biosorption processes depend on mass transfer rate and solid–liquid equilibrium. If the mass transfer occurs in the opposite direction, the process is called "desorption". A highly porous material is selected as the bio-adsorbent, and biosorption occurs primarily at specific locations on the walls of the pores or within the particles. The difference in shape, molecular weight, or polarity makes the molecule attach strongly to the surface of other materials, which makes separation easier (Blessy et al. 2016).

There are essentially many sources of waste biomass with biosorption properties, such as, rice husks, sawdust, tea and coffee waste, orange peel, peanut shells, dried leaves and bark, etc. (Kishore et al., 2008). The biosorption of heavy metal ions is the result of physicochemical interactions, mainly ion-exchange or complex formation between metal ions and functional groups present on the cell surface (Swati and Barkha, 2016).

16.7 CHARACTERIZATION OF LIGNOCELLULOSE-BASED GREEN BIO-ADSORBENT AND THEIR INTERACTION WITH HEAVY METALS AND DYES

Characterization of the bio-adsorbents is essential for understanding their physical, chemical, and mechanical properties, and provide the explanation of the biosorption mechanism. The bio-adsorbents selected for investigation were characterized by quantitative and qualitative analysis to determine the particle size, specific surface area, pore size and volume, surface morphology, and functional groups; the process included elemental analysis, crystallite analysis, magnetic intensity analysis, magnetic property analysis, etc., and provides brief descriptions of their principles and operation, selection of an appropriate strategy, and advantages based on physicochemical properties.

16.7.1 Infrared Spectroscopy

Infrared spectroscopy (IR) is a technique based on the vibrations of the atoms in a given molecule and is a result of the molecular vibration mechanism, which refers to energy on matter. Usually, an infrared spectrum is obtained by passing infrared radiation through a sample and quantifying the fraction of incident radiation of each wavelength that has been absorbed. The energy at which a peak appears in an absorption (or transmission) spectrum corresponds to the characteristic energy of the vibration in a part of the molecule. The selection rule for IR spectroscopy is that an electric dipole moment in the molecule has to change during the vibration

16.7.2 Scanning Electron Microscopy–Energy Dispersive X-ray Analysis (SEM–EDX)

Scanning electron microscopy (SEM), coupled to energy dispersive X-ray analysis (EDX), is a powerful instrumental combination to assess the mechanisms governing heavy metal removal by lignocellulosic biomass. SEM provides a considerable magnification of the surfaces and allows the gathering of high-quality images, from which the morphology and topography of the materials can be assessed. The additional use of the detection of backscattered electrons (BSE) helps finding target regions where metals may have been selectively accumulated in lignocellulosic biomaterials.

16.7.3 X-ray Photoelectron Spectroscopy

X-ray photoelectron spectroscopy (XPS) analysis, also known as electron spectroscopy, is a quantitative spectroscopic technique which allows analysis of the surface chemistry of materials. This technique provides valuable information about elemental composition, empirical formula, and the electronic state of the element present in a material. This technique was often used in biosorption studies to obtain information about the biosorption mechanism and the oxidation state of sorbed elements on the surface.

16.7.4 TEM (Transmission Electron Microscopy)

The ability of transmission electron microscopy (TEM) to provide information on crystalline structures, as well as density maps that reach subatomic resolution, is of great interest. With regards to its subatomic resolution, this technique can offer clearer evidence on the internal structure of the biomass. However, since TEM is very costly, its ability to probe the internal structure of biomass is not commonly utilized.

16.7.5 X-Ray Absorption–Fine Structure Spectroscopy (XAFS)

X-ray absorption–fine structure spectroscopy (XAFS) is a spectroscopic technique that utilizes X-rays to probe the physical and chemical structure of material at an atomic scale. XAFS is element-specific, in which X-rays are selected to be at and above the binding energy of a particular core electronic level of a certain atomic species.

16.7.6 X-Ray Diffraction (XRD) Spectroscopy

X-ray diffraction (XRD) is a non-destructive technique used to provide detailed information on the crystallographic structure of materials. This method offers several advantages, e.g., it is non-destructive, highly accurate, capable of detecting single crystals, polycrystalline or amorphous materials. Moreover, standards are readily accessible for thousands of material systems. Due to its versatility, XRD has been widely employed to assist in the characterization of biosorbents.

16.8 BEHAVIOUR EVOLUTION OF LIGNOCELLULOSIC BIO- ADSORBENT FACTORS

Biosorption process is mainly influenced by the nature of solution in which the contaminants are dispersed and the nature of bio-adsorbent used.

Some parameters affect biosorption process between the bio-sorbent and the adsorbate which are given below (Vieira and Volesky, 2010).

1. **pH of solution**. pH determines the adsorption in terms of hydrogen ion concentration in the solution. The adsorptive capacity of metal cations increases with increasing pH of the sorption system, but not in a linear relationship. Biosorption does not occur at highly acidic or alkaline conditions because hydrogen ions and hydroxyl ions compete for active sites on adsorbent surfaces, respectively.
2. **Degree of ionization of the adsorbate**. More highly ionized molecules are adsorbed to a smaller degree than neutral molecules. The degree of ionization of a species is affected by the pH.
3. **Pore and particle size.** It is the size of the molecule with respect to the size of the pores that affects the adsorpion behavior. Larger molecules may

be too large to enter small pores. This may reduce biosorption independent of other causes. Smaller particle sizes reduce internal diffusion and mass transfer limitation to the penetration of the adsorbate inside the adsorbent, i.e., equilibrium is more easily achieved, and nearly full adsorption capacity can be attained.

4. **Surface area of the bio-adsorbent**. Large surface area provides greater biosorption capacity. The smaller the particle size and the greater the surface area of the adsorbent, the more biosorption there is.
5. **Temperature**. Temperature is one of the most important parameters of the biosorption process which excites the metals and binding sites of bio-adsorbents for adsorption. Physisorption is quite fast at low temperatures, whereas chemisorption occurs at high temperatures. In general, both types of biosorption decrease with increasing temperature since the processes are exothermic. Generally, adsorption mechanisms are exothermic, that is to say, the equilibrium constant decreases with increasing temperature.
6. **Effect of adsorbent dose**. The percentage removal of heavy metals increases rapidly with increasing dose of the bio-adsorbent. But after a certain concentration of the adsorbent is reached, further increases in biomass concentration have no effect on percentage removal because there are no metal ions left to be adsorbed on the empty active sites after the equilibrium point is reached.
7. **Effect of initial concentration of metal ion**. The initial concentration of a metal ion provides an important driving force to overcome all mass transfer resistance of the metal between the aqueous and solid phases. There is a decrease in resistance for the uptake of the solute from the solution with increased metal ion concentration.
8. **Contact time**. The longer the contact time or retention, the more complete biosorption will be achieved. But after the equilibrium has been achieved, there is no increase in percentage removal of the metal ion.
9. **Solubility of solute**. Substances slightly soluble in water will be more easily removed from water than substances with high solubility.
10. **Effect of agitation rate**. Agitation means proper mixing of adsorbate and bioadsorbent. The adsorption removal efficiency is increased weakly with increasing agitation rate.

16.9 MODELING OF THE BIOSORPTION PROCESS FOR A LIGNOCELLULOSE-BASED BIO-ADSORBENT

16.9.1 Equilibrium Modeling

A precise mathematical description of the equilibrium isotherms is of paramount importance for the effective design of biosorption systems. The most widely used biosorption isotherms found in the literature to describe the amount of solute adsorbed as a function of the equilibrium concentration in solution are summarized in the differential equations in Table 16.3. Of these models, by far the most widely employed

TABLE 16.3
Modeling Differential Equations of Isotherms, Kinetics, and Thermodynamics

Biosorption Modeling Equations	Term Description
Biosorption isotherms **Langmuir** $\frac{1}{q_e} = \frac{1}{q_{max} K_L C_e} + \frac{1}{q_{max}}$	q_e is adsorption capacity (mg/g) *Ce* is the equilibrium concentration in solution (mg/L) K_L are the Langmuir constants, representing the maximum (mg/L) biosorption capacity for the solid phase loading and the energy constant related to the heat of biosorption.
Separation factor $R_L = \frac{1}{1 + K_L C_0}$	C_0 (mg/L) is initial concentration The value of R_L indicates the shape of the isotherm to be linear ($R_L = 1$) The R_L values were in the range of 0.05 to 0.005
Freundlich isotherm $log q_e = log K_f + \frac{1}{n_f} log C_e$	K_f and $1/n_f$ are the Freundlich constants, the characteristics of the system K_f and n_f are the indicator of the biosorption capacity and biosorption, respectively
Temkin isotherm $q_e = \left(\frac{RT}{b_T}\right) ln K_T + \left(\frac{RT}{b_T}\right) ln C_e$	R is common gas constant (0.008314 KJ /mol K) and T is the absolute temperature (K), 1/bT is the Temkin constant related to the heat of biosorption (KJ/mol) which indicates the biosorption potential. The biosorption capacity $(RT/b_T) ln K_T = A$ and $(RT/b_T) = B$, A and B values depend on K_T and b_T
Dubinin–Radushkevich (D–R) isotherm model $ln q_e = ln q_m - {}^2\mu^2$ $\mu = RT ln\left(1 + \frac{1}{C_e}\right)$ $E = \frac{1}{(2^2)^{1/2}}$	β is a constant related to the mean free energy of biosorption (mol ² K/J²) q_m (mg/g) is the theoretical saturation capacity based on the D-R isotherm, ε is the Polanyi potential R (J/mol/K) is the gas constant and T (K) is the absolute temperature. The mean energy of biosorption (Ea) is calculated from the value of β
Brunauer–Emmett–Teller (BET) isotherm $\frac{C_e}{q_e}\left[C_s - C_e\right] = \frac{1}{q_s C_B} + \left[\frac{C_B - 1}{q_s C_B}\right]\frac{C_e}{C_S}$	C_B (L/mg) is the BET adsorption isotherm Cs (mg/L) is the adsorbate monolayer saturation concentration q_s (mg/g) is the theoretical isotherm saturation capacity
Biosorption kinetics **Pseudo-first-order model** $log(q_e - q_t) = log q_e - \frac{K_1 t}{2.303}$	q_e (mq/g) and q_t (mg/g) are the biosorption capacity at equilibrium and time t, respectively, and K_1 is the rate of the pseudo-first-order biosorption
Pseudo-second-order model $\frac{t}{q_t} = \frac{1}{K_2 q_e^2} + \frac{t}{q_e}$ $h = K_2 q_e^2$	q_e and q_t (mg/g) are the biosorption capacities at equilibrium reaction and time t, respectively, and k_2 (g / mg min) is the rate constant for the pseudo-second-order biosorption *h* is the initial biosorption rate and is expressed in mg/g/min

(Continued)

TABLE 16.3 (CONTINUED)
Modeling Differential Equations of Isotherms, Kinetics, and Thermodynamics

Biosorption Modeling Equations	Term Description
Elovich kinetic model $q = \frac{ln(\alpha \acute{E})}{\acute{E}} + \frac{1}{\acute{E}} \ln t$	q (mg/g) is the adsorbate in solid phase at time t (min) a is the initial biosorption rate (mg/g min) for chemisorption ω is related to the extent of surface coverage and activation energy
Intra-particle diffusion model $q_t = K_{ip} t^{1/2} + I$	q (mg/g) is the concentration of adsorbate in solid phase at time t (min) kip (mg.g' min) is the intra-particle diffusion rate constant
Biosorption thermodynamics $\Delta G^0 = -RTlnK_C$ $K_C = \frac{q_e}{C_e}$ $lnK_c = \frac{\Delta S^0}{R} - \frac{\Delta H^0}{RT}$ $\theta = \left(1 - \frac{C_e}{C_i}\right)$ $S^* = (1-\theta)e^{-E_a/RT}$	R is the gas constant (8.314 J/mol K) T is the temperature (K) Kc is the equilibrium constant free energy (ΔG^0) KJ/mol, enthalpy (ΔH^0), KJ/mol, and entropy (ΔS^0), KJ/mol K Θ is the surface coverage fraction q_e and C_e are the equilibrium concentrations of metal ions on the bio-sorbent, mg/L The sticking probability, S*, is a function of the adsorbate/bio-adsorbent system

are the Langmuir and Freundlich isotherms. While the Langmuir isotherm model relies on the adsorption theory and assumes the formation of a sorbate monolayer, the Freundlich isotherm is an empirical model to correlate the concentration of a sorbate on the solid phase with the concentration of the sorbate in the fluid

16.9.2 Kinetic Studies

The effect of contact time on the biosorption of pollutants (heavy metals and dyes) onto lignocellulosic materials was determined by the technique, when a sample of the bio-adsorbent was added with stirring to a volume of the dye solution at an appropriate pH and knowing the initial concentration. The temperature of solutions was kept constant, in general at room temperature. After different contact times, volumes of supernatant were taken for spectrophotometric measurements of the dye and heavy metal contents.

The extent of biosorption was expressed by the fractional attainment of equilibrium:

$$\left(F = q_t / q\right)$$

where q_t and q (mg/g) are the amounts of dye or heavy metal adsorbed at time t and at equilibrium (24 h), respectively.

To analyze the bio-sorption of dyes onto the lignocellulosic sorbents and also to understand the dynamic of the sorption process in terms of the order of the rate:

- Pseudo-first-order model (Lagergreen 1898), applied when the sorption is preceded by diffusion through a boundary.
- Pseudo-second-order kinetic model, which assumes that the biosorption follows a second-order mechanism and the rate-limiting step may be chemical sorption involving valence forces or covalent forces between sorbent and sorbate.
- Intraparticle diffusion kinetic model (Webber and Morris, 1963), which assumes that adsorption is a multi-step process involving transport of sorbate from an aqueous solution to the sorption sites of the sorbent, with the diffusion into pores being the slowest rate-determining process.

16.9.3 Thermodynamic of Biosorption in Bio-adsorbents

The thermodynamic behavior of the sorption of pollutant ions onto green adsorbents from an aqueous solution shows whether the sorption process follows physisorption or chemisorption. In physisorption, a weak Van der Waals attraction is observed between an adsorbate (metal ion) and a surface. However, chemisorption occurs by the formation of chemical bonds between the surfaces of the solid (adsorbent) and the metal ion. In this case, it is difficult to remove heavy metals adsorbed from the adsorbent; therefore, chemisorption is of an irreversible nature. In contrast, the process of physisorption is reversible, i.e., desorption of the adsorbate occurs by increasing the temperature and does not require any activation energy.

16.10 DESORPTION AND REGENERATION OF LIGNOCELLULOSIC BIO-ADSORBENT.

Polluted biomass can either be directly disposed of or incinerated, delivering the ashes to a regular hazardous waste landfill. Incineration reduces waste volume, but enhances the metal content per mass unit and may cause environmental issues due to potential toxic metal leaching. Another alternative is regeneration of the biomass by elution of the loaded metal using desorbing agents. The selection criteria that guide selection of the appropriate desorbing agent are:

1. Small volume of eluent should yield high metal concentration in the resulting solution.
2. The structural integrity of the biomass must not be severely affected.
3. The eluent should be economically and environmentally friendly.

Most of the desorbing agents used for sorbents regeneration and metal recovery are based on strong mineral acids (HCl, HNO, and H_2SO_4), short-chain organic acids (HCOOH, CH_3COOH), bases (NaOH, $NaHCO_3$, Na_2CO_3, KOH, and K_2CO_3), salts (NaCl, KCl, $CaCl_2$, KNO_3), chelating agents (ethylenediaminetetraacetic acid (EDTA), diethylene-di-amine-pentacetic acid (DTPA), nitrilotri-acetic acid (NTA) or buffer solutions (phosphate, bicarbonate).

The main conclusion retrieved from the literature survey is that the recovery of metals from exhausted lignocellulosic materials and other low-cost sorbent materials, and the subsequent regeneration of the sorbent is not nowadays the focus of researchers.

16.11 CONCLUSION AND FUTURE RESEARCH

Lignocellulosic biomass based on green bio-adsorbent has a huge potential as a low cost, renewable source and environmentally friendly alternative to conventional methods for the removal of toxic pollutants from aqueous polluted solution as well as industrial effluent. This biosorption technology is a cost-effective method for the treatment of complex industrial wastewater containing high volumes of low concentrations of toxic pollutants. Many natural bio-sorbents, both from cellulosic-based and microbial origin, with efficient biosorption characteristics, have been identified. A thorough screening and selection of the most effective low-cost sorbents with sufficiently high metal-binding or pollutant-binding capacity, with selectivity for heavy metal ions and pollutants, are prerequisites for full-scale implementation in industrial processes.

Despite further research, efforts toward a full understanding of biosorption mechanisms and the development of more accurate mathematical models might be required for when the technology can be considered mature enough as to face scale-up scenarios to large scale. Industrial stakeholders, policymakers, and regulators have nowadays a challenging and exciting opportunity to take a step forward toward environmental sustainability, considering sorption onto biomaterials on their wastewater treatment schemes.

REFERENCES

Akar, T., Ozcan, A. S., Tunali, S., & Ozcan, A. (2008) Biosorption of a textile dye (Acid Blue 40) by cone biomass of Thuja orientalis: Estimation of equilibrium, thermodynamic and kinetic parameters. *Bioresource Technology*, 99, 3057–3065.

Al-Degs, Y., Khraisheh, M. A. M., Allen, S. J., & Ahmad, M. N. (2000) Effect of carbon surface chemistry on the removal of reactive dyes from textile effluent. *Water Research*, 34, 927–935.

Babel, S., & Kurniawan, T. A. (2003) Low-cost bio-adsorbents for heavy metals uptake from contaminated water: A review. *Journal of Hazardous Materials*, B97, 219–243.

Beckham, G. T., Johnson, C. W., Karp, E. M., Salvachúa, D., & Vardon, D. R. (2016) Opportunities and challenges in biological lignin valorization. *Current Opinion in Biotechnology*, 42(1), 40–53.

Castro-Gonzalez, M. I., & Méndez-Armenta, M. (2008) Heavy metals: Implications associated to fish consumption. *Environmental Toxicology and Pharmacology*, 26, 263–271.

Cazacu, G., Capraru, M., & Popa, V. I. (2013) Advances concerning lignin utilization in new materials. In S. Thomas, P. M. Visakh, & A. P. Mathew (Eds.), *Advances in Natural Polymers: Composites and Nanocomposites* (pp. 255–312). Berlin: Springer.

Cengiz, S., & Cavas, L. (2008) Removal of methylene blue by invasive marine seaweed: Caulerpa racemosa var. cylindracea. *Bioresource Technology*, 99, 2357–2363.

Garg, V. K., Gupta, R., Yadav, A. B., & Kumar, R. (2003) Dye removal from aqueous solution by adsorption on treated sawdust. *Bioresource Technology*, 89, 121–124.

Khlifi, R., & Hamza-Chaffai, A. (2010) Head and neck cancer due to heavy metal exposure via tobacco smoking and professional exposure: A review. *Toxicology and Applied Pharmacology*, 248, 71–88.

Khurma, M. P., Mudliar, S., & Bharati, A. V. (2013) Barks as biosorbent for exclusion of heavy metals: A review. *Journal of Applicable Chemistry*, 2(4), 850–862.

Krishnani, K. K., Meng, X. G., & Boddu, V. M. (2008) Biosorption mechanism of nine different heavy metals onto biomatrix from rice husk. *Journal of Hazardous Material*, 153(3), 1222–1234.

Lagergren, S., & Sensual, B. K. (1898) Zur theorie der sogenannten biosorption geloester stoffe. *Kunglinga Svenska Vetenskapsakademiens. Handlingar*, 24, 1–39.

Liao, J. J., Latif, N. H. A., Trache, D., Brosse, N., & Hussin, M. H. (2020) Current advancement on the isolation characterization and application of lignin. *International Journal of Biological Macromolecules*, 162, 985–1024.

Mathew, B. B., Jaishankar, M., & Biju, V. G. (2016) Role of bio bio-adsorbents in reducing toxic metals. *Journal of Toxicology*, 2016, 1–14.

Mosier, N., Wyman, C., Dale, B., Elander, R., Lee, Y. Y., Holtzapple, M., & Ladisch, M. (2005) Features of promising technologies for pretreatment of lignocellulosic biomass. *Bioresource Technology*, 96, 673–86.

Mueller, S. C., & Brown, Jr RM. (1980) Evidence for an intramembrane component associated with a cellulose microfibril-synthesizing complex in higher plants. *The Journal of Cell Biology*, 84(2), 315–326.

O'Connell, D. W., Birkinshaw, C., & O'Dwyer, T. F. (2008) Heavy metals adsorbents prepared from the modification of cellulose: A review. *Journal of BioresourceTechnology*, 99(15), 6709–6724.

Ragauskas, A. J., Beckham, G. T., Biddy, M. J., Chandra, R., Chen, F., Davis, M. F., Davison, B. H., Dixon, R. A., Gilna, P., Keller, M., & Langan, P. (2014) Lignin valorization: Improving lignin processing in the biorefinery. *Science*, 344(6185), 1246843.

Robinson, T., Chandran, B., & Nigam, P. (2002) Removal of dyes from a synthetic textile dye effluent by biosorption on apple pomace and wheat straw. *Water Research*, 36, 2824–2830.

Sud, D., Mahajan, G., & Kaur, M. (2008) Agricultural waste material as potential bio-adsorbent for sequestering heavy metal ions from aqueous solutions: A review. *Bioresource Technology*, 99, 6017–6027.

Swati, M., & Barkha, G. (2016) Heavy metals: Impact on human health and their biosorption. *Research & Reviews: Journal of Pharmacology and Toxicological Studies*, 4(2), 350–356.

Taherzadeh, M. J., & Karimi, K. (2008) Pretreatment of lignocellulosic wastes to improve ethanol and biogas production: A review. *International Journal of Molecular Sciences*, 9, 1621–1651.

Vieira, R. H., & Volesky, B. (2010) Biosorption: A solution to pollution? *International Microbiology*, 3(1), 17–24.

Vilvanathan, T., & Shanthakumar, S. (2017) Column adsorption studies on Nickel and Cobalt removal from aqueous solution using native and biochar form of tectona grandis. *Environmental Progress & Sustainable Energy*, 36(4), 1030–1038.

Wang, Z., Chen, L., & Huang, X. (2009) Research on advanced materials for Li-ion batteries. *Advanced Materials*, 21, 4593–4607.

Weber, Jr. W. J., & Morris, J. C. (1963) Kinetics of biosorption on carbon from solutions. *Journal of Sanitary Engineering Division, ASCE*, 89, 31–60.
Yu, X., Wei, Z., Lu, Z., Pei, H., & Wang, H. (2019) Activation of lignin by selective oxidation: An emerging strategy for boosting lignin depolymerization to aromatics. *Bioresource Technology*, 291, 121885.
Zheng, Y., Zhao, J., Xu, F., & Li, Y. (2014) Pretreatment of lignocellulosic biomass for enhanced biogas production. *Progress in Energy and Combustion Science*, 42, 35–53.

17 Carbon Nanotube Membranes from Waste Materials for Water Treatment

Trash to Treasure

Meenakshi Dhanawat
Amity Institute of Pharmacy, Amity University Haryana, Gurugram, India

Kashish Wilson
M.M. College of Pharmacy, Maharishi Markandeshwar (Deemed to be University), Mullana, Ambala, Haryana, India

Garima
M.M. College of Pharmacy, Maharishi Markandeshwar (Deemed to be University), Mullana, Ambala, Haryana, India

Sumeet Gupta
M.M. College of Pharmacy, Maharishi Markandeshwar (Deemed to be University), Mullana, Ambala, Haryana, India

Satish Sardana
Amity Institute of Pharmacy, Amity University Haryana, Gurugram, India

17.1 INTRODUCTION

Plastics and other non-biodegradable man-made materials have become incredibly common over the past 50 years. In America and Europe, precisely thirty million tons of plastic garbage are reportedly produced each year, with polyethylene (PE) making up more than 60% of those totals [1, 2]. Due to their mass production and low cost, plastics are frequently discarded, causing a significant problem with solid trash along

DOI: 10.1201/9781003432869-17

with pollution that contaminates both land and waterbodies. Programs employed for recycling have tried to solve this issue in the past, although the economics of this practice are not always sound because recycled plastics are typically considered to be inferior items. Even though the recycling rate in the USA climbed from around 6% to 33% of total municipal solid waste (MSW) during the previous 50 years, an additional rise for recycling is objectionable but also necessary because of the continual decrease in the amount of landfill area availability. However, it has been constrained by the recent global economic downturn's dramatic impact on the recycling market. For instance, over the past year, the cost of used plastic bottles has decreased from US 25 cents per pound to 2 cents per pound [3]. This poses a threat to the infrastructure needed for recycling, the markets for recycling products, and maybe even people's attitude toward the recycling process. Thus, the development of a modern market in recycled plastics, especially one that produces high-end goods, would be advantageous.

Producing high-end goods from recycled waste plastics, which includes combustion-generated carbon nanofibers (CGCF), notably nanofibers and carbon nanotubes, may halt this tendency and revive or even strengthen recycling's motivations. On the contrary, the price of carbonaceous precursors is the main obstacle to the practical implementation of either CNTs or CNFs. Twenty years after the discovery of CNT [4], those facilities with the capacity for extensive manufacturing are still forced to use high-quality, large-volume low-MW hydrocarbons as a source of carbon [5], such as ethane, ethylene, and methane, etc. Although the cost of these chemicals is high and there is a huge requirement across a wide range of industries, waste plastics offer a clear substitute for carbon sources due to their high carbon and hydrogen levels along with their enhanced energy level. The cost of nanostructured carbonaceous materials might be reduced with the application of corresponding cheap raw material for the production of precise CGCFs, and their widespread application in more commercial goods like composites and batteries could be accelerated. Waste polyethylene (PE) was used as a feedstock in this project, while additional waste plastics are currently being investigated.

The management, recycling, and disposal of plastic trash is a significant problem today. These changes place a significant strain on the ecosystem, so it is crucial to dig deeper into the impact of plastic waste and products on human health and the environment.

17.2 PLASTIC WASTE AND ITS COMBUSTION

Because of its fragility, low weight, and cost-effectiveness, plastic is both highly desirable and difficult to dispose of. Due to their fragility and frequent disposal after use, plastics endure in the environment. In India, like elsewhere, plastic is now pervasive. When sufficiently heated, both natural and manmade polymers will break down, or "pyrolyze", creating flammable volatiles [6]. Solid fuels go through a fairly intricate combustion process that includes desiccation, dehumidification, and volatile combustion, along with char combustion [7]. Thus, if the temperature is high enough, this plastic part as well as a mixture with air undergoes an ignition process. Combustible volatiles are often produced from the pyrolysis process of different polymers, in which released volatile materials undergo oxidation phenomena in polymer combustion, which involves the interaction of chemical and physical events in various gas and condensed phases [8].

Typical stages include thermal disintegration of combustible liquid and solid elements, ignition of the gas phase as well as petrol product combustion, heterogeneous ignition in embers, and burning of the residue of carbonaceous fuel. [9]. There are fundamentally four stages to the combustion of thermoplastics: softening, melting, thermal degradation, and combustion [10]. The material's physical characteristics change during the melting process, changing from a solid to a liquid. When the energy required to break the bond is less than the thermal energy, smaller molecular structures emerge instead in the substance's molecular structure decomposing at higher temperatures [11]. These binary phenomena are endothermic reactions. Formations of, among other things, aldehyde groups, carbonyl, hydroxyl, peroxides, or hydroperoxides, occur as well as the origin or termination of the polymer chain as a combined result of heat breakdown and oxidation [12]. As a result, newly created smaller molecules undergo additional oxidation through an exothermic reaction with oxygen, which mostly produces CO_2 and H_2O at its conclusion. The polymer structure, the thermolysis method, and the circumstances for oxidation, as well as the pyrolysis phenomenon, all play a role in smoke generation during diffusion combustion of polymers [13].

Burning plastic trash outside is a major source of air pollution. Dioxins, furans, mercury, and polychlorinated biphenyls are most frequently released into the atmosphere, when the 12% of plastics in the municipal solid trash are burned. Burning of polyvinyl chloride tends to pollute the environment by releasing harmful halogens into the air, which have an adverse effect on climate changes as shown in Figure 17.1. Because of the toxic substances released, the ecosystem, plant material, and the people's health and animals are under threat. Polystyrene is regarded as a main toxin

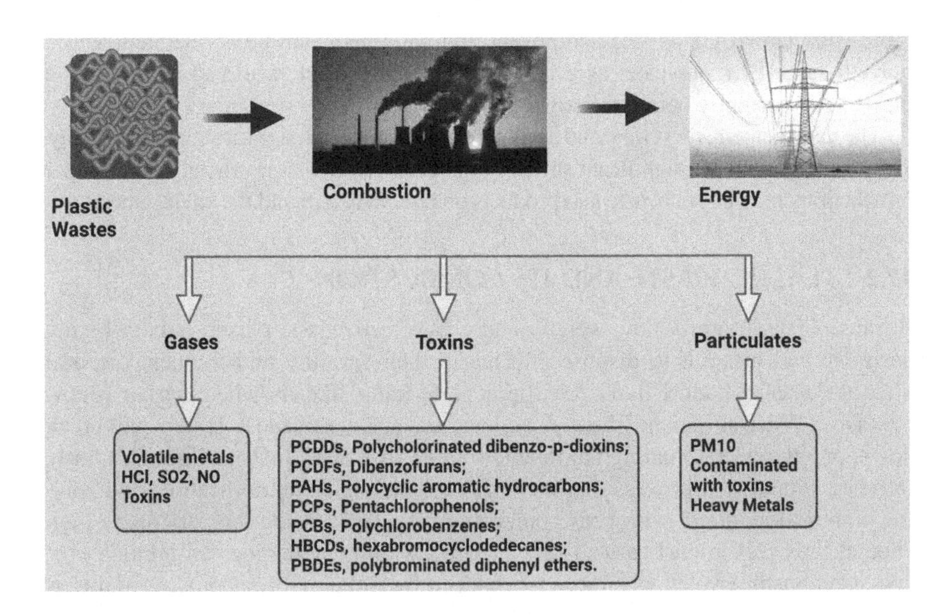

FIGURE 17.1 Major pollutants generated during the incineration of waste plastics.

for the central nervous system. The harmful brominated compounds are carcinogenic and mutagenic. Dioxins build up in crops and inside waterways before entering our food supply and subsequently our bodies. The worst component of these dioxins, 2,3,7,8-tetrachlorodibenzo-*p*-dioxin (TCDD), commonly known as Agent Orange, is a poisonous chemical that damages the neurological, reproductive, thyroid, and respiratory systems in addition to causing cancer, which is regarded as one of the deadliest persistent organic pollutant (POPs).

Consequently, in addition to damaging the neurological system, burning plastic garbage raises heart disease risk, aggravates respiratory disorders, such as asthma and emphysema, and produces irritation, sickness, or migraines. Therefore, environmentalists and scientists must take immediate action to take a sustainable step toward a future in which the environment is cleaner and healthier [14]. There hasn't been much research published with regard to the effects of such harmful gases in India. Fossil fuels come in second with a contribution of about 20% of the GHGs emanating from landfills. Currently, there are too many waste dumps in landfills, and this, combined with the burning of waste and plastic bags, poses serious health risks. The requirement for prompt action to solve this is imperative, using modern disposal techniques. Waste disposal inside landfills indicates unrecoverable energy dissipation and recovery of beneficial raw resources. Incomplete combustion of polyethylene (PE), polypropylene (PP), and polystyrene (PS) under thermal exploitation can result in significant levels of carbon monoxide (CO) and toxic emissions, while PVC burning creates dioxins, carbon black, and aromatics like pyrene and chrysene. Examples of dangerous pollutants include bromide and colored pigments which contain heavy metals such as copper, cobalt, chromium, cadmium, selenium, and lead. Every year, landfill fires and open burning of MSW will release 10,000 g of dioxins as well as furans inside the reduced atmosphere of Mumbai [15]. The waste plastic eventually makes its way into rivers, beaches, rivers, and open spaces. Despite waste management attempts, in excess of 91% of the MSW gathered remains discarded or discharged on open spaces [14, 16].

17.2.1 Strategies of Treatment of Plastic Waste

Four treatment strategies are typically taken into consideration when managing municipal solid waste (MSW), particularly when managing plastic solid waste (PSW), as shown in Figure 17.2 [17, 18]. For clean or semi-clean trash, a closed-loop process or primary recycling is used. This option allows the usage of thermoplastics. Typically, the same industry that produces scrap plastic employs this technique. Secondary recycling, called mechanical recycling, shows its importance because it entails physical processes and enables the use of plastic as a raw material in other plastic goods [18].

The variety in PSW makes it challenging to deal with by both main and secondary techniques. In tertiary recycling, chemical procedures are used to modify the structure of the plastic. Some chemical processes, such as viscosity breaking, pyrolysis/thermal cracking, steam or catalytic cracking, hydrolysis, glycolysis, aminolysis, gasification, and PSW use in the form of a reducing agent for blast furnaces, can be used in this recycling [19].

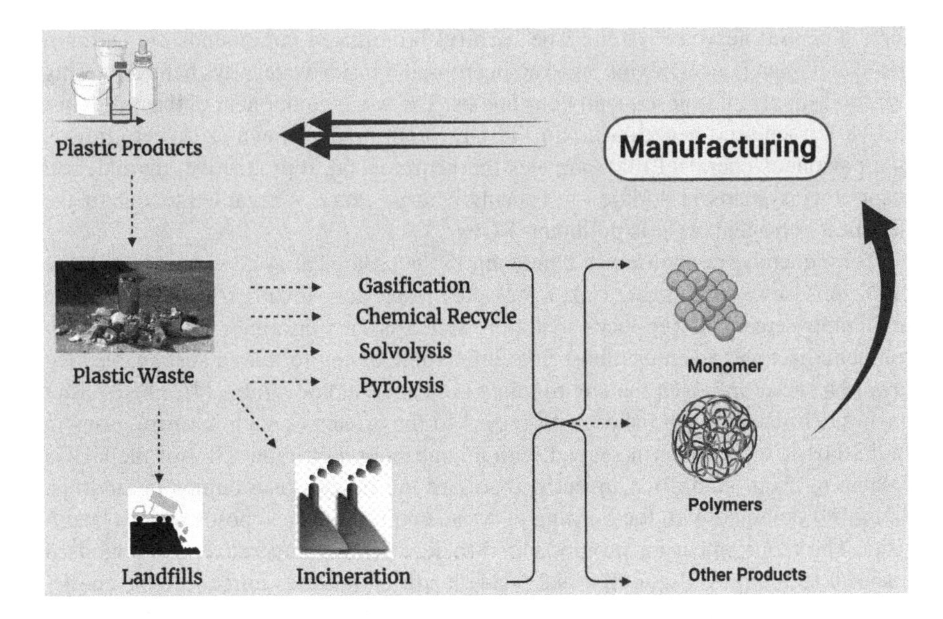

FIGURE 17.2 Approaches for recycling.

Based on the high calorific value of PSWs, quaternary recycling uses incineration for recovering energy through waste steam and heat along with the generation of electricity. Techniques include co-incineration using a single-stage direct combustion process of trash and a fluidized bed, and a two-stage incineration, rotary combustion, and cement kiln incineration [19]. Due to the emissions into the atmosphere, these options are not appealing.

In order to fully utilize waste plastic recycling and address environmental issues associated with waste plastics and hydrocarbons, thermal treatments of PSW in synthetic products prove to be possible solutions. Gasification and pyrolysis have been utilized to transform PSW into gases, liquids, and carbon nanotubes, i.e., developing materials through the deposition of gaseous hydrocarbons onto solid surfaces [20, 21]. This procedure not only reduces waste plastics and upcycles them, but it also offers a green option [22]. Pyrolysis is the most effective MSW treatment procedure when compared with the alternatives. Following European regulations for a more circular economy and producing monomers that may be utilized in a variety of ways is also more forgiving of dirty or mixed plastic garbage than mechanical recycling [23].

17.3 CARBON NANOTUBE MEMBRANE PROPERTIES

In relation to carbon nanotubes, there are numerous molecular forms, or allotropes, of carbon. CNTs, regarded as allotropes of carbon, are made of spherical graphite layers that have been expanded into a duct arrangement [24]. Single-walled carbon nanotubes (SWCNTs or SWNTs) are CNTs made up of just one sheet of graphene. On the other hand, many layers of graphene sheets are referred to as MWCNTs or MWNTs (multi-walled carbon nanotubes). The structure of MWCNTs and SWCNTs

TABLE 17.1
Comparison and Structure of SWCNTs and MWCNTs

Sr.no	Property	SWCNTs	MWCNTs
1	Graphene layer	One	Numerous
2	Bulk synthesis	Hard	Simple
3	Purity	Less	Excessive
4	Specific gravity	0.8	1.8
5	Chance of defect during functionalization	More	Less
6	Characterization and evaluation	Easy	Complex
7	Accumulation in body	Less	More

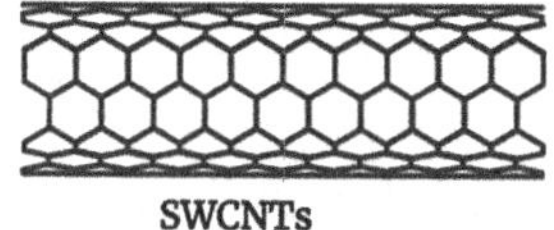

SWCNTs

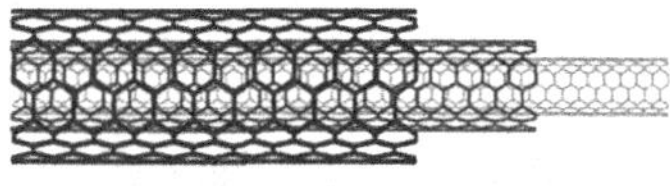

MWCNTs

structure are shown in Table 17.1 [25], which describes representations of different structures of SWCNTs on the basis of graphene wrapping sheets [26]. In Table 17.1, a comparison of SWCNTs and MWCNTs is also shown.

Due to their previously unheard-of mechanical, electrical, and thermal capabilities, carbon nanotubes (CNTs) have attracted awareness from academic institutions as well as industrial hubs since Sumio Iijima's (re)discovery of them in 1991 [27]. All of the chemical bonds in nanotubes are sp^2 bonds, exerting a more powerful nature than sp^3 bonds present in alkanes, and give nanotubes their distinctive robustness [28]. The most common techniques for producing CNTs are laser ablation, arc discharge, and chemical vapor deposition (CVD) [29]. Because of their exceptional mechanical, electrical, and optical characteristics, there are many kinds of research and products made of CNTs. Some of those applications are:

1. Improved electrical conductivity or mechanical qualities of composite materials.
2. Certain coatings that absorb microwaves and electromagnetic fields.
3. Materials present at thermal interfaces.
4. Transporters for ions and electrons, including in actuators, supercapacitors, batteries, fibers, and sensors.
5. Equipment for energy conversion and storage.
6. Emission from radiation sources.
7. Semiconductor gadgets of a nanometer scale.
8. For microscopy and lithography, there are emission displays, electron sources, scanning probe tips, X-ray tubes, gas discharge tubes, and vacuum microwave amplifiers.
9. To modify how they respond electronically, as a transistor or logic component.
10. To interconnect applications.

11. To be used as gas separation and water purification membranes.
12. To incorporate delivery of biomolecules and medication the intended organs, as well as biosensor diagnostics and analysis [30].

CNTs have been widely used to remove a variety of contaminants from aqueous solutions [31, 32]. Recent developments in the synthesis of new membranes required for water purification have drawn a lot of interest in CNTs [33, 34]. CNTs have been proposed for their application in numerous varieties of products, such as sensors, hydrogen storage devices, field emission displays, and high-strength conductive composites [35]. The use of CNTs in producing next-generation membranes with extreme selectivity, shift, and utilizing resistance is also gaining popularity [36]. The concept of size-based restriction and the division of chemical compounds has been influenced by the observation that the inner diameters of CNTs are identical in dimension to those of innumerable tiny molecules [37]. Due to the extremely smooth, defect-free walls of individual CNTs, early research using molecular dynamics models anticipated that gas and water molecules combination would travel through them extremely rapidly. [38]. Certain projections have been confirmed by research, showing that gases and liquids flowed very quickly through membranes consisting of aligned CNTs [39]. Various research was aimed at determining what causes the high permeability of aligned CNT membranes [40, 41]. It was found that the uniformity of interior nanotube walls was regarded as the principal key since it causes low resistance with surrounding molecules of water. One more point was the incredibly small nanotube diameter. Elongated water molecule chains, linked through extremely robust hydrogen bonds, form more readily, allowing the molecules to freely flow through the interior holes.

This provided the catalyst for more investigations into this innovative family of membrane components. Research on the cytotoxic properties of CNT membranes has provided evidence to support this, showing that these materials display longer membrane lifetimes by eradicating bacterial and viral infections and exhibiting lower susceptibility for contamination than most traditional polymeric membranes [42].

17.4 TYPES OF CNT MEMBRANES

As a consequence of the fabrication processes, carbon nanotube membranes undergo grouping into several classifications, but the two main types are shown in Figure 17.3:

1. Free-standing CNT membranes.
2. Mixed (nanocomposite) CNT membranes.

The major two categories for freestanding CNT membranes, bucky-paper membranes and vertically aligned CNT (VA-CNT) membranes, are commonly used in detoxification for water purification [43].

The cylindrical arrangement of the CNTs in VA-CNT membranes prevents fluid from passing outside of the empty central layer of CNT or the spaces among the bundles of CNTs. On the other hand, buckypaper CNT membrane have CNTs arranged

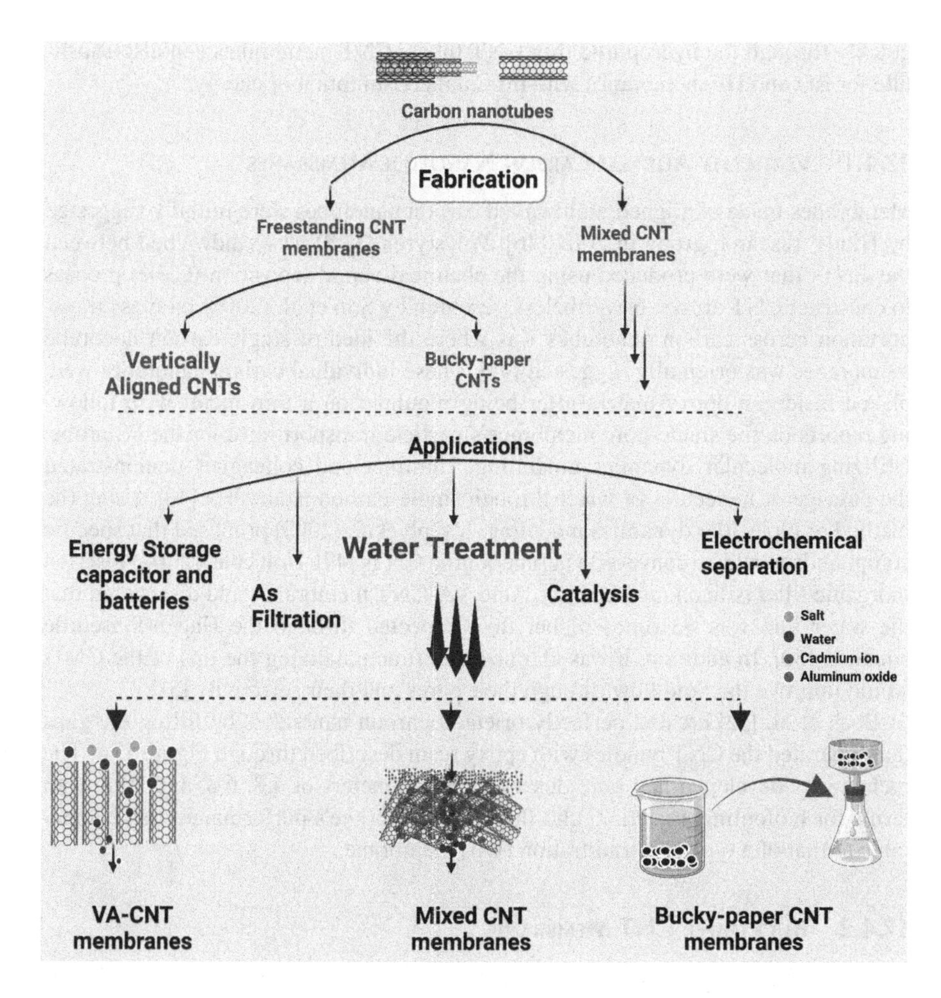

FIGURE 17.3 Categories and applications for CNTs.

at random within a sizable porous 3D network that has enhanced specific surface area [44]. Mixed-CNT membranes possess a top layer made of CNTs and a polymer, like thin-film composite reverse osmosis (RO) membranes [45].

There are benefits and drawbacks to both membrane types. For instance, the VA-CNT membranes benefit from enhanced flux of water, thanks to their dense nanotube junctions, along with low nanochannel expansion. Nevertheless, intricate manufacturing procedures make it challenging to produce these membranes for widespread use. The flux across these membranes, however, is not on a par with what was anticipated in VA-CNT membranes. On the other hand, the blended (nanocomposite) CNT membranes benefit from easy production processes. In the field of water purification therapy, CNT-based membranes have an ability to replace or outperform membrane filtration (MF), nanofiltration (NF), reverse osmosis (RO), ultrafiltration (UF) and front osmosis (FO) membranes. The polar molecules of water can move quickly through the hydrophilic thin CNT tubes. CNT membranes can also substitute for RO and UF membranes with minimal consumption of energy.

17.4.1 Vertically Aligned Carbon Nanotube Membranes

Membranes made of aligned multiwalled carbon nanotubes were initially suggested by Hinds' research group in 2004 [46]. Polystyrene (PS) was sandwiched between the CNTs that were produced using the chemical vapor deposition (CVD) process to construct CNT arrays. Nevertheless, research by Sun et al. (2000) on mass transportation across carbon nanotubes was where the idea of single-carbon nanotube membranes was originally suggested [47]. These individual carbon nanotubes were placed inside an epoxy matrix after being mounted on a thin metal wire following reports on the single-pore membrane's particle transport rates for the PS probe. Utilizing molecular dynamics modelling, Hummer and colleagues demonstrated the passage of molecules of water through single-carbon nanotubes [48]. Using the method of molecular dynamics modeling, Joseph et al. (2003) proposed that specific compounds would be conveyed via functioning CNTs [49]. Holt et al. (2004) inserted inorganic filler (silicon nitride, Si_3N_4) into VA-CNT membranes and discovered that the water flux was >3 times higher than expected through the Hagen-Poiseuille equation [50]. In addition, it was claimed that functionalizing the tips of the CNTs would improve the fluid flow through their pores and their selectivity [51].

Baek et al. [52] created perfectly oriented carbon nanotubes by filling the gaps that separated the CNT bundles with epoxy resin described through Figure 17.4. The membranes developed had pore densities and diameters of 4.8, 6.8, and 10 nm. In terms for biofouling, rejection, and flux, the membrane's performance was comparable to that of a typical ultrafiltration (UF) membrane.

17.4.2 Buckypaper CNT Membranes

Small CNT sheets are arbitrarily assembled and glued together via van der Waals interactions resulting in buckypaper barriers. Van der Waals interactions drive the sturdy CNT accumulation that result in viscous buckypaper networks. This particular sort of CNT membrane benefits from a substantial permeable 3D matrix and a huge specific surface area. Vacuum-aided purification, electrospinning methods, and layer-by-layer (LBL) accumulation are a few techniques that have been documented for making buckypaper CNT membranes.

The purification of CNTs represents one of the more crucial phases in the creation of buckypaper CNT membranes. In general, carbonaceous contaminants are eliminated through oxidative therapy. But these purifying techniques carry a threat of damaging and shortening CNTs. However, during this process, the CNTs are functionalized through carboxyl as well as hydroxyl groups, so that they are hydrophilic. This characteristic might help CNTs disperse better in polar liquids like water [53]. By employing chemically altered CNTs that had undergone UV/ozone therapy and coupling using alkoxysilane base categories, Dumée et al. [54] created buckypaper CNT membranes. These membranes were said to be extremely hydrophobic and to have an improved lifespan [53]. The fundamental steps require purification of initial

CNTs, and then distributing them in a suitable solvent. The CNTs become lodged in a permeable substrate after the dispersion solutions has been circulated across it, resulting in an optically transparent CNT buckypaper.

17.4.3 Mixed (Nanocomposite) CNT Membranes

The primary goal of the mixed (nanocomposite) CNT membranes is to enhance the functionality of current membranes (mostly polymeric) by incorporating CNTs. CNTs have a number of desirable properties, including surface hydrophilic nature, thermal and mechanical rigidity, antimicrobial and anti-fouling characteristics, and improved rejection of salt potential. Van der Waals interactions cause CNTs to cluster, which restricts their use in particular circumstances [55]. However, compared with unmodified CNTs, CNTs that had been subjected to acid showed enhanced dispersal in the matrix of polymers [56]. There are various techniques to introduce CNTs into a polymer matrix. The most popular methods are layer-by-layer, solution casting, interfacial polymerization, and phase inversion [57, 58].

17.5 CARBON NANOTUBE MEMBRANE SYNTHESIS FROM PLASTIC WASTE

CNTs can be made via a variety of processes, including electrolysis, electric discharge, CVD, laser blasting, flame synthesis, high-pressure carbon monoxide, and hydrothermal process, among others. These production techniques demand expensive processing conditions, such as gas feedstock, high vacuum, high temperature, inert atmosphere, and lengthy processing durations, which increase the cost of production. However, over the past two decades, the CVD approach gained popularity for CNT synthesis due to its continuous mass production, ease of use, and low cost [59].

Given that commercial uses of CNTs are yet to reach their maximum capability, there is still a great deal of scope for the invention of CNT fabrication from waste plastics as an additional method for large-scale CNT production [60]. The methods for making CNTs that are currently available, such as the catalytic chemical vapour deposition (CVD), electric arc discharge method, flame synthesis, laser ablation method, and the solar energy route, are resource- and energy-intensive.

Regarding these methods, CVD seems to be currently the most likely and popular method for large-scale manufacturing [61]. Plastic polymers have been suggested as one of the possible solutions for the carbonaceous feed. By simultaneously addressing the issues of plastic waste and CNT manufacture, significant profits may be gained if a workable technology could convert widely available, carbon-rich plastic trash into extremely valuable products like CNTs. based on the different sorts of reactors that are employed in the nanoproduct manufacture.

17.5.1 Quartz Tube

A Russian research team did some of the first experiments to try and make CNTs from plastic polymer sources, and they created twisted carbon nanotubes at 800°C with diameters of 20 to 60 nm [62].

17.5.2 Autoclave

In an autoclave at 700°C, Kong and Zhang were successful in manufacturing CNTs with sizes between 20 and 60 nm at high yields. They have a tendency to create 5% helical CNTs during the process. Generally speaking, carbon nanotubes can take on a variety of shapes in addition to being straight, including bent, helical and constant spirals [63]. Helical CNTs are particularly interesting in the nanoelectromechanical domains because of their verified special mechanical nature, and magnetic and electrical characteristics [64]. Certain researchers hypothesized that the creation of helical CNTs in this case might be achieved through catalyst-driven acetylene pyrolysis, and it is created as a secondary by-product during the breakdown of maleic anhydride-polypropylene (MA-PP) and PE.

A comparable study was published by Zhang et al. [65] only a couple of weeks after the first publication. Throughout the two evaluations, the only noteworthy differences were that, at this point, Ni was utilized in place of ferrocene, and polypropylene was employed as a reagent instead of polyethylene. Compared from the previous experiment, the synthesized CNTs had a median wall depth of 45 nm and an average outer diameter of 160 nm. However, even under operating conditions of 800°C, the yield was still as high as 80%. Carbon spheres were regarded as minimum products created when no catalyst was used. Additionally, it is hypothesized that MA-PP may reduce the rate of distribution of carbon atoms over the surface of a catalyst, which would encourage the formation of graphite sheets along with CNTs. The method of Zhang et al. [66] of pyrolyzing PP, maleated PP, and ferrocene $[Fe(C_5H_5)_2]$ within an autoclave likewise resulted in the production of disordered CNTs in the shape of microspheres with dimensions of 5.5–7.1 nm.

Additionally, Pol and Thiyagarajan demonstrated how MWCNTs could be produced through thermal breakdown of HDPE, LDPE, or PS at a temperature of 700°C inside an autoclave with a cobalt acetate catalyst. Carbon spheres are produced with or without catalysts, in low or high volumes [67].

17.5.3 Crucible

Song et al. employed HZMS-5 or H-Beta zeolites as synergistically supplements in a straightforward reactor crucible to promote the formation of multi-walled carbon nanotubes from polypropylene (PP) via burning in the presence of Ni_2O. It was hypothesized that the combustion of clay would promote the creation of MWCNTs [68].

17.5.4 Muffle Furnace

By incorporating Mo and Mg into the catalyst for major current investigations, Song and Ji provided notable refinement over the prior use of Ni in the synthesis of CNTs from PP. The method of Li et al. was used to produce and burn these catalysts. According to additional research, helical products are produced when molybdenum-containing catalysts are used [69]. In another study, Chen et al. heated a mixture of nickel formate, PP, and organo-montmorillonite (OMMT) *in situ* to produce chestnut-like carbon nanotube spheres [70].

17.5.5 Moving Bed

Another recent work by Liu et al. focuses on creating hydrogen in the synthesis process as an advantageous consequence, in addition to the manufacturing of CNTs from PP. In a previous study, in a highly comparable binary-step process for treating both virgin and waste PP, Liu et al. investigated various Si/Al ratios for the zeolite catalyst [71].

17.5.6 Fluidized Beds

In order to mass generate multiwall carbon nanotubes, Arena and Mastellone put forward a series of studies on the economical pyrolysis of different polyolefins in an air-bubbling fluidized bed. Nitrogen gas was applied for various tests that were used to fluidize the bed. Sand that was high in quartz or alumina made up the bed. During the experiment, hydrocarbon concentrations in the gas phase were identified and measured using online gas analyzers. When the products' resistance to thermal oxidation was examined in one of their earlier works in atmosphere, it was found that at about 690–700°C, the major breakdown stage started to appear [72]. In general, this solid-gas fluidized approach allowed for reliable authority over resident instances, perpetual operation, and elevated temperatures, as well as exchange coefficients.

Waste PET (recyclable plastic bottles) were gathered for MWCNT synthesis. These bottles were initially fragmented, thoroughly cleaned, and dried. These plastic objects were subjected to heating for 15 minutes at 3000°C in a muffle furnace. Nickel metal powder that had been ball milled served as the catalyst for the production of MWCNTs. Two quartz tubes each held 10 g of plastic along with 1 g of nickel catalyst. These two tubes underwent positioning in zones 1 and 2 of the CVD chamber. In the tube furnace, the catalyst was heated to 10,000°C and the plastic to 8,000°C as part of the synthesis process. A regulated circulation of process gas, such as hydrogen H_2 and an inert gas, such as argon (90 L/min.), was used to conduct this heat treatment during an entire hour (Figure 17.1). In order to purify the synthesized MWCNTs, they were removed by filtration from the Whatman filter paper. The CNTs were additionally washed with dilute hydrochloric acid (HCl). to produce a greenish-colored solution, which shows that the nickel catalyst has been completely removed. The CNTs were air dried after being given another rinse with distilled water to get rid of the HCl [73].

17.6 APPLICATION OF CNTS TO WATER PURIFICATION

Membrane technology is already attracting more and more interest as a potential replacement for traditional separation techniques in a number of industries. Membranes are used for many different things such as water and wastewater treatment, food processing, chemical industries, biotechnology, pharmaceuticals, and energy conversion. The growing use of membrane technology is a result of its benefits, which include a more energy-efficient process, lower operating costs, less environmental impact, and simplicity of operation. When it comes to water treatment, reverse osmosis (RO) membranes took the place of thermal-based desalination methods in around 60% of all desalination plants [74].

However, CNT membranes also have a number of drawbacks that restrict their ability to operate. The selectivity and permeability of synthetic membranes made of ordinary materials (polymeric and inorganic) are typically compromised, and they also have a significant fouling propensity. To address these drawbacks, advanced research on membrane alterations has been developed. The majority of the modifications center on the incorporation of nanoparticles, such as zeolites and graphene, along with different nano-metal oxide fragments and CNTs, as fillers in a matrix of membranes [75]. Water treatment by various CNTs is depicted in Figure 17.3.

17.6.1 Applications of Buckypaper CNT Membranes for Water Treatment

Buckypaper CNT membranes were created, and the performance as well as the potential of this membrane is connected with direct contact membrane distillation (DCMD), as evaluated by Dumée et al. The buckypaper CNT membranes prone to be very porous, highly hydrophobic, and possess higher thermal efficiency. A demonstration of buckypaper CNT membranes shows a great deal of potential for use in DCMD desalination. About 99% of the salt was removed at an average water flow of 12 kg/m^2/h and a vapor of water relative pressure differential of 22.7 kPa [76].

Using a polymeric support, a buckypaper CNT membrane was successfully used to desalinate seawater using DCMD. Additionally, buckypapers were employed as a fine filter. A CNT buckypaper film (2 nm thick) positioned on a cellulose acetate (CA) disk, according to Viswanathan et al., was capable of effectively removing particulates with an average diameter of 100–500 nm to a level that was greater than that needed for high-efficiency particulate air (HEPA) filters. CNT membranes made of buckypaper also showed excellent antibacterial capabilities [55].

17.6.2 Applications of Vertically Aligned Carbon Nanotube (VA-CNT) Membranes for Water Treatment

In addition to a greater flow rate, the VA-CNT membranes demonstrated potent prophylactic properties as well as superior salt elimination abilities. A SWNT filter was suggested by Brady-Estevez et al. to remove microbiological pathogens from water [42]. A SWNT filter was shown to have a strong antibacterial activity toward *Escherichia coli* K12. Lee et al. created an ultrafiltration membrane that was millimeters thick and had a very high water permeability [77]. Utilizing the model bacterium, *Pseudomonas aeruginosa* PA01, the antibiofouling potential of the membranes was assessed. The formations of biofilms and bacterial adhesion were found to be inhibited by CNT membranes. Du et al. published a straightforward procedure for producing epoxy composite membranes with elongated, vertically aligned carbon nanotubes (SLVA-CNT). Water, dodecane, and hexane were only a few of the liquids for which the constructed membrane had an exceptionally high flow rate. Iron nanoparticles were drawn to and removed by the membrane following application of a magnetic field, which was then utilized for blocking as well as opening CNT membranes, utilizing currents [78].

Through simulation studies, Chan et al. predicted that the functional zwitterion substitutes present at the CNT terminals would provide a barrier to nearly every ion, permitting significant water flux. The ion rejection frequency is near 100% when twin zwitterions are linked between the two ends of CNTs with a dimension of around 15. According to Li et al., a novel idea for the production of an ultrafiltration membrane that has elevated flux was developed using a pre-aligned MWCNT array and polyethersulfone (PES). A straightforward drop-casting and phase-inversion procedure was used to manufacture the membrane. According to Figure 17.3, PES matrix contained a homogeneous distribution of vertically oriented CNTs [80]. The results showed that the water speed via the generated membranes was more than ten times that of a pure PES membrane and nearly three times that of a straightforward mixed CNT/PES membrane, under the same pressure load.

A brand-new *in-situ* bulk polymerization technique was used by Kim et al. to create VA-CNT/polymer composite films. A styrene monomer was added to a VA-CNT array together with a small quantity of the plasticizing polystyrene-polybutadiene (PS-PB) copolymer. Micro-indentation studies showed that the lengthening at breakage of the CNT/PS hybrid sheet was enhanced with the addition of PS-b-PB copolymer to the matrices. It was discovered that the manufactured CNT/polymer composite had enhanced water as well as gas permeability characteristics in comparison with prior VA-CNT hybrid membranes [81]. A VA-CNT membrane was synthesized using the technique of CVD, according to Park et al. The produced VA-CNT membranes possess elevated permeability for water, although, as a result of interactions among the membrane surface, along with foulants, they rapidly and irreversibly fouled. The VA-CNT membranes' rejection capacity were compared with commercialized polymeric UF membranes. Nanoparticles larger than 10 nm experienced high rejection rates. However, the properties of VA-CNTs and UF membranes properties could be altered by methacrylic acid (MA) surface modification through graft polymerization. The addition of carboxylic groups decreased the membrane contact angle. The BSA (bovine serum albumin) inhibition was increased, and the fouling tendency was decreased in the modified CNT membrane. The carbon nanotube membrane walls tended to prevent adherence of bacteria, along with prevention of biofilm production [82].

17.7 CURRENT CHALLENGES AND NEED FOR FURTHER RESEARCH

These CNT membranes possess a great chance of becoming the water filtration technology of the future. However, the creation of these membranes and their use in the treatment of water are still at the early stages, and many important problems have not yet been adequately resolved. Increasing the effectiveness of production, reducing the price of CNTs, and identifying the possible adverse effects of CNTs are the main challenges that must be overcome. The main barrier to commercial-scale

CNT use is still the large-scale synthesis of CNTs with appropriate permeability as well as dispersion. For the cost-effective manufacture of CNTs on a widespread scale, investigators might examine as well as develop innovative techniques. The price of CNTs, especially SWCNTs, is another barrier that prevents large-scale CNT adoption. Despite the fact that researchers are striving to develop a technology for making CNTs at a low cost in large quantities, their current cost prohibits their widespread application. However, it is projected that, as commercial production rises, the price of CNTs would fall considerably in the coming years. Furthermore, there are important questions that, given the potential risks, ought to be addressed, that CNTs may pose to the environment and human health. The prevailing consensus is that chemically functionalized CNTs are less hazardous than raw CNTs. This might be because raw CNTs contain the metal catalyst. Investigation on any potentially harmful impacts of CNTs on individuals and the natural environment is crucial in order to prescribe any necessary safety precautions. The challenges in CNT development for appropriate membrane orientation in VA-CNTs represent another barrier. To successfully address this problem and improve membrane performance, more research is required. It is also necessary to look at practical ways to functionalize CNT tips without sacrificing their qualities. Additionally, there is a need to standardize the language used to evaluate the effectiveness of differences described in the published material. In addition, the vast majority of CNT-based membranes are built on ceramic or polymeric membranes which raise the possibility of CNT characteristics being impacted or weakened. To progress the existing CNT-based membrane method, more work needs to be put into expanding the production process from laboratory-scale trials to pilot-plant trials and improving the fabrication processes.

17.8 CONCLUSION

Future membranes that purify water should be made with carbon nanotube membranes. Even though there are still many obstacles to overcome, current research trends indicate that numerous advancements in CNT membrane manufacturing methods and applications can be anticipated in the near future More study is needed to address these challenges and to position CNT membranes at the forefront in membrane technology for water purification. Undoubtedly, CNT-based membranes have a bright future in the desalination and water purification industries.

REFERENCES

1. R. Andrews, D. Jacques, A.M. Rao, F. Derbyshire, D. Qian, X. Fan, E.C. Dickey, J. Chen, Continuous production of aligned carbon nanotubes: A step closer to commercial realization, *Chem. Phys. Lett.* 303 (1999) 467–474. https://doi.org/10.1016/S0009-2614(99)00282-1.
2. R.L. Vander Wal, T.M. Ticich, V.E. Curtis, Diffusion flame synthesis of single-walled carbon nanotubes, *Chem. Phys. Lett.* 323 (2000) 217–223. https://doi.org/10.1016/S0009-2614(00)00522-4.
3. C.K. Gonçalves, J.A.S. Tenório, Y.A. Levendis, J.B. Carlson, Emissions from the premixed combustion of gasified polyethylene, *Energy and Fuels.* 22 (2007) 372–381. https://doi.org/10.1021/EF700379C.

4. J.A. Conesa, R. Font, A. Martilla, A.N. Garcia, Pyrolysis of polyethylene in a fluidized bed reactor, *Energy and Fuels.* 8 (1994) 1238–1246. https://doi.org/10.1021/EF00048A012/ASSET/EF00048A012.FP.PNG_V03.
5. H. Richter, W.J. Grieco, J.B. Howard, Formation mechanism of polycyclic aromatic hydrocarbons and fullerenes in premixed benzene flames, *Combust. Flame.* 119 (1999) 1–22. https://doi.org/10.1016/S0010-2180(99)00032-2.
6. D. Price, G. Anthony, P. Carty, Introduction: Polymer combustion, condensed phase pyrolysis and smoke formation, fire retard. *Mater.* (2001) 1–30. https://doi.org/10.1533/9781855737464.1.
7. K. Křůmal, P. Mikuška, J. Horák, F. Hopan, K. Krpec, Comparison of emissions of gaseous and particulate pollutants from the combustion of biomass and coal in modern and old-type boilers used for residential heating in the Czech Republic, Central Europe, *Chemosphere.* 229 (2019) 51–59. https://doi.org/10.1016/J.CHEMOSPHERE.2019.04.137.
8. B. Camino, G. Camino, The chemical kinetics of the polymer combustion allows for inherent fire retardant synergism, *Polym. Degrad. Stab.* 160 (2019) 142–147. https://doi.org/10.1016/J.POLYMDEGRADSTAB.2018.12.018.
9. K. Vershinina, P. Strizhak, V. Dorokhov, D. Romanov, Combustion and emission behavior of different waste fuel blends in a laboratory furnace, *Fuel.* 285 (2021) 119098. https://doi.org/10.1016/J.FUEL.2020.119098.
10. J. Baron, E.M. Bulewicz, S. Kandefer, M. Pilawska, W. Zukowski, A.N. Hayhurst, The combustion of polymer pellets in a bubbling fluidised bed, *Fuel.* 85 (2006) 2494–2508. https://doi.org/10.1016/J.FUEL.2006.05.004.
11. J. Izdebska, Aging and degradation of printed materials, *Print, Polym. Fundam. Appl.* (2015) 353–370. https://doi.org/10.1016/B978-0-323-37468-2.00022-1.
12. V. Kholodovych, W.J. Welsh, Thermal-oxidative stability and degradation of polymers, *Phys. Prop. Polym. Handb.* (2007) 927–938. https://doi.org/10.1007/978-0-387-69002-5_54.
13. S.V. Levchik, M.M. Hirschler, E.D. Weil, *Practical Guide to Smoke and Combustion Products from Burning Polymers: Generation, Assessment and Control* (2011) 240. Rapra Technology Ltd.
14. R. Verma, K.S. Vinoda, M. Papireddy, A.N.S. Gowda, Toxic pollutants from plastic waste: A review, *Procedia Environ. Sci.* 35 (2016) 701–708. https://doi.org/10.1016/J.PROENV.2016.07.069.
15. *Catalogue Indian Emission Inventory Reports* (2022). https://www.teriin.org/sites/default/files/files/Indian-Emission-Inventory-Report.pdf
16. The Solid Waste Management Sector in India. *Ministry of Finance Government of India* (2009). http://www.indiaenvironmentportal.org.in/files/ppp_position_paper_solid_waste_mgmt.pdf
17. S.M. Al-Salem, A. Antelava, A. Constantinou, G. Manos, A. Dutta, A review on thermal and catalytic pyrolysis of plastic solid waste (PSW), *J. Environ. Manage.* 197 (2017) 177–198. https://doi.org/10.1016/J.JENVMAN.2017.03.084.
18. N. Singh, D. Hui, R. Singh, I.P.S. Ahuja, L. Feo, F. Fraternali, Recycling of plastic solid waste: A state of art review and future applications, *Compos. Part B Eng.* 115 (2017) 409–422. https://doi.org/10.1016/J.COMPOSITESB.2016.09.013.
19. S.M. Al-Salem, P. Lettieri, J. Baeyens, Recycling and recovery routes of plastic solid waste (PSW): A review, *Waste Manag.* 29 (2009) 2625–2643. https://doi.org/10.1016/J.WASMAN.2009.06.004.
20. J. Wang, B. Shen, M. Lan, D. Kang, C. Wu, Carbon nanotubes (CNTs) production from catalytic pyrolysis of waste plastics: The influence of catalyst and reaction pressure, *Catal. Today.* 351 (2020) 50–57. https://doi.org/10.1016/J.CATTOD.2019.01.058.

21. C. Zhuo, Y.A. Levendis, Upcycling waste plastics into carbon nanomaterials: A review, *J. Appl. Polym. Sci.* 131 (2014) 39931. https://doi.org/10.1002/APP.39931.
22. S.S. Sharma, V.S. Batra, Production of hydrogen and carbon nanotubes via catalytic thermo-chemical conversion of plastic waste: review, *J. Chem. Technol. Biotechnol.* 95 (2020) 11–19. https://doi.org/10.1002/JCTB.6193.
23. R.X. Yang, S.L. Wu, K.H. Chuang, M.Y. Wey, Co-production of carbon nanotubes and hydrogen from waste plastic gasification in a two-stage fluidized catalytic bed, *Renew. Energy.* 159 (2020) 10–22. https://doi.org/10.1016/J.RENENE.2020.05.141.
24. H. Dai, Carbon nanotubes: Opportunities and challenges, *Surf. Sci.* 500 (2002) 218–241. https://doi.org/10.1016/S0039-6028(01)01558-8.
25. Y.L. Zhao, J.F. Stoddart, Noncovalent functionalization of single-walled carbon nanotubes, *Acc. Chem. Res.* 42 (2009) 1161–1171. https://doi.org/10.1021/AR900056Z/ASSET/IMAGES/MEDIUM/AR-2009-00056Z_0001.GIF.
26. K. Balasubramanian, M. Burghard, Chemically functionalized carbon nanotubes, *Small.* 1 (2005) 180–192. https://doi.org/10.1002/SMLL.200400118.
27. P. Calvert, Nanotube composites: A recipe for strength, *Nature.* 399 (1999) 210–211. https://doi.org/10.1038/20326.
28. X. Wang, Q. Li, J. Xie, Z. Jin, J. Wang, Y. Li, K. Jiang, S. Fan, Fabrication of ultralong and electrically uniform single-walled carbon nanotubes on clean substrates, *Nano Lett.* 9 (2009) 3137–3141. https://doi.org/10.1021/NL901260B/SUPPL_FILE/NL901260B_SI_001.PDF.
29. Y.M. Manawi, A.S. Ihsanullah, T. Al-Ansari, M.A. Atieh, A review of carbon nanomaterials' synthesis via the chemical vapor deposition (CVD) method, *Mater.* 11 (2018) 822. https://doi.org/10.3390/MA11050822.
30. N. Anzar, R. Hasan, M. Tyagi, N. Yadav, J. Narang, Carbon nanotube: A review on synthesis, properties and plethora of applications in the field of biomedical science, *Sensors Int.* 1 (2020) 100003. https://doi.org/10.1016/J.SINTL.2020.100003.
31. F.A. Ihsanullah, B. Al-Khaldi, M.K. Abusharkh, M.A. Atieh, M.S. Nasser, T. Laoui, T.A. Saleh, S. Agarwal, I. Tyagi, V.K. Gupta, Adsorptive removal of cadmium(II)ions from liquid phase using acid modified carbon-based adsorbents, *J. Mol. Liq.* 204 (2015) 255–263. https://doi.org/10.1016/J.MOLLIQ.2015.01.033.
32. F.A. Ihsanullah, B. Al-Khaldi, A.M. Abu-Sharkh, M.I. Abulkibash, T. Qureshi, M.A. Laoui Atieh, Effect of acid modification on adsorption of hexavalent chromium (Cr(VI)) from aqueous solution by activated carbon and carbon nanotubes, *New Pub. Balaban.* 57 (2015) 7232–7244. https://doi.org/10.1080/19443994.2015.1021847.
33. H.A. Asmaly, B. Abussaud, T.A.S. Ihsanullah, V.K. Gupta, M.A. Atieh, Ferric oxide nanoparticles decorated carbon nanotubes and carbon nanofibers: From synthesis to enhanced removal of phenol, *J. Saudi Chem. Soc.* 19 (2015) 511–520. https://doi.org/10.1016/J.JSCS.2015.06.002.
34. H.A. Asmaly, B. Abussaud, T.A.S. Ihsanullah, A.A. Bukhari, T. Laoui, A.M. Shemsi, V.K. Gupta, M.A. Atieh, Evaluation of micro- and nano-carbon-based adsorbents for the removal of phenol from aqueous solutions, *Toxicol. Environ. Chem.* 97 (2015) 1164–1179. https://doi.org/10.1080/02772248.2015.1092543.
35. S. Kim, J.R. Jinschek, H. Chen, D.S. Sholl, E. Marand, Scalable fabrication of carbon nanotube/polymer nanocomposite membranes for high flux gas transport, *Nano Lett.* 7 (2007) 2806–2811. https://doi.org/10.1021/NL071414U/SUPPL_FILE/NL071414USI20070723_122121.PDF.
36. P.W. Barone, S. Baik, D.A. Heller, M.S. Strano, Near-infrared optical sensors based on single-walled carbon nanotubes, *Nat. Mater.* 41 (2004) 86–92. https://doi.org/10.1038/nmat1276.

37. G.M. Spinks, V. Mottaghitalab, M. Bahrami-Samani, P.G. Whitten, G.G. Wallace, Carbon-nanotube-reinforced polyaniline fibers for high-strength artificial muscles, *Adv. Mater.* 18 (2006) 637–640. https://doi.org/10.1002/ADMA.200502366.
38. A.I. Skoulidas, D.M. Ackerman, J.K. Johnson, D.S. Sholl, Rapid transport of gases in carbon nanotubes, *Phys. Rev. Lett.* 89 (2002) 185901. https://doi.org/10.1103/PHYSREVLETT.89.185901/FIGURES/4/MEDIUM.
39. M. Majumder, N. Chopra, R. Andrews, B.J. Hinds, Enhanced flow in carbon nanotubes, *Nat.* 438 (2005) 44. https://doi.org/10.1038/438044a.
40. K. Falk, F. Sedlmeier, L. Joly, R.R. Netz, L. Bocquet, Molecular origin of fast water transport in carbon nanotube membranes: Superlubricity versus curvature dependent friction, *Nano Lett.* 10 (2010) 4067–4073. https://doi.org/10.1021/NL1021046/SUPPL_FILE/NL1021046_SI_001.PDF.
41. M. Whitby, N. Quirke, Fluid flow in carbon nanotubes and nanopipes, *Nat. Nanotechnol.* 22 (2007) 87–94. https://doi.org/10.1038/nnano.2006.175.
42. A.S. Brady-Estévez, S. Kang, M. Elimelech, A single-walled-carbon-nanotube filter for removal of viral and bacterial pathogens, *Small.* 4 (2008) 481–484. https://doi.org/10.1002/SMLL.200700863.
43. R. Das, M.E. Ali, S.B.A. Hamid, S. Ramakrishna, Z.Z. Chowdhury, Carbon nanotube membranes for water purification: A bright future in water desalination, *Desalination.* 336 (2014) 97–109. https://doi.org/10.1016/J.DESAL.2013.12.026.
44. S. Kar, R.C. Bindal, P.K. Tewari, Carbon nanotube membranes for desalination and water purification: Challenges and opportunities, *Nano Today.* 7 (2012) 385–389. https://doi.org/10.1016/J.NANTOD.2012.09.002.
45. C.H. Ahn, Y. Baek, C. Lee, S.O. Kim, S. Kim, S. Lee, S.H. Kim, S.S. Bae, J. Park, J. Yoon, Carbon nanotube-based membranes: Fabrication and application to desalination, *J. Ind. Eng. Chem.* 18 (2012) 1551–1559. https://doi.org/10.1016/J.JIEC.2012.04.005.
46. B.J. Hinds, N. Chopra, T. Rantell, R. Andrews, V. Gavalas, L.G. Bachas, Aligned multiwalled carbon nanotube membranes, *Science (80-).* 303 (2004) 62–65. https://doi.org/10.1126/SCIENCE.1092048/SUPPL_FILE/HINDS.SOM.PDF.
47. L. Sun, R.M. Crooks, Single carbon nanotube membranes: A well-defined model for studying mass transport through nanoporous materials, *J. Am. Chem. Soc.* 122 (2000) 12340–12345. https://doi.org/10.1021/JA002429W/SUPPL_FILE/JA002429W_S.PDF.
48. G. Hummer, J.C. Rasaiah, J.P. Noworyta, Water conduction through the hydrophobic channel of a carbon nanotube, *Nat.* 414 (2001) 188–190. https://doi.org/10.1038/35102535.
49. S. Joseph, R.J. Mashl, E. Jakobsson, N.R. Aluru, Electrolytic transport in modified carbon nanotubes, *Nano Lett.* 3 (2003) 1399–1403. https://doi.org/10.1021/NL0346326/ASSET/IMAGES/MEDIUM/NL0346326N00001.GIF.
50. J.K. Holt, A. Noy, T. Huser, D. Eaglesham, O. Bakajin, Fabrication of a carbon nanotube-embedded silicon nitride membrane for studies of nanometer-scale mass transport, *Nano Lett.* 4 (2004) 2245–2250. https://doi.org/10.1021/NL048876H/ASSET/IMAGES/MEDIUM/NL048876HN00001.GIF.
51. Y. Manawi, V. Kochkodan, M.A. Hussein, M.A. Khaleel, M. Khraisheh, N. Hilal, Can carbon-based nanomaterials revolutionize membrane fabrication for water treatment and desalination?, *Desalination.* 391 (2016) 69–88. https://doi.org/10.1016/J.DESAL.2016.02.015.
52. Y. Baek, C. Kim, D.K. Seo, T. Kim, J.S. Lee, Y.H. Kim, K.H. Ahn, S.S. Bae, S.C. Lee, J. Lim, K. Lee, J. Yoon, High performance and antifouling vertically aligned carbon nanotube membrane for water purification, *J. Memb. Sci.* 460 (2014) 171–177. https://doi.org/10.1016/J.MEMSCI.2014.02.042.

53. K. Sears, L. Dumée, J. Schütz, M. She, C. Huynh, S. Hawkins, M. Duke, S. Gray, Recent developments in carbon nanotube membranes for water purification and gas separation, *Materials (Basel).* 3 (2010) 127. https://doi.org/10.3390/MA3010127.
54. L. Dumée, V. Germain, K. Sears, J. Schütz, N. Finn, M. Duke, S. Cerneaux, D. Cornu, S. Gray, Enhanced durability and hydrophobicity of carbon nanotube bucky paper membranes in membrane distillation, *J. Memb. Sci.* 376 (2011) 241–246. https://doi.org/10.1016/J.MEMSCI.2011.04.024.
55. S. Kang, M. Pinault, L.D. Pfefferle, M. Elimelech, Single-walled carbon nanotubes exhibit strong antimicrobial activity, *Langmuir.* 23 (2007) 8670–8673. https://doi.org/10.1021/LA701067R/SUPPL_FILE/LA701067R-FILE002.PDF.
56. J.H. Choi, J. Jegal, W.N. Kim, Fabrication and characterization of multi-walled carbon nanotubes/polymer blend membranes, *J. Memb. Sci.* 284 (2006) 406–415. https://doi.org/10.1016/J.MEMSCI.2006.08.013.
57. C.C. Yu, H.W. Yu, Y.X. Chu, H.M. Ruan, J.N. Shen, Preparation thin film nanocomposite membrane incorporating PMMA modified MWNT for nanofiltration, *Key Eng. Mater.* 562–565 (2013) 882–886. https://doi.org/10.4028/WWW.SCIENTIFIC.NET/KEM.562-565.882.
58. L. Liu, M. Son, S. Chakraborty, C. Bhattacharjee, H. Choi, Fabrication of ultra-thin polyelectrolyte/carbon nanotube membrane by spray-assisted layer-by-layer technique: Characterization and its anti-protein fouling properties for water treatment, *New Pub. Balaban.* 51 (2013) 6194–6200. https://doi.org/10.1080/19443994.2013.780767.
59. Y. Zhang, G. Ji, D. Ma, C. Chen, Y. Wang, W. Wang, A. Li, Exergy and energy analysis of pyrolysis of plastic wastes in rotary kiln with heat carrier, *Process Saf. Environ. Prot.* 142 (2020) 203–211. https://doi.org/10.1016/J.PSEP.2020.06.021.
60. R.K. Singh, B. Ruj, A.K. Sadhukhan, P. Gupta, Conventional pyrolysis of plastic waste for product recovery and utilization of pyrolytic gases for carbon nanotubes production, *Environ. Sci. Pollut. Res.* 29 (2022) 20007–20016. https://doi.org/10.1007/S11356-020-11204-1/METRICS.
61. R. Miandad, M.A. Barakat, A.S. Aburiazaiza, M. Rehan, I.M.I. Ismail, A.S. Nizami, Effect of plastic waste types on pyrolysis liquid oil, *Int. Biodeterior. Biodegradation.* 119 (2017) 239–252. https://doi.org/10.1016/J.IBIOD.2016.09.017.
62. C. Okolo, R. Rafique, S.S. Iqbal, M.S. Saharudin, F. Inam, Carbon nanotube reinforced high density polyethylene materials for offshore sheathing applications, *Molecules.* 25 (2020). https://doi.org/10.3390/MOLECULES25132960.
63. J.F. Colomer, L. Henrard, E. Flahaut, G. Van Tendeloo, A.A. Lucas, P. Lambin, Rings of double-walled carbon nanotube bundles, *Nano Lett.* 3 (2003) 685–689. https://doi.org/10.1021/NL034159W/ASSET/IMAGES/MEDIUM/NL034159WN00001.GIF.
64. A. Volodin, D. Buntinx, M. Ahlskog, A. Fonseca, J.B. Nagy, C. Van Haesendonck, Coiled carbon nanotubes as self-sensing mechanical resonators, *Nano Lett.* 4 (2004) 1775–1779. https://doi.org/10.1021/NL0491576/ASSET/IMAGES/MEDIUM/NL0491576N00001.GIF.
65. J. Zhang, J. Li, J. Cao, Y. Qian, Synthesis and characterization of larger diameter carbon nanotubes from catalytic pyrolysis of polypropylene, *Mater. Lett.* 62 (2008) 1839–1842. https://doi.org/10.1016/J.MATLET.2007.10.015.
66. J. Zhang, J. Du, Y. Qian, S. Xiong, Synthesis, characterization and properties of carbon nanotubes microspheres from pyrolysis of polypropylene and maleated polypropylene, *Mater. Res. Bull.* 45 (2010) 15–20. https://doi.org/10.1016/J.MATERRESBULL.2009.09.007.
67. V.G. Pol, P. Thiyagarajan, Remediating plastic waste into carbon nanotubes, *J. Environ. Monit.* 12 (2010) 455–459. https://doi.org/10.1039/B914648B.

68. R. Song, B. Li, S. Zhao, L. Li, Transferring polypropylene into carbon nanotubes via combustion of PP/zeolites (H-ZSM-5 or H-beta)/Ni2O3, *J. Appl. Polym. Sci.* 112 (2009) 3423–3428. https://doi.org/10.1002/APP.29754.
69. T. Somanathan, A. Pandurangan, Helical multiwalled carbon nanotubes (h-MWCNTs) synthesized by catalytic chemical vapor deposition, *New Carbon Mater.* 25 (2010) 175–180. https://doi.org/10.1016/S1872-5805(09)60024-X.
70. X. Chen, J. He, C. Yan, H. Tang, Novel in situ fabrication of chestnut-like carbon nanotube spheres from polypropylene and nickel formate, *J. Phys. Chem. B.* 110 (2006) 21684–21689. https://doi.org/10.1021/JP064682.
71. J. Liu, Z. Jiang, H. Yu, T. Tang, Production of hydrogen and carbon nanotubes by catalytic pyrolysis of waste polypropylene in a two-step process, *Polym. Degrad. Stab.* 96 (2011) 1711–1719. https://doi.org/10.1016/j.polymdegradstab.2011.08.008.
72. U. Arena, M.L. Mastellone, G. Camino, E. Boccaleri, An innovative process for mass production of multi-wall carbon nanotubes by means of low-cost pyrolysis of polyolefins, *Polym. Degrad. Stab.* 91 (2006) 763–768. https://doi.org/10.1016/J.POLYMDEGRADSTAB.2005.05.029.
73. A. Pattanshetti, N. Pradeep, V. Chaitra, V. Uma, Synthesis of multi-walled carbon nanotubes (MWCNTs) from plastic waste & analysis of garlic coated gelatin/MWCNTs nanocomposite films as food packaging material, *SN Appl. Sci.* 2 (2020) 1–7. https://doi.org/10.1007/S42452-020-2442-8/TABLES/4.
74. H. Strathmann, A. Grabowski, G. Eigenberger, Ion-exchange membranes in the chemical process industry, *Ind. Eng. Chem. Res.* 52 (2013) 10364–10379. https://doi.org/10.1021/IE4002102.
75. A.K. Wardani, A.N. Hakim, I.G. Wenten, Combined ultrafiltration-electrodeionization technique for production of high purity water, *Water Sci. Technol.* 75 (2017) 2891–2899. https://doi.org/10.2166/WST.2017.173.
76. L.F. Dumée, K. Sears, J. Schütz, N. Finn, C. Huynh, S. Hawkins, M. Duke, S. Gray, Characterization and evaluation of carbon nanotube Bucky-Paper membranes for direct contact membrane distillation, *J. Memb. Sci.* 351 (2010) 36–43. https://doi.org/10.1016/J.MEMSCI.2010.01.025.
77. B. Lee, Y. Baek, M. Lee, D.H. Jeong, H.H. Lee, J. Yoon, Y.H. Kim, A carbon nanotube wall membrane for water treatment, *Nat. Commun.* 61 (2015) 1–7. https://doi.org/10.1038/ncomms8109.
78. F. Du, L. Qu, Z. Xia, L. Feng, L. Dai, Membranes of vertically aligned superlong carbon nanotubes, *Langmuir.* 27 (2011) 8437–8443. https://doi.org/10.1021/LA200995R/ASSET/IMAGES/MEDIUM/LA-2011-00995R_0001.GIF.
80. S. Li, G. Liao, Z. Liu, Y. Pan, Q. Wu, Y. Weng, X. Zhang, Z. Yang, O.K.C. Tsui, Enhanced water flux in vertically aligned carbon nanotube arrays and polyethersulfone composite membranes, *J. Mater. Chem. A.* 2 (2014) 12171–12176. https://doi.org/10.1039/C4TA02119C.
81. S. Kim, F. Fornasiero, H.G. Park, J. B. In, E. Meshot, G. Giraldo, M. Stadermann, M. Fireman, J. Shan, C.P. Grigoropoulos, O. Bakajin, Fabrication of flexible, aligned carbon nanotube/polymer composite membranes by in-situ polymerization, *J. Memb. Sci.* 460 (2014) 91–98. https://doi.org/10.1016/J.MEMSCI.2014.02.016.
82. S.M. Park, J. Jung, S. Lee, Y. Baek, J. Yoon, D.K. Seo, Y.H. Kim, Fouling and rejection behavior of carbon nanotube membranes, *Desalination.* 343 (2014) 180–186. https://doi.org/10.1016/J.DESAL.2013.10.005.

18 Emerging Techniques for Wastewater Treatment using Biomass

Indrani Das
Department of Environmental Science, Asutosh College, 92, Shyama Prasad Mukherjee Rd, Jatin Das Park, Bhowanipore, Kolkata, West Bengal, India

Shramana Roy Barman
Department of Environmental Science, Asutosh College, 92, Shyama Prasad Mukherjee Rd, Jatin Das Park, Bhowanipore, Kolkata, West Bengal, India

18.1 INTRODUCTION

Water is an essential component of the Earth's biosphere, and is a finite resource. Though the total Earth's surface consists of approximately 71% water, only a small portion of this can be accessible for human use. This very little proportion of the total water supports all life systems on our Earth (Barman et al., 2019). Water is present in various forms like solid, liquid, and gaseous forms in different locations such as the ocean, underground, atmosphere and the Earth's surface, etc. Groundwater, like water in soil pores and cracks in rocks, and surface water, such as rivers, lakes, and wetlands, are the main sources of water that have the potential to be used for various purposes, like industrial, agricultural and municipal applications (Khodakarami and Bagheri, 2021).

18.1.1 Water Distribution on Earth's Surface

Water is one of the most essential components of living organisms on Earth. Across the world, water is commonly recognised as an economic and social capital. The overall water content on Earth is approximately 71%, of which only 2.5% can be considered to be pure water. Freshwater is absolutely necessary to sustain life forms as well as keep the whole ecosystem afloat. Freshwater consists of lakes, groundwater, streams, rivers and ponds. Around 1% of freshwater is accessible for human use, with the remainder being inaccessible in the form of glaciers, snow cover, or inaccessible groundwater sources. Due to rapid industrialization and population growth, these limited water resources are being depleted constantly. Climate change and weather variability

DOI: 10.1201/9781003432869-18

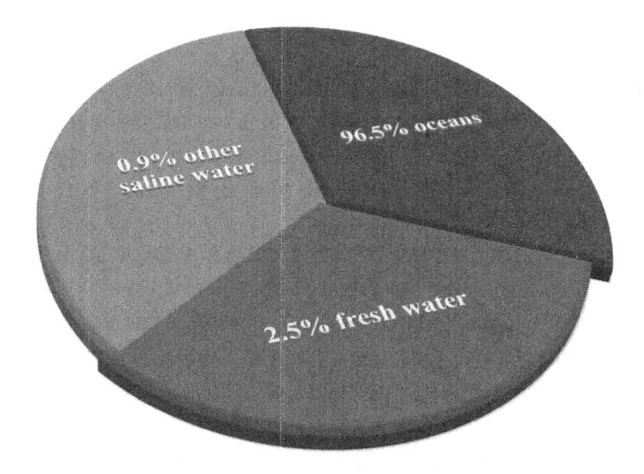

FIGURE 18.1 A pie chart showing the water distribution globally.

hamper the water cycle as well as water usage for energy generation purposes, etc., leading to freshwater depletion and scarcity (Karmakar and Barman, 2023).

18.1.2 Water as a Resource

Freshwater is necessary to maintain human health and to ensure prosperity and security (Hussam, 2013). Approximately over 70% of the Earth is covered with water but only a small proportion of it is viable for human consumption. Furthermore, it plays a vital role in economic activities such as transportation, industrial production, agriculture, fishing, trade, and food processing, etc. Water scarcity is due to the spatial distribution of ever-increasing population growth in comparison with the availability of water resources. In the past few decades, rapid population growth together with increased economic activities like industrialization, urbanization, and agricultural production, etc. has led to water scarcity by polluting existing water (Oki and Quiocho, 2020). The emergence of water contaminants has rapidly increased not only water quality degradation but also harmful health impacts on living organisms and persistence in the ecosystem (Hussam, 2013).

18.1.3 Water Pollution

Drinking water is essential for our survival. More than 1 billion people have limited access to potable water in the world. In addition, approximately 2.6 billion people have poor sanitation accessibility, resulting in degraded water quality. Over 1 million people die each year due to sickness through the consumption of unsafe water. Drinking water gets polluted because of careless disposal of various kinds of chemicals like fertilisers, pesticides, and pharmaceuticals. Thus, freshwater bodies get contaminated with plastics, heavy metals, organic and inorganic contaminants, as well as pesticides, radionuclides, and pharmaceuticals. Contaminated drinking water is a main source of various diseases in human beings and other living organisms. To overcome this serious issue it is necessary to purify polluted water before consumption (Ahuja, 2021).

18.1.4 Water Crisis

Various international organizations reveal in statistical reports that 70% of global freshwater usage is related to food production and crop irrigation purposes which will eventually increase by 19% due to global population growth over the next 40 years. In addition, various contaminants eventually find their way to contaminating underground freshwater sources, thus reducing the quality of water and making it unsafe for drinking purposes. Approximately 70% of industrial waste is discarded into water streams without any treatment in developing countries, whereas 90% of contaminated water finds its way into lakes, rivers, and coastal zones. Thus water quality degrades, which hampers food security, and makes the water unsafe for drinking and bathing. Ever-increasing water demand together with reduced sources of clean freshwater is expected to lead toward a global water crisis in the coming decades. If no action is taken to control water pollution, it is expected that two-thirds of the world will face water stress by 2025. Freshwater sources fail to meet the increasing water demand due to their poor quality resulting from contamination (Wong et al., 2018).

18.1.5 Water Quality and Sustainable Development

The UN Sustainable Development Goals (SDGs) include Target 6.3: By 2030, improve water quality by:

- Reducing pollution.
- Eliminating dumping and minimizing the release of hazardous chemicals and materials.
- Halving the proportion of untreated wastewater.

Substantially increasing recycling and safe reuse globally (Alcamo, 2019).

18.2 MAJOR SOURCES OF WATER POLLUTION

Domestic discharge, livestock and aquaculture effluent, industrial wastewaters, landfill leachates, hospital effluents, and runoff from agriculture act as major sources of emerging contaminants. Specifically, the main contributor to emerging contaminants is the effluent discharge from treatment plants of municipal wastewater. Various emerging contaminants include pesticides, drugs, cosmetics, pharmaceuticals, personal care products, surfactants, food additives, food packaging, cleaning products, industrial formulations and chemicals, metalloids, nanomaterials, rare earth elements, pathogens, and microplastics (Morin-Crini et al., 2022).

18.2.1 Organic Pollutants

With the fast-growing chemical industry, different types of organic contaminants, like phenol, manufactured drugs, pesticides from agricultural residue, dyes, and aromatic hydrocarbons get released into water bodies which negatively impacts aquatic animals as well as human beings and also degrade the water quality (Dehgani et al., 2020).

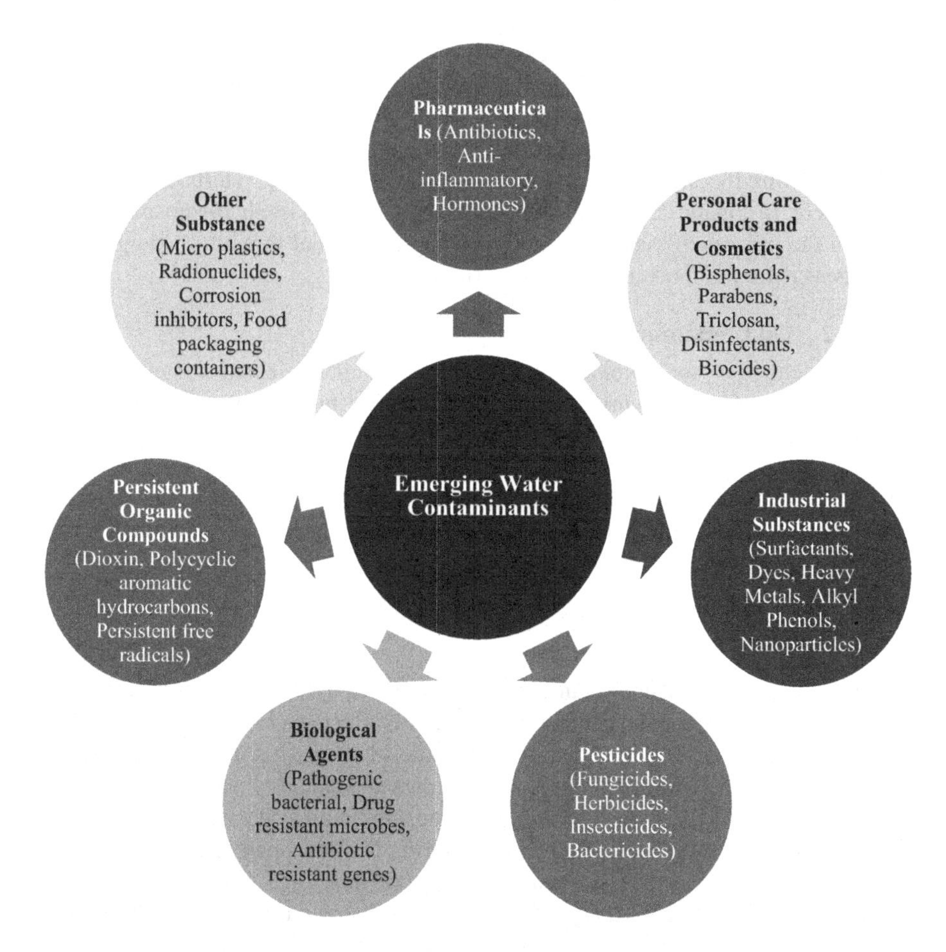

FIGURE 18.2 The circular figure shows the emerging water contaminants.

Organic pollutants like polyaromatic hydrocarbons are lipophilic in nature and are less soluble in water. Phenol can be found in various wastewater including refineries, polymer resin, the coal industry, the printing industry, pesticides, pharmaceuticals, and the petrochemical industry stream. Organic pollutants exhibit toxicity in low concentrations, making them the most dangerous chemical class. Phenol causes convulsions, and damage to the liver, kidney, vascular system, and lungs, also causing coma and cyanosis when it enters our body through various routes (Soffian et al., 2022).

18.2.1.1 Persistent Organic Pollutants

Persistent organic pollutants can be found in water, soil, atmosphere, and every environmental compartment. Their lipophilic nature helps them to accumulate in the tissues of organisms; along the food chain, they become biomagnified and pose serious threats to human beings and other living organisms. They also show resistance to environmental degradation. Cancer, immune and reproductive dysfunctions, nervous system-related issues, birth defects, and other health impacts are related to exposure to persistent organic pollutants (PoPs). The major sources of PoPs are the industrial sector, agriculture, etc., and POPs can be transported long distances by atmospheric and oceanic processes. To eliminate PoPs, many international treaties have been set up, like the Stockholm Convention. The major sinks of PoPs are lake ecosystems, including the organisms, surrounding soils, sediments, waterbodies, and the atmosphere. In aquatic ecosystems, PoPs can accumulate in living organisms' tissues through the process of bioaccumulation, posing a potential threat to humans if the organisms enter the food chain (Morin-Crini et al., 2022).

18.2.1.2 Oil Contaminants

Another dangerous source of water pollution is oil-contaminated wastewater. The major sources of oil contamination are oil refineries, metal processing, the petrochemical industry, crude oil production, vehicle washing, cooling agents, compressor condensates, and lubricants. Phenols, polyaromatic hydrocarbons and petroleum products containing oily wastes which can be categorized as high-risk wastewater due to their toxicity, and carcinogenic, and mutagenic natures (Al-Anzi and Siang, 2017).

18.2.2 Dyes

The most deleterious contaminants in wastewater are dyes because of their immunogenic, carcinogenic, mutagenic and teratogenic effects (Azari et al., 2019). They are organic compounds with a complex aromatic structure that imparts bright color to other materials. However, this structure provides stability and makes it non-biodegradable. The excessive use of dyes creates pollution as it causes the discharge of colored wastewater. Dyes are used for coloring textiles, fibers, paper, and wools; the use of natural dyes dates from as long as 5,000 years ago. From the study conducted by Soffian et al. (2022) it can be concluded that more than 10,000 dyes of various chemical structures are commercially available. They can be classified as (a) anionic, (b) cationic or (c) nonionic. Synthetic and natural dyes are also found, the latter being largely replaced by the former due to advantages like low cost and vast color range. Their cheap cost of production, high resistance to biodegradation, and brightness enables them to be widely used in a broad range of sectors like the paper, plastic, leather, cosmetics, printing, textiles, and cosmetics industries (Soffian et al., 2022). Dyes are discharged from various sources like food processing, textile and ink production factories, and can cause serious threats to the ecosystem due to their toxicity (Bello et al., 2015). Dyes cause harmful health-related risks to human beings as well as to aquatic organisms. Moreover, organic cationic dyes like methylene blue (MB) are more toxic than anionic dyes.

18.2.3 Pharmaceuticals (Pc)

Production of pharmaceuticals adds physiologically active contaminants in significant amounts if effluent is released without any treatment into the waterbodies. These pollutants can not be removed by usual treatment procedures. Paracetamol, ibuprofen, and ketoprofen, etc. are non-steroid medicines which are used for anti-inflammatory purposes which are frequently found in water sources (Soffian et al., 2022). Pharmaceuticals show harmful effects on ecosystems as well as on human beings in low concentrations too (ng/L to μg/L). Pharmaceuticals are mainly used to improve the life quality of human beings and other animals but when they (or their metabolites) are released untreated into waterbodies, they can severely damage the whole ecosystem in numerous ways as they can harm human beings, slow down the cell renewal process, and affect lipid metabolism-related issues. In fish and other aquatic species, various biochemical, enzymatic, and hematological changes have been observed. In humans, allergies, fetal developmental issues, reproductive malfunction, carcinogenicity, bacterial resistance, genotoxicity, etc. have been observed (Michelon et al., 2022).

18.2.4 Heavy Metals

Heavy metals are one of the most common industrial pollutants, consisting of inorganic elements which exhibit high toxicity at low concentrations. In industrial processes, various harmful heavy metals are released into water bodies without necessary treatment. They include Cu(II), Cd(II), As(III), Pb(II), Zn(II), Cr(VI), and Ni(II). The heavy metals may cause ecological disturbance by imparting adverse effects on human beings and aquatic organisms when they are disposed of in wastewater. In human beings, they show carcinogenic effects, respiratory issues, and skin conditions. Furthermore, they severely affect aquatic animals, which can lead to huge economic loss on commercially important species, and also creates challenges to recycle heavy metals from wastewater (Soffian et al., 2022).

18.2.5 Microplastics

Another pollutant class, namely microplastics (MPs), are also found in industrial wastewater and may turn out to be more harmful than pharmaceuticals. A vast amount of plastic particles (for example, plastic beads in regular chemical products, fibers within laundry wastewater, car tire wear debris, and other fragmented plastic waste) enter into the effluent system and then are released into waterbodies and consequently into the sea. According to their source, MPs are classified into primary and secondary microplastics. The primary source consists of the direct introduction of microbeads that come from cosmetics and pellets that are pre-manufactured as well as in used-product erosion of particles and fibers, while secondary MPs are released into the environment by the breakdown of larger plastic items, which consist of manufactured products, vehicle tires, household items, color flakes, and plastic fabrication (Soffian et al., 2022). Microplastics refer to plastics having a size less than 5 mm, which are used utilized in various industrial sectors, such as the textiles, pharmaceuticals, and electronics industries. Anthropogenic activities or the weathering processes of nature have resulted in the breakdown of plastic into smaller-sized units (microplastics) (Parsai et al., 2022).

18.3 EXISTING CONVENTIONAL TECHNIQUES FOR WASTEWATER TREATMENT

The existing conventional wastewater treatment process consists of several techniques including chemical, biological and physical methods to eliminate organic matter and nutrients as well as suspended and dissolved solids from wastewater. Mainly, there are various steps to be followed in wastewater treatment to get desired results by increasing the degree of treatment level. The levels are pretreatment, primary treatment, secondary treatment, and tertiary or advanced treatment (Ranjit et al., 2021).

There are many conventional techniques of wastewater treatment available, namely physical methods (screening, sedimentation, and filtration), chemical methods (coagulation and flocculation, photocatalysis, ion exchange, and chemical oxidation) and biological methods (aerobic and anaerobic treatments, phytoremediation, and bioremediation).

18.3.1 Physical Methods

In the wastewater treatment process, physical methods include screening, filtration, sedimentation, and aeration. For oil- and grease-contaminated wastewater, sand filters are used for oil removal purposes (Saravanan et al., 2018).

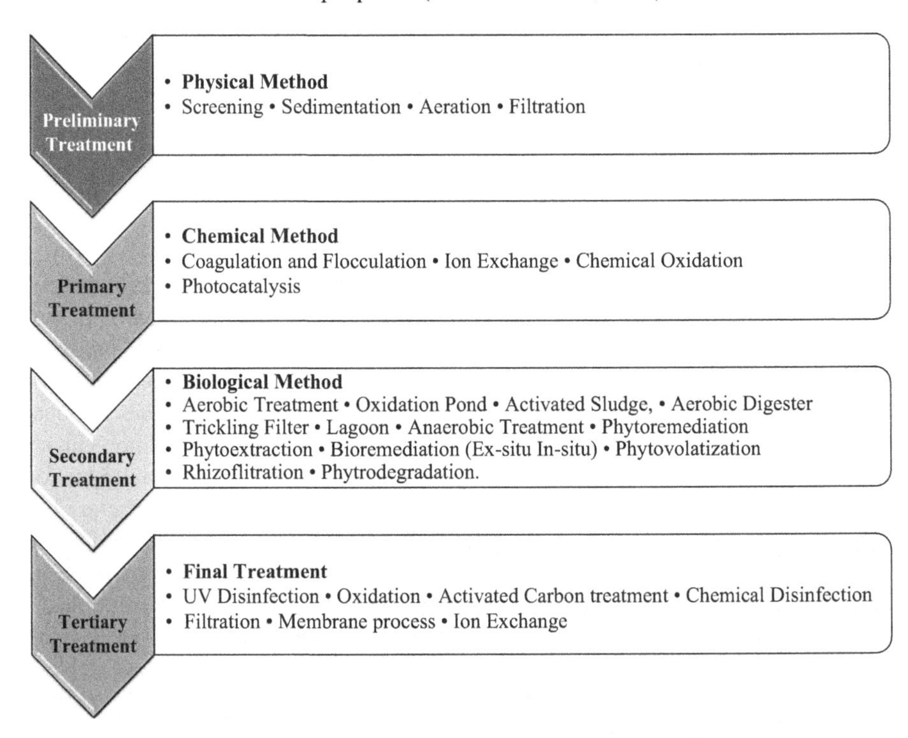

FIGURE 18.3 The figure shows various steps of the conventional wastewater treatment process.

18.3.1.1 Screening

Screening is the first step in the wastewater treatment process, which is sub-divided into two steps, namely coarse and fine screening. The screening process achieves the elimination of large, non-biodegradable particles which are floating, such as tin, paper, rag, wood, plastics, etc. (Santoro et al., 2017). Efficient removal of these particles helps to maintain the safety of downstream equipment from potential damage (Karmakar and Barman, 2023).

18.3.1.2 Sedimentation

Particle size is the most vital factor for the sedimentation process as the large particles easily settle out due to gravitational force acting on them. Through this process, two distinct zones are created: a clear zone and a sludge zone. Furthermore, this technique also reduces the amount of coagulant to be needed. Chemical coagulants are added for floc formation which adds more sediments at the bottom (Karmakar and Barman, 2023).

18.3.1.3 Filtration

The filtration process helps to remove particulate matter, suspended solids, and chemical compounds, as well as microorganisms from wastewater. Removal of contaminants according to their size can be done by the process of filtration. Different types of filters are used, depending on the types of impurities. Two types of filters are mainly used, namely particle filters and membrane filters (Ahmadi et al., 2020). Particle filters help to remove coarse particles which are larger than one micron in size, whereas membrane filters or membrane barriers are used for the particle removal process. Different types of membrane filters are used according to the particle size to be removed (Karmakar and Barman, 2023).

18.3.2 Chemical Methods

Various types of chemical processes are used in coupling with biological and physical processes to get the end product, which is clean freshwater. The most commonly used chemical methods are coagulation and flocculation, photocatalysis, oxidation, and ion exchange (Karmakar and Barman, 2023).

18.3.2.1 Coagulation and Flocculation

This is the most essential process for industrial wastewater cleanup for separation of solids from the liquid. Chemical coagulants are added to destabilise the colloidal particles and neutralise charges to form an aggregation of small particles (Ganesh et al., 2020). In addition, flocculants are used to increase the efficiency of the settlement process. Different flocculants/coagulants, with different structural properties, such as molecular weight, functional group, charge, etc. are used in the wastewater treatment process (Kolya and Tripathy, 2013).

To remove suspended solids, organic, inorganic, or combined coagulants are used as they help to neutralise the negative charge on particles. Flocculants accumulate

TABLE 18.1
Classification of Membrane Filtration Techniques Concerning Their Particle Retention Size and Cross-membrane Pressure Requirement (Barman et al., 2021)

Particle Size (nm)	Filtration Technique	Required Pressure across the Membrane (psi)
<1	Reverse osmosis	35–100
1–10	Nanofiltration	10–30
1–100	Ultrafiltration	1–10
100–10000	Microfiltration	1–10

the destabilized particles and help them to clump together. Through this process, particles get separated from the solution (Karmakar and Barman, 2023).

18.3.2.2 Chemical Oxidation

The chemical oxidation process is mainly used for disinfection and it also helps to reduce the biological oxygen demand (BOD) of the wastewater. Furthermore, it helps to reduce toxicity by converting the contaminants into carbon dioxide and water (Karmakar and Barman, 2023).

18.3.2.3 Photocatalysis

This is a process of advanced oxidation for eliminating POPs and microorganisms, particularly bacteria. Active oxygen forms are generated by the solid and semiconductive catalyst on thesurface during the process of photocatalysis. This process happens when catalysts are exposed to the appropriate wavelength of light (Barman et al., 2022; Ben-Sasson et al., 2014).

18.3.2.4 Ion Exchange

This is the process of replacing ions from hard water, such as magnesium and calcium, by adding sodium ions. Through this process, sodium chloride salt is added to soften the water as it replaces calcium and magnesium ions present in water. This process is used in wastewater treatment to replace the electric charges from contaminated waters (Kanakaraju et al., 2018).

18.3.3 Biological Methods

This is the process of the biodegradation of dissolved and suspended organic pollutants in wastewater with the help of microorganisms. Through the process of biological oxidation and biosynthesis, microorganisms can break down organic pollutants in wastewater. Biological methods include aerobic and anaerobic treatment, phytoremediation, and bioremediation processes (Karmakar and Barman, 2023).

18.3.3.1 Aerobic Treatment

This treatment involves an aerobic tank, oxidation pond, activated sludges, trickling filters, aerobic digester, lagoon-based treatment, oxidation ditch, spray aeration, etc. In this process, a wetland is constructed for wastewater treatment with an attached aeration system for oxygen supply and odor reduction. In the biological treatment system, a diffused oxygen supply is provided to enhance oxygen transport for microbial growth. As a result, microbes degrade organic compounds, forming carbon dioxide and biomass in the presence of dissolved oxygen. Furthermore, microbes convert organic nitrogen into nitrate by the process of the oxidation of ammonium and nitrite. Certain environmental conditions should be maintained to complete the aerobic process, such as appropriate pH, temperature, freedom of toxic substances, etc. (Karmakar and Barman, 2023).

18.3.3.2 Anaerobic Treatment

This is a comparatively slow process due to the decay of large amounts of organic material and it also forms an unpleasant odor. Anaerobic bacteria are used for decomposing organic substances in anaerobic digesters, lagoons, or septic tanks. Energy can be recovered by this process as it produces biogas, thus waste-to-energy conversion is possible (Karmakar and Barman, 2023).

18.3.3.3 Phytoremediation

Phytoremediation is an eco-friendly, cheaper, and more sustainable method for the elimination of contaminants as, in this process, plants are used for cleaning. The process of remediation is carried out either by degradation or elimination of pollutants from soil, water, or air (Lakshmi et al., 2017). Water-tolerant trees are mostly used for waste stream cleaning by the processes of rhizofiltration, phytovolatilization, phytodegradation, phytoextraction, and phytotransformation. For the phytoremediation process, the exposure period, pollutant concentration, pH, temperature, plant type, etc. all play a significant role in the pollutant degradation process (Karmakar and Barman, 2023). The ability of plants to take up or bind various metals is called phytoextraction which is an excellent technique for the removal of hazardous metals from wastewater. Furthermore, aquatic plants can reduce the level of total suspended solids as well as BOD from wastewater. Plants like *Eichhornia crassipes* can bioaccumulate heavy metals such as Zn, Cd, Pb, Cr, Cu, etc. (Lakshmi et al., 2017).

18.3.3.4 Bioremediation

In this process, naturally occurring microbes are used to eliminate nutrients from wastewaters in an eco-friendly way. It is a cost-effective method and also helps to restore ecological damage by improving the efficiency of microorganisms. Through this process, microbes break down, degrade, remove or convert hazardous organic substances to harmless or less harmful products by metabolism. Environmental conditions, type of pollutants, geographical factors, and microbial ecology play a vital role in this process. *In-situ* and *ex-situ* processes of bioremediation are the two main types. In the *in-situ* process, contaminants are treated at the site of their origin whereas, in the *ex-situ*, process, pollutants are treated elsewhere. Various technologies like bioreactors, bioaugmentation, biostimulation, phytoremediation, fungal remediation, leaching, precipitation, etc. are used in the bioremediation process (Karmakar and Barman, 2023).

18.4 ADSORPTION FOR WASTEWATER TREATMENT

This is a physical method of separation by using a solid material to remove certain liquid or gaseous compounds. The absorbent binds compounds to their surface by multiple mechanisms (Soffian et al., 2022). Adsorption is a process of binding of various types of molecules to a particular surface thereby forming a film of absorbent. It is a process of mass transfer and it particularly helps to remove particulate contaminants from wastewater. The advantage of absorbents is that toxicity and salinity do not affect them, so that they can be used for multiple cycles of the wastewater treatment process. In industrial wastewater treatment, this process is used as a part of tertiary treatment to treat dissolved contaminants that escape biological and chemical oxidation stages (Barman et al., 2019). It is a cost-effective process by which various kinds of pollutants from wastewater can be removed, such as heavy metals, pesticides, toxins, pharmaceuticals, and refractory wastes, etc. by transferring them to other phases (Soffian et al., 2022). Easy to operate, versatile, economical, with minimum energy requirement and no by-product, this process is an effective one. Three types of adsorption are there, such as physical absorption, electrostatic, and chemisorption. In addition, important parameters regarding this process which help to enhance the effectiveness of adsorbent are temperature, ion concentration, pH, ionic strength, coexisting ions, adsorbent dosage, stirring speed, contact period, etc. (Sadegh and Ali, 2021).

The compound being absorbed is called the adsorbate, while the material used for the process is called the adsorbent. They have specific properties depending upon their constituents. In the process of physical adsorption, the acting forces are the van der Waals forces. As it is a weak force, the attraction process shows reversibility, while, in the chemisorption process, the attraction force is due to chemical bonding to the solid surface. Chemical bonding is irreversible in nature due to its strong nature in comparison with the van der Waals forces. A study conducted by De Gisi et al. (2016) recognised activated carbon to be the most widely used adsorbent nowadays. It also has some disadvantages like high cost, its regeneration, and disposal-related issues. To solve this problem, various types of low-cost adsorbents have been scrutinized for their potential ability to remove pollutants from wastewater.

Low-cost adsorbents can be divided into the following categories according to Ali et al. (2012): (a) agricultural and household by-products, (b) industrial by-products, (c) sludge, (d) soil and ore materials, (e) sea materials, and (f) novel low-cost adsorbents.

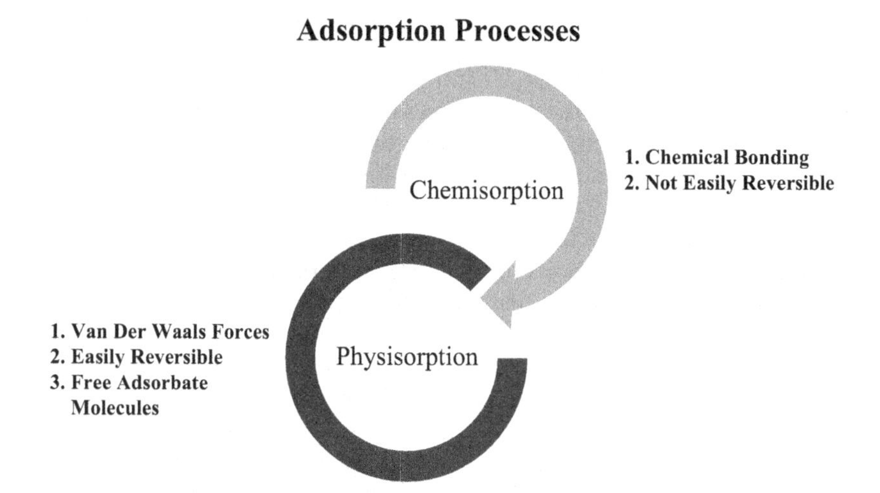

FIGURE 18.4 The graphical representation shows the process of physisorption and chemisorption.

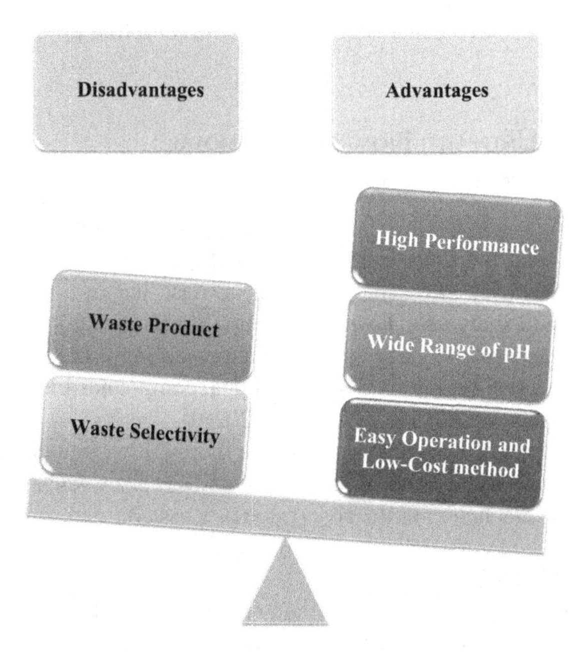

FIGURE 18.5 The figure of a balance shows the advantages and disadvantages of the adsorption process.

18.5 EMERGING TECHNIQUES FOR WASTEWATER TREATMENT USING BIOMASS

Biomass materials are low-cost, easily accessible, carbon-rich, ubiquitous, and recyclable, making them efficient materials for carbon supply. Carbon from biomass has a porous and organised nature which makes it suitable for electrolyte penetration and also reduces the distance of ion diffusion. Most of the biomass material consists of carbon, nitrogen, boron, and other elements which act as heteroatoms to provide excess active sites. In addition, biomass from agricultural residues like straws, bamboo, etc. is a cost-effective and environment-friendly solution for wastewater treatment. Activated carbon can be made from various sources of biomass including coconut, sewage sludge, and ash residue from industrial processes, rice husk, etc. Agricultural waste and forest residue are the major sources of natural biomass which are used for biochar fabrication due to their ubiquitous nature and low cost. Every year, a huge amount of biomass waste is generated due to human activities like rice husk, wood chips, sawdust, straws, etc. Large amounts of lignin and cellulose are present in biomass waste which can be used as a low-cost source of biochar. Furthermore, biomass can be used as adsorbents due to its low cost, low energy investment, and renewable and eco-friendly nature. The use of waste materials in wastewater treatment has become popular, as it is a part of the circular economy through waste-to-resource conversion by reuse and recycling processes (Hossain et al., 2020).

18.5.1 Use of Untreated Biomass for Removal of Pollutants from Wastewater

A study conducted by Hossain et al. (2020) found that naturally available low-cost adsorbents like coal, mud, waste from industrial sources (sugar, paper mills, tea), agricultural residue (rice husk, shells, straw), sawdust, bagasse, unburned carbonaceous material, and sludge etc. have the potential to remove 60–100% of contaminants found in wastewaters. They can eliminate emerging contaminants like heavy metals (Cu, Pb, Ni, Cr, Zn, Cd) as well as chemical compounds like chloride, nitrate, fluoride, phenol, acetone, etc. (Hossain et al., 2020).

Unmodified forms of agricultural waste are thoroughly cleaned, oven dried, and ground up appropriately to bring out particular particle sizes which could be further used as adsorbent precursors in the adsorption process (Barman et al., 2019).

18.5.1.1 Algal Biomass

The ability of microalgal cells to assimilate nutrients can help to eliminate nitrogen and phosphorus from wastewater. They are also used as fertilisers by which they can return absorbed nutrients to the environment. They can be used for various wastewater treatment purposes like municipal wastewater, aquaculture wastewater, livestock and industrial wastewater, etc. Algal biomass has the potential to remove greater than 80% of nutrients from wastewater (Tao et al., 2017).

18.5.1.2 Rice Starch

Unmodified rice starch coupled with an alum can be used as a potential coagulant in the coagulation-flocculation process of wastewater treatment. A study conducted by Teh et al. (2014) showed that unmodified rice starch together with alum improves the coagulation process by reducing the amount of alum (−47.95%), settling time (−58.66%), and chemical oxygen demand (COD), as well as enhancing the total suspended solids (TSS) of agro-industrial effluent from a palm mill adsorption.

18.5.1.3 Wood Chips

Another study, Li et al. (2019) showed that wood chips can be used for diclofenac removal by crushing and sieving the materials, thus increasing the surface area and pore volume.

18.5.1.4 Bagasse and Mushroom Residue

Diclofenac elimination can also be achieved using unmodified 'Isabel' grape bagasse which is used as a biosorbent material. Removal of acetaminophen using shiitake and champignon mushroom stem residue was achieved by the adsorption process. Samples went through the process of drying, grinding, and sieving, after which they showed a high percentage of removal (98%) efficiency (Michelon et al., 2022).

18.5.1.5 Water Hyacinth

The application of water hyacinth (WH) biomass for the elimination of pollutants like metals and organics dyes from wastewater have been reported. Water hyacinth is an effective absorber of organic and inorganic contaminants released from the textile industry. Furthermore, a blooming water hyacinth plant can adsorb organophosphate and organochlorine pesticides through the process of phytoremediation. Another study also found that the adsorption of pharmaceuticals has been achieved from contaminated waters by WH via roots, making WH a potential component for phytoremediation (Amalina et al., 2022).

18.5.2 Use of Modified Biomass for Removal of Pollutants from Wastewater

In the process of adsorption, carbon-based adsorbents are excellent due to their porous nature, consisting of various types of atoms such as oxygen and hydrogen. They show superior kinetic properties and they are renewable in nature. Also, they can adsorb non-polar organic compounds found in wastewater. Agricultural residue, as well as industrial waste, including rice husk, modified sewage sludge, plant leaves, vegetable peel, and tree bark, are used as adsorbents for the removal of various contaminants from wastewater. A modified form of biomass is thermally or chemically treated to improve its adsorption capacity (Barman et al., 2019). After the use of adsorbent loses its effectiveness, it could then be utilized as boiler fuel, in landfill, in construction raw materials, or in the brick-making process (Mohanta and Ahmaruzzaman, 2018).

18.5.2.1 Activated Carbon

Activated carbon (AC) is a carbonaceous solid with large micropores that has a great adsorption potential due to its enhanced internal surface area. It has an excellent adsorption capacity in the wastewater treatment process as well as in air pollution control. It has a high adsorption ability for polar organic compounds such as PAHs, PCBs, pesticides, and phenol. AC is usually used for drinking water and wastewater treatment and for purification purposes. Furthermore, AC also shows high effectiveness in the removal of heavy metals, organic matter, pesticides, microplastics, disinfection by-products and medicines (Soffian et al., 2022). ACis the most frequently used adsorbent and can be prepared from any carbon-containing materials by the physical or chemical activation process. They can be divided into the following categories based on their physical properties such as powder AC, AC pellets, and granular AC (Barman et al., 2019). Coal, lignite, and peat are the commercially available, non-renewable forms of AC which contain high carbon content and porosity. There is an ongoing research interest in finding out low-cost biomass waste as a potential adsorbent due to associated environmental and economic challenges regarding conventional ones (Jjagwe et al., 2021).

- **Sources**

Low-cost agricultural waste residues, such as coconut shells, apple, banana, and orange peel, waste tea, coffee grounds, straw, maize stalks, rice and barley husk, wood chips, sawdust, bagasse, aquatic plants, fungi, etc. with high carbon content have been utilized as a raw material for activated carbon preparation. They are easily available, inexpensive, and could also be used directly as adsorbent for dye removal from wastewater (Bello et al., 2015).

- **Activated Carbon Preparation**

The preparation of activated carbon consists of various steps including pretreatment, soaking, activation washing, sieving, and drying (Ahmed, 2017). The properties of AC depend on the nature and the way the raw materials are prepared, including activation steps. There are two types of activation stages, namely physical and chemical activation. In physical activation, two stages are followed, such as the conversion of the raw material into the carbonized form (char) and then an oxidizing gas like carbon dioxide helps to activate the material. In chemical activation, the raw material is soaked in dehydrating chemicals such as zinc chloride ($ZnCl_2$), phosphoric acid (H_3PO_4), or sulfuric acid (H_2SO_4) and carbonized at varying temperatures (Bello et al., 2015). For various lignocellulosic materials, phosphoric acid is used as the activating agent, as cellulose shows acid hydrolysis resistivity. So, in the H_3PO_4-lignocellulose interaction, the acid first reacts with cellulose and then lignin (Soffian et al., 2022). The chemical activation process is preferable to the physical one due to its high carbonization yield as well as the fact that it can be carried out at lower temperatures (Bello et al., 2015).

- **Modification of Activated Carbon**

Modification of activated carbon can be conducted by enhancing the hydrophilicity of AC by adding polar functional groups to its surface. By altering the surface chemistry, selective particles can be adsorbed by the modified AC, such as ammonia, nitrate, cyanide, sulfide, and chloride as well as Cu, Zn, Pb, etc. (Bjorklund and Li, 2017).

18.6 BIOCHAR

Biochar is a porous solid material with a rich source of carbon formed by the process of thermal decomposition of biomass in the absence of oxygen. It has high resistance and a high level of aromatization (Rangabhashiyam and Balasubramanian, 2019). Agricultural residues like crop waste, animal waste, sludge, algal biomass, energy crops, etc. are the main source of biomass feedstock. Many biochemical, physical, and thermochemical methods are used to produce high-value biochar. Biochar is formed by the thermochemical method of conversion of carbonaceous biomass material at high temperatures (300–900°C) in the presence of limited oxygen concentration. The following processes are used in biochar formation: pyrolysis, hydrothermal carbonization, gasification, and torrefaction. Different physical, mechanical, and chemical properties could be present in the biochars generated according to the pyrolysis process and the raw materials used (Amalina et al., 2022). The adsorption of pollutants onto biochar is achieved by various interaction processes like pore-filling, hydrogen bonding, diffusion, π–π interaction, and hydrophobic and electrostatic interaction mechanisms, or a combination of them. Thus, easy availability, eco-friendly nature, and low cost of the biomass feedstock and the excellent adsorption

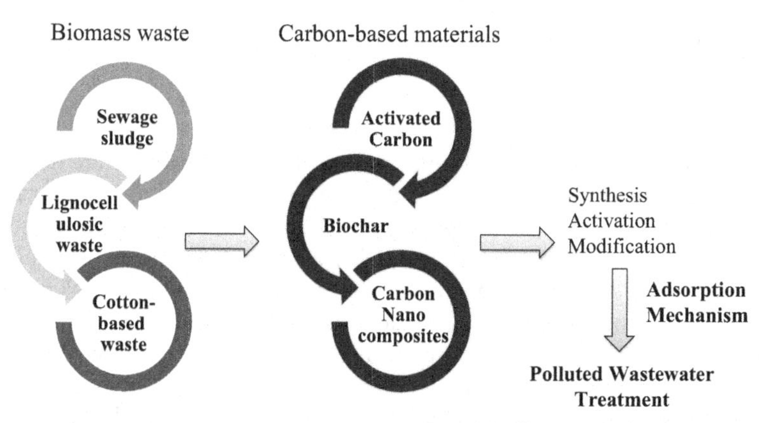

FIGURE 18.6 The figure shows the application of carbon-based biomass waste in wastewater treatment by adsorption mechanism.

capacity of organic contaminants make biochar an excellent alternative for the remediation process of wastewater contaminants (Barman et al., 2019).

- **Sources**

The main feedstocks of biochar are biomass, animal manure, agricultural residue, municipal waste, etc. Some lignocellulosic biomass, including crop and wood waste, plant and animal residue, energy crops etc., are excellent sources of biochar. Non-lignocellulosic biomass sources, including sewage sludges, algal biomass, animal fur, and skeletal and other wastes, need critical management during usage due to their complexity and diverse nature (Amalina et al., 2022).

- **Production of Biochar**

For the production of biochar, various methods could be used like hydrothermal carbonization (HTC), pyrolysis, gasification, and torrefaction, among which pyrolysis is the most commonly used method. The properties of biochars depend on the types of raw material used, the pyrolysis process, duration, temperature, oxygen concentration, and heating rates. In oxygen-free or oxygen-limited conditions, biochar is produced by pyrolysis (300–700°C) of a variety of biomass materials (Soffian et al., 2022).

- **Modification**

To enhance the pollutant removal efficiency and increase the effectiveness of degradation of pollutants, biochars can be modified by chemical and physical methods. The chemical modification method consists of acid or alkaline modifications as well as modification with oxidizing agents. Physical modification is carried out by gas purging (Soffian et al., 2022).

18.6.4 Hydrochar

In the pyrolysis process, high energy input and expensive pretreatment are required for biomass feedstock. To make it more convenient, hydrothermal carbonization (HTC) is preferred as a thermochemical process in place of pyrolysis for some high-moisture-content biomass waste. In HTC, a lower temperature is required (80–240°C) to form solid hydrochar under subcritical water pressure (Nadarajah et al., 2021). Hydrochar became popular due to its low cost and environmentally sustainable nature as well as its effectiveness in the process of adsorption to remove pollutants from water. Hydrochar can be obtained from a wide range of biomass, such as agricultural waste, manure, microalgal biomass, ans sewage sludge through HTC. This process ensures sustainability not only in sources of raw materials but also in terms of energy consumption and management of waste. The hydrochar surface contains many chemically reactive functional groups. These acidic functional

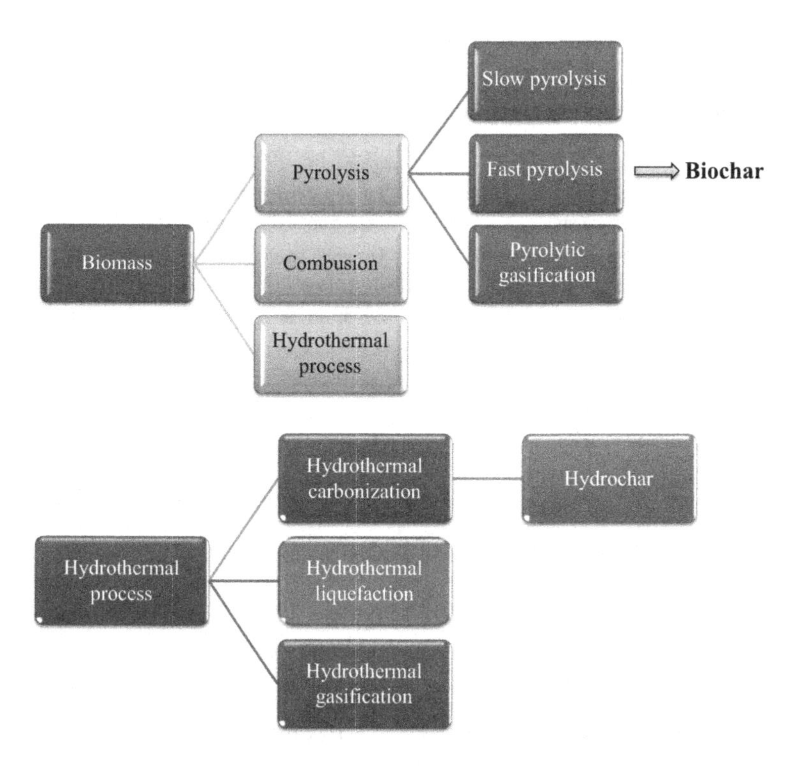

FIGURE 18.7 Figure showing various pathways for conversion of biomass into biochar and hydrochar.

groups drive the adsorption process and help to remove contaminants from wastewater through electrostatic interactions.

Hydrochar produced by the HTC method has a low surface area; to increase its surface area and porosity activation, steps need to be added for remediation purposes. The chemisorption of water pollutants by hydrochar is controlled by the properties of contaminants and the surface chemistry of the hydrochar. Functional groups of hydrochar can be optimized for better adsorption capacity for removing contaminants. The activation process of hydrochar is achieved using $ZnCl_2$, KOH, K_2CO_3, NaOH, H_3PO_4, and other bases, acid-oxidizing agents, and salts. Low temperature and residence time of the HTC process helps to enhance hydrochar efficiency to remediate pollutants by increasing surface area (Padhye et al., 2022).

18.7 Carbon-based Nanocomposites

To overcome the challenges associated with typical adsorbents, nanoadsorbents have been formed with large surface areas, multiple active sites, pore sizes that can be adjusted, and better surface chemistry with more rapid kinetics. Nanocomposites show electron affinity and flexibility as well as mechanical strength. Thus, they

TABLE 18.2
Comparison between Hydrochar and Biochar

Properties	Hydrochar	Biochar	Reference
Specific surface area and porosity	Non-porous, low specific surface area	Porous and depends on the reaction temperature	Masoumi et al. (2021)
Morphology	Spherical shape	Graphite-like layers	Masoumi et al. (2021)
Total carbon content	58–64 wt. %	60–80 wt. %	Masoumi et al. (2021)
H/C molar ratio	>2.3	>1.5	Masoumi et al. (2021)
O/C molar ratio	>1.7	>0.7	Masoumi et al. (2021)
pH	Mostly acidic	Mostly alkaline	Masoumi et al. (2021)
Aromacity	Contains alkyl moieties	Contains aromatic groups	Masoumi et al. (2021)
Techniques	Hydrothermal carbonisation **Advantages**: • Eliminates energy-intensive dry methods • High moisture biomass is suitable for this method	Pyrolysis **Advantages:** • Simple and low-residence time • High biochar yield • Any organic feedstock can be utilized • Large surface area, high conductivity and a porous structure	Amalina et al. (2022)

could be used for water treatment and desalination processes also (Soffian et al., 2022). Although the porous nature of carbon is an excellent adsorbent for the removal of pollutants, in recent times carbon-based nanocomposites became popular due to their greater adsorption capacity as well as antimicrobial properties. Nanocomposites have unique properties like large surface area which enhance the adsorption capacity for inorganic and organic contaminants in wastewater. A huge vacant surface area allows a high rate of mass transfer, as various nanoparticles efficiently remove pollutants. Furthermore, they also have metallic or semi-metallic characteristics to target particular pollutants for interaction. Many low-cost and readily available metal particles are produced from bacteria, algae, plants, etc. Metal nanocomposites consisting of zinc, gold, lead, and palladium have also been biosynthesized and are cheap and widely available. In addition, biochar and activated carbon are both used in nanocomposite preparation. Biochar-based nanocomposites have been used to eliminate heavy metals like arsenic, copper, cadmium, chromium, lead, and platinum from wastewater. Nanoscale zerovalent ions, as well as graphite, have been used to remove organic chemicals, while AC composites have been used to eliminate various metals. AC have reactive centers like -COOH, -OH, and $-NH_2$ able to attract nanoparticles through interactions like hydrogen bonding, electrostatic forces, and π–π interactions. Furthermore, CuO nanoparticles easily attach to AC because of the presence of a reactive oxygen center and hard/soft interactions with the metal center. AC-based nanocomposites are used for the removal of various petroleum products, dyes, and aromatic compounds (Barman et al., 2019). Nanomaterials with high surface area, renewable nature, nanosorbent capacity, and chemically modified forms play a significant role in the removal of contaminants from wastewater. Carbon-based nanocomposites, like carbon nanotubes, graphene oxide, graphene, and activated carbon, are widely used because of their reliable nature in the water treatment process. The size range of nanomaterials is greater than 1 nm to less than 100 nm, they have high surface area, and there exists strong and weak interactions between them. Due to these characteristics, nanocomposites have unique features like small size effect, interfacial effect, and surface effect (Arun et al., 2022).

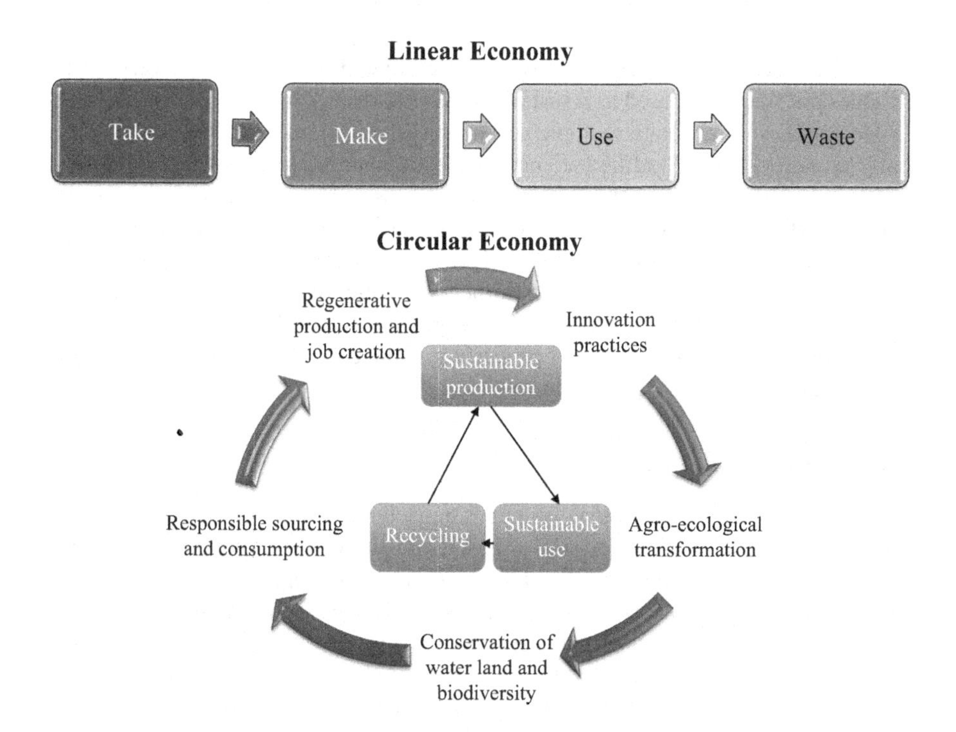

FIGURE 18.8 Graphical representations showing linear and circular economy concepts.

18.8 CIRCULAR ECONOMY AND SUSTAINABILITY OF BIOMASS

The United Nations formulated 17 Sustainable Development Goals (SDGs) that are a global call for action to protect the environment, for better prospects of lives for everyone as well as to end poverty and ensure global peace. Currently, the progress of SDGs is dragging behind its 2030 target. Reaching the target of adequate water quality and quantity maintenance is a significant challenge. Of the 17 goals, SDG 6 is focused on "clean water and sanitation for all" as water is considered a crucial factor in achieving other goals due to its connections with water, food, energy, and the economy. Hence, it is necessary to address the growing challenges related to water quality and scarcity to achieve UN SDGs by 2030. However, the latest data on SDG 6 shows very alarming trends in the safe management of drinking water resources as well as sanitary services. The "circular economy" concept has become an excellent solution with regard to accelerating the progress rate of achieving SDGs. One of the most promising solutions is the application of modified biomass as an emerging technique in the wastewater treatment process (Padhye et al., 2022). The concept of the circular economy is the process by which products and materials remain in circulation throughout the economy for as long as possible. Wastes act as a secondary raw material which can be further used by the recycling and reuse process. A linear economy is different from the circular economy in the way that it consists of a "take-make-use-dispose" process. In this process, wastes are generated at the last stage of production. A circular economy promotes sustainability of energy and material management by reducing the amount of waste generated and using the waste as an asset, thus keeping the process circulating (Neczaj and Grosser, 2018). Agricultural residues are frequently used as a sustainable source. Hence, using these waste value-added products to generate multiple sustainable applications is an example of a circular economy in use. Modified biomass like hydrochar is a low-cost material as well as being easily available in the form of biomass waste (like agricultural and forestry waste). Furthermore, a less energy-intensive process is operated during biomass activation as biomass has vast numbers of chemically reactive functional groups on its surface. Thus, it acts as an excellent adsorbent of contaminants through various interactions. Functional groups can be easily optimized to adsorb a particular pollutant by modifying the physicochemical properties of the adsorbent to remove particular contaminants from wastewater (Padhye et al., 2022).

18.9 CONCLUSION

The existence of living organisms depends entirely on water resources. Though the Earth is 70% covered by water, only 3% of it constitutes freshwater which is accessible for human consumption. In addition, this small proportion of freshwater is contaminated due to various anthropogenic activities like the discharge of toxic contaminants, including dyes, pharmaceuticals, heavy metals, organic pollutants like phenol, and pesticides (Ighalo et al., 2022). The rapid increase in distribution of various pollutants in different waterbodies is currently a major threat to human beings as well as all living organisms (Ajala et al., 2022). The main cause of the hike in water pollutant concentrations is mainly due to the advancement of technologies resulting in the production of various products in different industries (Ewis and Hameed, 2021). Contaminated water is highly toxic in nature and causes harmful effects on living beings as well as damage to the overall ecosystem. To overcome these negative impacts of polluted waters on human health and the global environment, researchers are seeking sustainable solutions to perform high-performance water purification (Ighalo et al., 2022).

The typical methods for the elimination of contaminants from water are adsorption, ion exchange, sedimentation, coagulation, flocculation, reverse osmosis, precipitation, biological treatment, complexation, and electrochemical operations (Ighalo et al., 2022).

For the elimination of water pollutants from wastewater, various treatment techniques are available, based on biological, chemical, and physical methods. Among these conventional techniques, adsorption is extensively used for the elimination of a variety of water contaminants (Ighalo et al., 2022). Emerging new techniques are developing over time to improve efficiency as well as to resolve challenges associated with sustainability. Adsorption associated with activated carbon has been extensively used due to its low cost, ease of use, and high efficiency in the removal of organic substances or heavy metals from wastewater streams (Bello et al., 2015). Furthermore, various types of adsorbents have been selected by researchers for the removal of contaminants from water, including nanocomposites, magnetic porous carbon, activated biocarbon, hydrochar, activated carbon from sewage sludge, carbon nanotubes, and other organic wastes (Ighalo et al., 2022).

Various emerging concepts and technologies have fast been replacing conventional wastewater treatment methods. Nanocomposites with high adsorption capacity, high reactivity, large surface area, high degree of functionality, and size-dependent properties make them an excellent material for wastewater treatment as well as water purification (Bora and Dutta, 2014). Carbon-based materials (CBMs) provide efficient water disinfection and purification; the study conducted by Soffian et al. (2022) shows that CBMs have the capacity for high-level adsorption and efficient elimination of pollutants by forming bonds with elements. CBMs like biochar, activated carbon, carbon nanotubes, and graphene are promising technologies for wastewater treatment; by altering their properties through various synthesis and modification methods, their adsorption capacity can be further enhanced. The progress on SDGs to reach their targets by 2030 has been slow to date. The universal accessibility of sufficient quality and quantity of freshwater is considered to be the most significant challenge in contemporary times. The potential solution to increasing the rate of progress to achieve SDG6 is the circular economy concept. It is an engineered solution designed to apply to the water treatment process with the help of modified biomass (Padhye et al., 2022).

REFERENCES

Ahmadi, F., Zinatizadeh, A.A., Asadi, A., McKay, T. and Azizi, S., 2020. Simultaneous carbon and nutrients removal and PHA production in a novel single air lift bioreactor treating an industrial wastewater. *Environmental Technology & Innovation*, *18*, p.100776.

Ahmed, M.J., 2017. Adsorption of non-steroidal anti-inflammatory drugs from aqueous solution using activated carbons. *Journal of Environmental Management*, *190*, pp.274–282.

Ahuja, S., 2021. Select applications of nanomaterials for water purification. In Ahuja, S., (Ed.), *Handbook of Water Purity and Quality* (pp. 339–357). Academic Press.

Ajala, O.J., Khadir, A., Ighalo, J.O. and Umenweke, G.C., 2022. Cellulose-based nano-biosorbents in water purification. In Denizli, A., Bilal, M., Nguyen, T.A., Ali, N. and Khan, A., (Eds.), *Nano-biosorbents for Decontamination of Water, Air, and Soil Pollution* (pp. 395–415). Elsevier.

Al-Anzi, B.S. and Siang, O.C., 2017. Recent developments of carbon based nanomaterials and membranes for oily wastewater treatment. *RSC Advances*, *7*(34), pp.20981–20994.

Alcamo, J., 2019. Water quality and its interlinkages with the sustainable development goals. *Current Opinion in Environmental Sustainability*, *36*, pp.126–140.

Ali, I., Asim, M. and Khan, T.A., 2012. Low cost adsorbents for the removal of organic pollutants from wastewater. *Journal of Environmental Management*, *113*, pp.170–183.

Amalina, F., Abd Razak, A.S., Krishnan, S., Sulaiman, H., Zularisam, A.W. and Nasrullah, M., 2022. Biochar production techniques utilizing biomass waste-derived materials and environmental applications: A review. *Journal of Hazardous Materials Advances*, *7*, p.100134.

Amalina, F., Abd Razak, A.S., Krishnan, S., Zularisam, A.W. and Nasrullah, M., 2022. Water hyacinth (Eichhornia crassipes) for organic contaminants removal in water: A review. *Journal of Hazardous Materials Advances*, *7*, p.100092.

Arun, J., Nirmala, N., Priyadharsini, P., Dawn, S.S., Santhosh, A., Gopinath, K.P. and Govarthanan, M., 2022. A mini-review on bioderived carbon and its nanocomposites for removal of organic pollutants from wastewater. *Materials Letters*, *310*, p.131476.

Azari, A., Noorisepehr, M., Dehghanifard, E., Karimyan, K., Hashemi, S.Y., Kalhori, E.M., Norouzi, R., Agarwal, S. and Gupta, V.K., 2019. Experimental design, modeling and mechanism of cationic dyes biosorption on to magnetic chitosan-lutaraldehyde composite. *International Journal of Biological Macromolecules*, *131*, pp.633–645.

Barman, S.R., Banerjee, P., Mukhopadhayay, A. and Das, P., 2022. Biopolymer linked activated carbon-nano-bentonite composite membrane for efficient elimination of PAH mixture from aqueous solutions. *Biomass Conversion and Biorefinery*, *14*, pp.359–373.

Barman, S.R., Roy, U., Das, P. and Mukhopadhayay, A., 2021. Membrane processes for removal of polyaromatic hydrocarbons from wastewater. In Sharma, S.K., (Ed.), *Green Chemistry and Water Remediation: Research and Applications* (pp. 189–207). Elsevier.

Barman, S.R., Mukhopadhyay, A. and Das, P., 2019. Green synthesis of carbonaceous adsorbents and their application for removal of polyaromatic hydrocarbons (PAHs) from water. In Sharma, S.K., (Ed.), *Bioremediation* (pp. 229–258). CRC Press.

Bello, O.S., Adegoke, K.A., Olaniyan, A.A. and Abdulazeez, H., 2015. Dye adsorption using biomass wastes and natural adsorbents: Overview and future prospects. *Desalination and Water Treatment*, *53*(5), pp.1292–1315.

Ben-Sasson, M., Lu, X., Bar-Zeev, E., Zodrow, K.R., Nejati, S., Qi, G., Giannelis, E.P. and Elimelech, M., 2014. In situ formation of silver nanoparticles on thin-film composite reverse osmosis membranes for biofouling mitigation. *Water Research*, *62*, pp.260–270.

Björklund, K. and Li, L.Y., 2017. Adsorption of organic stormwater pollutants onto activated carbon from sewage sludge. *Journal of Environmental Management*, *197*, pp.490–497.

Bora, T. and Dutta, J., 2014. Applications of nanotechnology in wastewater treatment: A review. *Journal of Nanoscience and Nanotechnology*, *14*(1), pp.613–626.

De Gisi, S., Lofrano, G., Grassi, M. and Notarnicola, M., 2016. Characteristics and adsorption capacities of low-cost sorbents for wastewater treatment: A review. *Sustainable Materials and Technologies*, *9*, pp.10–40.

Dehgani, Z., Ghaedi, M., Sabzehmeidani, M.M. and Adhami, E., 2020. Removal of paraquat from aqueous solutions by a bentonite modified zero-valent iron adsorbent. *New Journal of Chemistry*, *44*(31), pp.13368–13376.

Ewis, D. and Hameed, B.H., 2021. A review on microwave-assisted synthesis of adsorbents and its application in the removal of water pollutants. *Journal of Water Process Engineering*, *41*, p.102006.

Hossain, N., Bhuiyan, M.A., Pramanik, B.K., Nizamuddin, S. and Griffin, G., 2020. Waste materials for wastewater treatment and waste adsorbents for biofuel and cement supplement applications: A critical review. *Journal of Cleaner Production*, *255*, p.120261.

Hussam, A., 2013. Potable water: Nature and purification. In Ahuja, S. (Ed.), *Monitoring Water Quality*, pp.261–283. Elsevier. https://doi.org/10.1016/B978-0-444-59395-5.00011-X

Ganesh Kumar, A., Anjana, K., Hinduja, M., Sujitha, K. and Dharani, G., 2020. Review on plastic wastes in marine environment-Biodegradation and biotechnological solutions. *Marine Pollution Bulletin*, *150*, p.110733.

Ighalo, J.O., Rangabhashiyam, S., Dulta, K., Umeh, C.T., Iwuozor, K.O., Aniagor, C.O., Eshiemogie, S.O., Iwuchukwu, F.U. and Igwegbe, C.A., 2022. Recent advances in hydrochar application for the adsorptive removal of wastewater pollutants. *Chemical Engineering Research and Design*, *184*, pp.419–456.

Jjagwe, J., Olupot, P.W., Menya, E. and Kalibbala, H.M., 2021. Synthesis and application of Granular activated carbon from biomass waste materials for water treatment: A review. *Journal of Bioresources and Bioproducts*, *6*(4), pp.292–322.

Kanakaraju, D., Glass, B.D. and Oelgemöller, M., 2018. Advanced oxidation process-mediated removal of pharmaceuticals from water: A review. *Journal of Environmental Management*, *219*, pp.189–207.

Karmakar, S. and Barman, S.R., 2023. Sustainable wastewater treatment using membrane technology. In Nadda, A.K., Banerjee, P., Sharma, S., and Nguyen-Tri, P., (Eds.), *Membranes for Water Treatment and Remediation* (pp. 23–53). Springer Nature.

Khodakarami, M. and Bagheri, M., 2021. Recent advances in synthesis and application of polymer nanocomposites for water and wastewater treatment. *Journal of Cleaner Production*, *296*, p.126404.

Kolya, H. and Tripathy, T., 2013. Preparation, investigation of metal ion removal and flocculation performances of grafted hydroxyethyl starch. *International Journal of Biological Macromolecules*, *62*, pp.557–564.

Lakshmi, K.S., Sailaja, V.H. and Reddy, M.A., 2017. Phytoremediation-a promising technique in waste water treatment. *International Journal of Scientific Research and Management (IJSRM)*, *5*(6), pp.5480–5489.

Li, Y., Taggart, M.A., McKenzie, C., Zhang, Z., Lu, Y., Pap, S. and Gibb, S., 2019. Utilizing low-cost natural waste for the removal of pharmaceuticals from water: Mechanisms, isotherms and kinetics at low concentrations. *Journal of Cleaner Production*, *227*, pp.88–97.

Masoumi, S., Borugadda, V.B., Nanda, S. and Dalai, A.K., 2021. Hydrochar: A review on its production technologies and applications. *Catalysts*, *11*(8), p.939.

Michelon, A., Bortoluz, J., Raota, C.S. and Giovanela, M., 2022. Agro-industrial residues as biosorbents for the removal of anti-inflammatories from aqueous matrices: An overview. *Environmental Advances*, *9*, p.100261.

Mohanta, D. and Ahmaruzzaman, M., 2018. Bio-inspired adsorption of arsenite and fluoride from aqueous solutions using activated carbon@ SnO2 nanocomposites: Isotherms, kinetics, thermodynamics, cost estimation and regeneration studies. *Journal of Environmental Chemical Engineering*, *6*(1), pp.356–366.

Morin-Crini, N., Lichtfouse, E., Liu, G., Balaram, V., Ribeiro, A.R.L., Lu, Z., Stock, F., Carmona, E., Teixeira, M.R., Picos-Corrales, L.A. and Moreno-Pirajan, J.C., 2022. Worldwide cases of water pollution by emerging contaminants: A review. *Environmental Chemistry Letters*, *20*(4), pp.2311–2338.

Nadarajah, K., Bandala, E.R., Zhang, Z., Mundree, S. and Goonetilleke, A., 2021. Removal of heavy metals from water using engineered hydrochar: Kinetics and mechanistic approach. *Journal of Water Process Engineering*, *40*, p.101929.

Neczaj, E. and Grosser, A., 2018. Circular economy in wastewater treatment plant–challenges and barriers. *Proceedings* (Vol. 2, No. 11, p. 614). https://doi.org/10.3390/proceedings2110614

Oki, T. and Quiocho, R.E., 2020. Economically challenged and water scarce: Identification of global populations most vulnerable to water crises. *International Journal of Water Resources Development*, *36*(2–3), pp.416–428.

Padhye, L.P., Bandala, E.R., Wijesiri, B., Goonetilleke, A. and Bolan, N., 2022. Hydrochar: A promising step towards achieving a circular economy and sustainable development goals. *Frontiers in Chemical Engineering*, *4*, 867228.

Parsai, T., Figueiredo, N., Dalvi, V., Martins, M., Malik, A. and Kumar, A., 2022. Implication of microplastic toxicity on functioning of microalgae in aquatic system. *Environmental Pollution*, *38*, p.119626.

Rangabhashiyam, S. and Balasubramanian, P., 2019. The potential of lignocellulosic biomass precursors for biochar production: Performance, mechanism and wastewater application-a review. *Industrial Crops and Products*, *128*, pp.405–423.

Ranjit, P., Jhansi, V. and Reddy, K.V., 2021. Conventional wastewater treatment processes. In Maddela, N.R., Cruzatty, G., and Chakraborty, S., (Eds.), *Advances in the Domain of Environmental Biotechnology: Microbiological Developments in Industries, Wastewater Treatment and Agriculture* (pp. 455–479). Springer.

Sadegh, H. and Ali, G.A.M., 2021. Potential applications of nanomaterials in wastewater treatment. In *Research Anthology on Synthesis, Characterization, and Applications of Nanomaterials* (pp. 1230–1240). Information R Management Association

Santoro, C., Arbizzani, C., Erable, B. and Ieropoulos, I., 2017. Microbial fuel cells: From fundamentals to applications: A review. *Journal of Power Sources*, *356*, pp.225–244.

Saravanan, A., Kumar, P.S. and Renita, A.A., 2018. Hybrid synthesis of novel material through acid modification followed ultrasonication to improve adsorption capacity for zinc removal. *Journal of Cleaner Production*, *172*, pp.92–105.

Soffian, M.S., Halim, F.Z.A., Aziz, F., Rahman, M.A., Amin, M.A.M. and Chee, D.N.A., 2022. Carbon-based material derived from biomass waste for wastewater treatment. *Environmental Advances*, *16*, p.100259.

Tao, Q., Gao, F., Qian, C.Y., Guo, X.Z., Zheng, Z. and Yang, Z.H., 2017. Enhanced biomass/biofuel production and nutrient removal in an algal biofilm airlift photobioreactor. *Algal Research*, *21*, pp.9–15.

Teh, C.Y., Wu, T.Y. and Juan, J.C., 2014. Optimization of agro-industrial wastewater treatment using unmodified rice starch as a natural coagulant. *Industrial Crops and Products*, *56*, pp.17–26.

Wong, S., Ngadi, N., Inuwa, I.M. and Hassan, O., 2018. Recent advances in applications of activated carbon from biowaste for wastewater treatment: A short review. *Journal of Cleaner Production*, *175*, pp.361–375.

19 Potential Effects and Removal Strategies for Contaminants of Emerging Concern (CECs) in the Circular Economy of Water

Serin Zachariah
Department of Bioscience and Engineering,
National Institute of Technology, Calicut, India

Reshmi Sasi
Department of Bioscience and Engineering,
National Institute of Technology, Calicut, India

T. V. Suchithra
Department of Bioscience and Engineering,
National Institute of Technology, Calicut, India

19.1 INTRODUCTION

Fresh water is a valuable resource for which we have no substitutes. The circular economy concept aims to preserve material goods within closed loops, minimizing or eliminating the linear resource extraction, transformation, and disposal processes. In contrast, the linear economy's consumption patterns result in wasteful energy and resource usage, potentially leading to environmental contamination (Figure 19.1). Over time, the term "circular economy" has changed. Initially, it was known as the "service economy" (Stahel, 1997) or the "performance economy" (Stahel, 2010), but its meaning has become intertwined with the economy of sustainable development (Sauvé et al., 2016).

While the biogeochemical cycle of water may give the impression of a straightforward circular process, it is essential to recognize that clean water undergoes a complex journey. It evaporates from the planet's surface, forms clouds, and returns

DOI: 10.1201/9781003432869-19

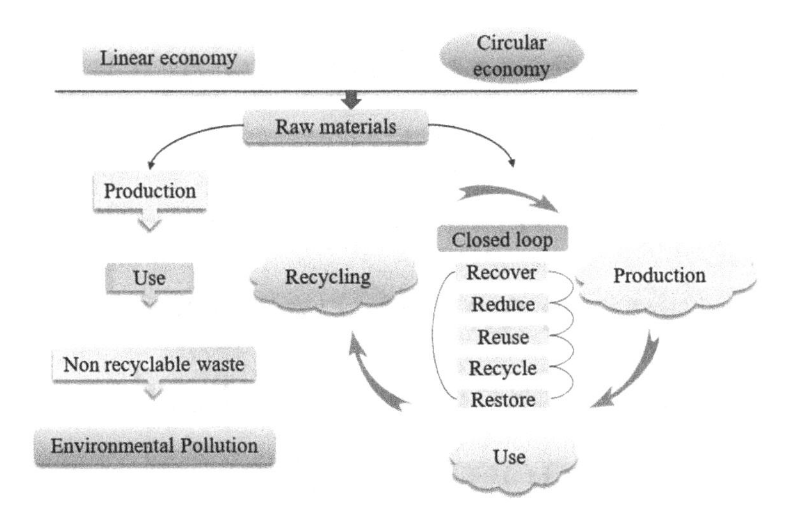

FIGURE 19.1 Comparison of linear and circular economy concepts.

to Earth as rain. At a global level, water follows a circular path within a dynamic and self-regulated system. Large-scale water movements encompass the entire water cycle, including precipitation, surface flow, evaporation, and subsequent rainfall. However, localized water usage and dispersion on a smaller scale contribute to evapotranspiration, increasing air humidity, and leading to localized rain. The water must be sourced from rainfall, surface waterbodies, groundwater, or through recycling efforts to sustain agriculture, food production, and various domestic or commercial activities. It is essential to monitor and manage the water balance closely and have a thorough understanding of the circularity of the system. However, it is worth noting that the water balance is a constantly shifting target, as climate change and land use practices significantly impact water availability. These factors alter runoff patterns, recharge rates, and overall water availability. Localized land modifications can also disrupt the natural water cycle by reducing soil permeability, promoting water evaporation, hindering infiltration and groundwater replenishment, and accelerating soil erosion. Therefore, comprehensive management approaches are necessary to maintain the circularity and sustainability of water resources at different scales. The natural circularity of water in the environment is no longer a perfect process, as it can lead to water contamination through various means. Water traverses agricultural fields and can introduce nutrients, pesticides, and pharmaceuticals into the water supply. Additionally, it can contribute to soil erosion, wash pollutants from roads, increase waterbody temperatures, and release harmful substances into the environment when utilized in agri-food and industrial activities. It is essential to recognize that the water we consume rarely maintains its pristine state as it undergoes some level of reuse. Unfortunately, this aspect is often disregarded or overlooked (Sauvé et al., 2021).

Water itself cannot be classified as sustainable or unsustainable; instead, how we use it can. We look for the finest methods and technologies that may be used to monitor water usage together with its immediate and long-term effects. Extensive attempts have been undertaken to assess the water footprint, aiming to classify water appropriation into three distinct categories: blue, green, and gray, allowing for a comprehensive quantification of the sustainability of water utilization (Hoekstra and Mekonnen, 2012). Bluewater consumption refers to the quantity of water, whether from surface or groundwater sources, that is evaporated from or incorporated into the production of a particular commodity. On the other hand, green water consumption represents the volume of rainwater utilized. Graywater production is closely linked to the levels of water contamination. It signifies the importance of the water required to dilute wastewater and reduce the contaminants to acceptable quality thresholds. The quality of wastewater treatment determines the extent to which "circularity" is achieved. The actual methods and technologies for treating water will also affect the circular economy. Water circularity is evaluated, not based on the water's mass balance, but on the amount of energy, chemicals, and other consumables consumed.

Circularity concepts are used across the entire water management cycle to establish a sustainable water future. This entails adopting cutting-edge water treatment and purification technologies and optimizing water usage efficiency in households, businesses, and agriculture. The safeguarding of water quality is a concern of circularity (Mohamed et al., 2023). Pollutants in wastewater can be effectively eliminated through a range of approaches, including physical methods such as membrane filtration and adsorption, chemical techniques like precipitation and oxidation, and biological processes such as sludge-based activated carbon, membrane bioreactors, and biological aerated filters, as well as a combination of these techniques in hybrid systems (Englande et al., 2015). Innovating and integrating energy- and resource-efficient anaerobic effluent treatment plants, as well as increased carbon captured to be diverted to energy recovery schemes are the two primary approaches to establishing more rational and circular-economy-based effluent treatment approaches (Ghimire et al., 2021).

19.2 CONTAMINANTS OF EMERGING CONCERN

Contaminants of emerging concern (CECs) are diverse pollutants that pose significant environmental and public health risks. This category contains various chemical components, including pharmaceuticals, personal care products, insecticides, microplastics, and endocrine-disrupting chemicals. These substances represent just a few examples of the broad range of pollutants that fall under the classification of CECs and warrant attention due to their potential adverse impacts. The term "emerging" indicates that these hazards have recently been recognized, and ongoing research is being conducted to understand their actions, outcomes, and effects. The development of awareness of these contaminants is credited mainly to improvements in detection methods and an increase in public awareness of their presence in various environmental compartments. Traditional pollutants still pose a threat, but CECs pose particular problems because of their persistence, bioaccumulation, and potential for adverse effects, even at low doses. The prevalence of CECs in waterbodies is one of their most significant challenges (Figure 19.2). Agricultural runoff, wastewater effluents, and poor disposal techniques release many of these environmental pollutants. Once discharged, they can stay in the ecosystem and contaminate ecological systems and water sources that supply drinking water. According to research, these pollutants can potentially disrupt endocrine systems, impair reproductive health, destroy

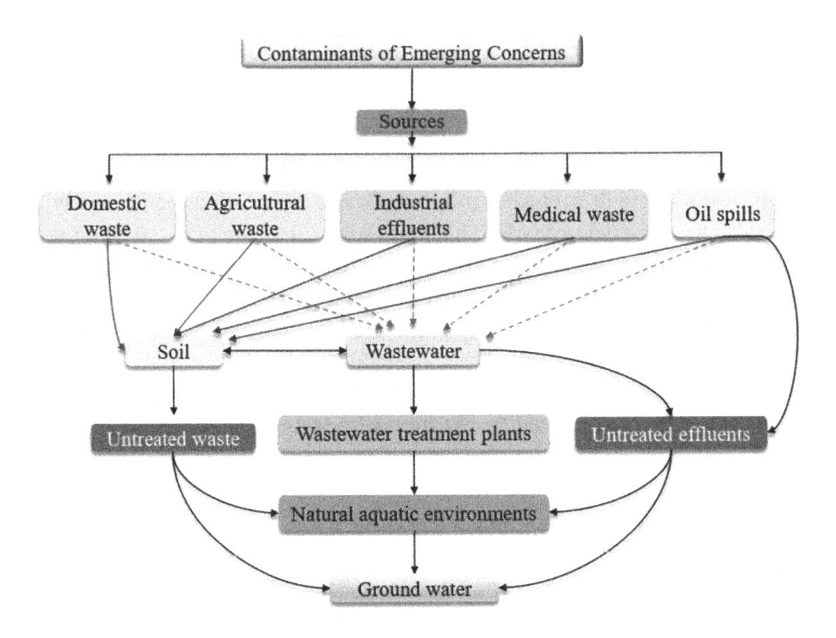

FIGURE 19.2 Routes of emerging pollutants from sources to different parts of the environment.

aquatic life, and perhaps enter the food chain, raising questions about their effects on humans. A multidisciplinary approach is necessary to completely understand the fate, transit, and ecological impact of CECs due to their complexity.

19.2.1 Persistent Organic Pollutants

Persistent organic pollutants (POPs) are highly hazardous chemicals that can travel long distances through air and water currents. These pollutants can persist in the environment for extended periods and tend to accumulate within organisms through bioaccumulation. Due to these pollutants' resilience to degradation, they stay in the background for long periods, even after their production and usage have been interrupted. POPs encompass a range of organic compounds, including Per- and polyfluoroalkyl substances (PFASs), polycyclic aromatic hydrocarbons (PAHs), polybrominated diphenyl ethers (PBDEs), certain pesticides like DDT and chlordane, as well as unintended by-products of industrial activities like dioxins and furans. These substances are of global concern due to their detrimental effects on the environment and human health (Matei et al., 2023).

Polycyclic aromatic hydrocarbons are dangerous compounds, identified as being mutagenic and carcinogenic. These compounds, consisting of two or more fused benzene rings, are formed as by-products of incomplete combustion processes like burning fossil fuels, wood, or organic matter (Gaurav et al., 2021). Although PAHs are poorly soluble in water, they can adhere to particulate matter and other particles, which allows them to be transported and deposited in aquatic systems. Due to their hydrophobicity and endurance, they are susceptible to bioaccumulation in organisms and subsequent biomagnification along the food chain. As a result, PAHs can endanger aquatic and terrestrial organisms, such as fish, birds, mammals, and humans.

Per- and poly-fluoroalkyl substances are persistent chemicals in all environmental matrices worldwide. The global concerns surrounding PFASs are escalating due to their ability to bioaccumulate and threaten human health and the environment. These pollutants comprise various anthropogenic aliphatic compounds used in non-stick cookware, food additives, water repellents, and diverse manufacturing processes (Gagliano et al., 2020). The contamination of the environment by PFASs can occur through multiple pathways, including the immediate release of treated sludge or wastewater effluents from treatment plants. Traditional water treatment methods, like filtration, flocculation, coagulation, sedimentation, and ultraviolet disinfection, are inadequate for removing PFASs from drinking water. However, using activated carbon (AC) and anion-exchange treatments has proven effective in PFAS removal. Extensive medical research has linked PFASs to hyperthyroidism, carcinogenesis, adverse impacts on birth weight, and poor growth. Consequently, regulatory measures have been implemented by the European Union, the Stockholm Convention, and the US Environmental Protection Agency to restrict exposure to PFASs.

Polybrominated diphenyl ethers are a class of intolerant and bioaccumulative halogenated chemicals that have emerged as another significant environmental pollutant. In consumer products, such as electrical equipment, building materials, coatings, textiles, and polyurethane foam in furniture cushioning, PBDEs are employed as a flame-retardant. PBDEs resemble polychlorinated biphenyls in structure and are resistant to environmental deterioration. Tetra-, penta-, and hexa-brominated PBDEs have a strong affinity for lipids and can accumulate in the bodies of both humans and animals (Siddiqi and Clinic, 2003). Studies have revealed that PBDEs may be harmful to human health. Some PBDE congeners have been linked to developmental neurotoxicity, reproductive problems, and perhaps carcinogenic implications. They are also thought to interfere with the endocrine system. PBDE exposure, particularly at critical periods like the prenatal and early infancy stages, is of concern due to its possible impact on neurological development. Most conventional wastewater treatment systems cannot effectively eliminate PBDEs due to their high hydrophobicity, poor solubility, and limited biodegradability. As a result, approximately 60 to 90% of the PBDEs that enter these systems accumulate in sewage sludge (Kim et al., 2017).

19.2.2 Pharmaceuticals

Over the past 15 years, pharmaceuticals have attracted increasing attention as potential bioactive compounds present in the environment (Kümmerer, 2009). They are regarded as emerging contaminants in waterbodies since they are either not yet regulated or are in the process of being so, even if the regulations and legal frameworks have not been established. Pharmaceuticals are widely used and frequently released into the environment, impacting water quality, water supplies, ecosystems, and human health (Sirés and Brillas, 2012). Only in recent years have efforts been made to quantify and acknowledge the environmental presence of pharmaceuticals in water, despite their detection for many years, recognizing their potentially detrimental effects on ecosystems. Antibiotics, antacids, steroids, antidepressants,

TABLE 19.1
Different Therapeutic Groups as Contaminants of Emerging Concern (Sirés and Brillas, 2012)

Type of Pharmaceuticals	Drugs Detected
Analgesics and anti-inflammatory medications	Acetylsalicylic acid, diclofenac, ibuprofen, paracetamol, ketoprofen, and naproxen
Antidepressants	Benzodiazepines
Antiepileptics	Carbamazepine
Lipid-lowering drugs	Fibrates
Antibiotics	Tetracyclines, macrolides, β-lactams, penicillins, quinolones, sulfonamides, fluoroquinolones, chloramphenicol, imidazole derivatives
Antacids	Ranitidine
β-blockers	Atenolol, sotalol, propranolol
Antimicrobial	Triclosan
Non-steroidal anti-inflammatories	Mefenamic acid

analgesics, anti-inflammatories, antipyretics, beta-blockers, lipid-lowering medications, tranquilizers, and stimulants are the therapeutic categories most frequently identified in water treatment effluents (Table 19.1).

Pharmaceuticals, the majority of which have molecular masses around 500 Da, vary from typical industrial chemical pollutants in a number of ways. These molecules possess multiple ionizable groups and exhibit polarity, with their ionization and properties being influenced by the pH of the surrounding medium. Additionally, they have a lipophilic nature, and some demonstrate moderate solubility in water; their formation involves large and chemically intricate compounds that display considerable variation in molecular weight, structure, functionality, and shape (Lipinski et al., 1997). Following administration, the molecules are absorbed, distributed, and subject to metabolic reactions that can modify their chemical structure.

The inadequate elimination of many of these compounds by conventional treatments is evidenced by the presence of various pharmaceuticals in the inflow and outflow water of different wastewater treatment plants. More effective and targeted approaches are needed to reduce these contaminants' possible environmental effects. In recent times, researchers have begun utilizing activated carbon as a means to adsorb pollutants, particularly aromatic chemicals, from pharmaceuticals. The significant advantage of employing activated carbon for drug removal is that it does not generate hazardous or pharmaceutically active by-products. Existing literature suggests that activated carbons generally exhibit a high capacity for adsorbing pharmaceuticals, with effectiveness dependent on factors such as the specific type of activated carbon used, the composition of the pharmaceutical substances, and the characteristics of the solution in which they are present (Kümmerer, 2009).

Electrochemical techniques have also emerged as an effective and targeted approach for the remediation of waterbodies contaminated with pharmaceutical micropollutants. While implementing large-scale wastewater collection may not be a practical solution, concerns regarding the high costs associated with advanced treatment methods, such as source separation and separate urine collection, are valid related to energy consumption, cost, and elimination strategies. Further research is essential to develop sustainable solutions that can meet the needs of future generations. These solutions should address the challenges of pharmaceutical pollutants in water systems while considering environmental impact, cost-effectiveness, and long-term viability (Rivera-Utrilla et al., 2013).

19.2.3 Microplastics

In recent years, microplastics, which are minute plastic particles measuring less than five millimeters, have emerged as contaminants of increasing concern. The disintegration of larger plastic waste, the release of microbeads from personal care items, and the release of fibers from synthetic textiles during washing are a few of the sources of these minute particles. Even though they are tiny, microplastics provide significant risks to the environment, ecosystems, and possibly even human health. There are numerous routes for microplastics to get into the atmosphere. They are frequently discovered in freshwater and marine ecosystems, where currents and wind can carry them. Furthermore, microplastics have been found in soil, sediments, and the atmosphere. Their extensive distribution draws attention to the pervasiveness of plastic pollution and its possible adverse effects on many ecosystems.

Microplastics have the potential to accumulate within aquatic organisms, such as fish and shellfish, either by ingestion or entanglement. This characteristic, coupled with their small size and persistence, can result in their movement through the food chain, eventually reaching higher trophic levels, including humans. The most commonly used polymers for manufacturing microplastics include polyethylene, polypropylene, polyvinyl chloride, polystyrene, polyethylene terephthalate, and polyurethane. Medical face masks can comprise various nano- and microfiber polymeric materials, such as polypropylene, polyurethane, polyacrylonitrile, polystyrene, polycarbonate, polyethylene, and polyester. However, with the anticipated increase in their usage, future investigations should explore incorporating emerging novel materials like Tritan, which aim to address various technological challenges associated with medical face mask production and use (Rubio-Armendariz et al., 2022).

The effects of microplastics on marine life include physical harm, decreased feeding efficiency, impaired reproduction, and changes in the benthic flora and fauna. Additionally, heavy metals and persistent organic pollutants can adsorb and move around in the environment via microplastics. The threats to wildlife and human health can be further increased due to the transfer of these toxins to organisms following ingestion. Microplastic sources, distribution, and effects must be understood to resolve the problem. Researchers and government authorities are making dedicated efforts to gain a comprehensive understanding of the sources and destiny of microplastics within the environment while simultaneously striving to develop methods for their detection and quantification. In order to stop the intrusion of microplastics into the environment, there is also increasing attention on minimizing plastic waste, enhancing waste management systems, and encouraging sustainable practices.

19.2.4 Biocidal Products and Pesticides

Biocidal products and pesticides control pests, diseases, and unwanted organisms. Concerns about their possible environmental and human health effects have been raised because of their extensive and widespread use. Pesticides are chemicals that explicitly target pests and other undesirable organisms, whereas biocidal products cover a wide variety of chemicals such as disinfectants, insecticides, fungicides, and herbicides. Using these chemicals for pest management, applying them to crops, or using them in industrial, municipal, or domestic settings are just a few ways they might reach the environment. They may harm ecosystems and human populations by contaminating the soil, water, and air.

Pesticides and biocidal chemicals both have several adverse effects as developing pollutants. They can persist, absorb, and transfer between various environmental and ecological compartments. Harmful effects on non-target animals, such as beneficial insects, birds, fish, and other wildlife, can result from prolonged exposure to these toxins. Some effects of its widespread use include the disturbance of ecological equilibrium and biodiversity loss. In addition, environmental pesticides and biocidal agents put human health at risk. Direct exposure can happen while touching contaminated food or drink, using it, or consuming it. Prolonged exposure to low levels of these pollutants has been associated with various health issues, including neurotoxicity, disruption of hormonal balance, and an elevated likelihood of developing specific types of cancer.

Regulatory frameworks and risk assessment processes exist in many nations to address potential risks related to biocidal agents and pesticides. These seek to safeguard these compounds' sustainable and safe use, reduce their harmful environmental effects, and safeguard public health. Many practices are being adopted to lessen reliance on biocidal chemicals, promote alternatives to chemical pesticides, and increase the use of integrated pest management strategies. Our understanding of the environmental destiny, behavior, and toxicity of biocidal agents and pesticides is still being extended by research. This includes investigating the usage of natural products, developing more environmentally friendly and target-specific substitutes, and enhancing waste management techniques to stop trash from entering the environment.

19.2.5 Endocrine Disruptors

Endocrine disruptors can be found in various everyday products and substances, including pesticides, plastics, industrial chemicals, personal care products, and pharmaceuticals. They can enter the environment through agricultural practices, waste disposal, product consumption, and manufacturing processes. Once unleashed, they threaten wildlife and human populations since they can pollute air, water, soil, and food. The fact that endocrine disruptors can function in minute quantities is one of their troubling characteristics. Because they can disrupt hormone synthesis and metabolism, interfere with hormone signaling pathways, and affect interactions between hormone receptors, even low concentrations of these compounds can significantly impact the endocrine system. This can result in developmental anomalies, reproductive issues, immune system problems, and a higher risk of several cancers. Endocrine disruptors can have an impact on populations and ecosystems in addition to individuals. They can interfere with animal reproduction, diminish fertility, change sex ratios, and cause population collapse. Additionally, a few endocrine disruptors can potentially have transgenerational impacts, which means they can impact future generations by disrupting germ cells and changing the expression of genes that control hormones.

There have been initiatives to regulate and reduce exposure to endocrine disruptors in light of possible risks associated with these chemicals. Regulatory bodies have put screening programs and recommendations into place to evaluate the potential for endocrine disruption by chemicals and encourage using safer alternatives. Additionally, consumer education and awareness are essential for empowering consumers to make well-informed choices and promote environmentally friendly and sustainable practices. Our understanding of endocrine disruptors and their modes of action continues to evolve because of ongoing research. This entails researching the combined impacts of several disruptors, identifying vulnerable groups, and developing novel screening and assessment methods to detect possible endocrine disruptors. By making these steps, we can lessen the harmful effects of endocrine disruptors on human health, wildlife, and the environment (Kumar et al., 2022).

19.2.6 Heavy Metals

The occurrence of various heavy metals, including copper (Cu), nickel (Ni), zinc (Zn), cadmium (Cd), lead (Pb), chromium (Cr), mercury (Hg), and manganese (Mn, in activated sludge restricts its suitability for land application due to the potential for soil and groundwater contamination, posing risks to human and animals. The concentrations of these constituents vary considerably based on the origin of the sewage sludge. Among the identified metals, Cu and Zn were present in the highest concentrations in biosolids derived from wastewater treatment facilities (Mulchandani and Westerhoff, 2016). Supported liquid membranes have been employed to eliminate heavy metals from leached sewage sludge effluents. The extraction of Cu and Zn, two moderately volatile heavy metals, from sewage sludge was achieved by calcination in the presence of a Cl-donor additive ($MgCl_2$) under either an inert (N_2) or oxidizing environment (air). Electrokinetic therapy was also explored as a potential method for heavy metal removal, wherein significant removal efficiencies for Cu, Zn, Cr, Pb, Ni, and Mn were observed when combined with the addition of a chelating agent (tetrasodium of *N*, *N-bis* (carboxymethyl) glutamic acid) and a biodegradable biosurfactant (rhamnolipid) as an electrolyte.

19.3 ENVIRONMENTAL IMPACTS OF CECs

The environmental impacts of CECs are far-reaching and diverse. These substances have the potential to contaminate various environmental compartments, including waterbodies, soils, and the atmosphere. CECs can disrupt ecosystems, leading to reduced biodiversity, altered reproductive patterns, and impaired growth and development of organisms. They can also bioaccumulate in organisms over time, resulting in higher concentrations at the top of the food chain. Furthermore, the release of CECs into the environment can contribute to antibiotic resistance, disrupt hormonal systems, and increase the risk of cancer and neurological disorders in humans (Pesqueira et al., 2020). The major health risks and environmental impacts of CECs are discussed below.

19.3.1 Human Health Risk

CECs have raised significant concerns regarding their potential health effects on humans. The effects of these pollutants on various organismal systems are described in Table 19.2.

19.3.2 Ecological Risk

CECs can have profound ecological effects, impacting various components of ecosystems and posing risks to biodiversity and ecosystem function. Major ecological effects associated with CECs are explained below (Impellitteri et al., 2023).

19.3.2.1 Aquatic Toxicity

CECs that enter waterbodies, such as rivers, lakes, and oceans, can be highly toxic to aquatic organisms. These substances can disrupt the normal physiological functions of fish, amphibians, invertebrates, and other aquatic species. They can interfere with reproduction, growth, and development, impair behavior, and compromise immune systems, ultimately leading to population declines and reduced biodiversity (Pastorino and Ginebreda, 2021).

19.3.2.2 Bioaccumulation and Biomagnification

Many CECs can accumulate in organisms over time. When organisms at lower trophic levels are exposed to CECs, the contaminants can accumulate in their tissues. As predators consume these organisms, the CECs bio-magnify, resulting in higher concentrations in organisms higher up the food chain. This process can lead to amplified toxic effects, particularly in top predators, including birds, mammals, and even humans (Shi et al., 2015).

19.3.2.3 Reproductive and Developmental Disruption

CECs, particularly those exhibiting endocrine-disrupting properties, can interfere with reproductive processes in animals. They can cause abnormalities in gonadal development, disrupt hormone production, alter reproductive behaviors, and reduce

TABLE 19.2
The Major Health Effects of Contaminants of Emerging Concern on Human Health (Kasonga et al., 2021; Yadav et al., 2021)

Health Effects	Description
Carcinogenicity	Potential ability to cause cancer in humans and animals through DNA damage, mutations, or tumor promotion.
Neurotoxicity	Toxic effects on the nervous system result in neurological disorders, impaired cognitive functions, and behavioral abnormalities.
Respiratory system	Irritation of the respiratory system causes coughing, wheezing, shortness of breath, or more severe respiratory illnesses.
Cardiovascular system	Impacts on the cardiovascular system include changes in heart rate, blood pressure, and increased risk of cardiovascular diseases.
Developmental effects	Adverse effects on fetal development include congenital disabilities, impaired growth, cognitive deficits, and behavioral disorders.
Reproductive system	Interference with reproductive systems leading to infertility, reduced fertility, reproductive abnormalities, and hormonal disruptions.
Allergenic reactions	Induction of allergic reactions in susceptible individuals, such as skin rashes, itching, hives, respiratory distress, and anaphylaxis.
Gastrointestinal system	Irritation of the gastrointestinal tract causing nausea, vomiting, diarrhea, abdominal pain, and gastrointestinal disorders.
Renal and liver system	Impacts on the kidneys and liver including organ damage, impaired function, and increased risk of kidney or liver diseases.
Immune system	Modulation of the immune system leading to immunosuppression, increased susceptibility to infections, and autoimmune disorders.
Genetic and epigenetic effects	DNA damage, mutations, or epigenetic alterations which can affect gene expression, leading to long-term health implications.
Endocrine system disruptors	Disruption of hormonal regulation affecting various body functions, leads to hormonal imbalances, and developmental abnormalities. metabolism, growth, development, and reproductive processes.
Skin irritation and sensitization	Provoking skin irritation, rashes, dermatitis, or allergic contact dermatitis upon exposure.
Eye and ocular effects	Irritation, redness, itching, and other eye-related problems due to direct contact or airborne exposure to CECs.
Antibiotic resistance	Selection and proliferation of antibiotic-resistant bacteria, reducing the effectiveness of antibiotics in treating infections.

fertility. These disruptions can have significant consequences for population dynamics and the long-term sustainability of species (Guillette Jr. et al., 2000).

19.3.2.4 Impacts on Wildlife Behavior and Ecology

CEC exposure can alter the behavior and ecology of wildlife. Organisms may exhibit abnormal feeding patterns, migration disruptions, changes in foraging behaviors, and reduced ability to adapt to changing environmental conditions. Such behavioral changes can have cascading effects on species interactions, predator–prey dynamics, and overall ecosystem stability (Gwenzi et al., 2018).

19.3.2.5 Soil and Terrestrial Effects

CECs that enter soil through various pathways, such as agricultural runoff or the application of contaminated bio-solids, can impact soil health and terrestrial ecosystems. These substances can accumulate in the soil, potentially affecting soil microorganisms, nutrient cycling, and plant growth. Contaminated soil can also lead to the uptake of CECs by plants, which may enter the food chain and indirectly impact wildlife and human health (Maddela et al., 2022).

19.3.2.6 Pollinator and Beneficial Organism Disturbance

CECs can have detrimental effects on pollinators, such as bees and butterflies, and other beneficial organisms, like earthworms. Exposure to CECs can impair pollination processes, affecting plant reproduction and crop yields. The decline of pollinators and beneficial organisms can disrupt ecosystem services, such as pollination and soil health, with cascading effects on plant communities and food webs (Shah et al., 2023).

19.3.2.7 Genetic and Evolutionary Impacts

CECs can induce genetic and epigenetic alterations in exposed organisms, affecting their ability to adapt and survive. These alterations can be passed on to subsequent generations, potentially leading to reduced genetic diversity and impaired evolutionary potential. Such effects may impact the resilience and long-term viability of populations and ecosystems (Vassallo et al., 2021).

Addressing and mitigating the ecological effects of CECs requires comprehensive monitoring, risk assessment, and management strategies. Effective wastewater treatment, source control measures, and environmentally conscious practices in agriculture, industry, and consumer behavior can help minimize the release and exposure of CECs. Additionally, promoting sustainable alternatives, responsible chemical use, and conservation efforts can contribute to safeguarding ecosystems and preserving biodiversity in the face of CEC contamination.

19.4 TREATMENT OF CECs

CECs pose significant environmental and human health challenges, requiring effective treatment methods to mitigate their impacts (Figure 19.3). Various approaches have been developed to address the removal and degradation of CECs, aiming to ensure ecosystem protection and clean water resources. This section explores several treatment methods for CECs, including the sludge-based activated carbon method (SBAC), membrane bioreactors (MBR), advanced oxidation processes (AOP), Fenton-based processes (FBP), phytoremediation, electrokinetic remediation, photodegradation by nano-scale TiO_2, and hybrid systems. These methods offer diverse strategies for treating CECs, targeting different types of contaminants and employing unique mechanisms to achieve effective removal or degradation. By examining

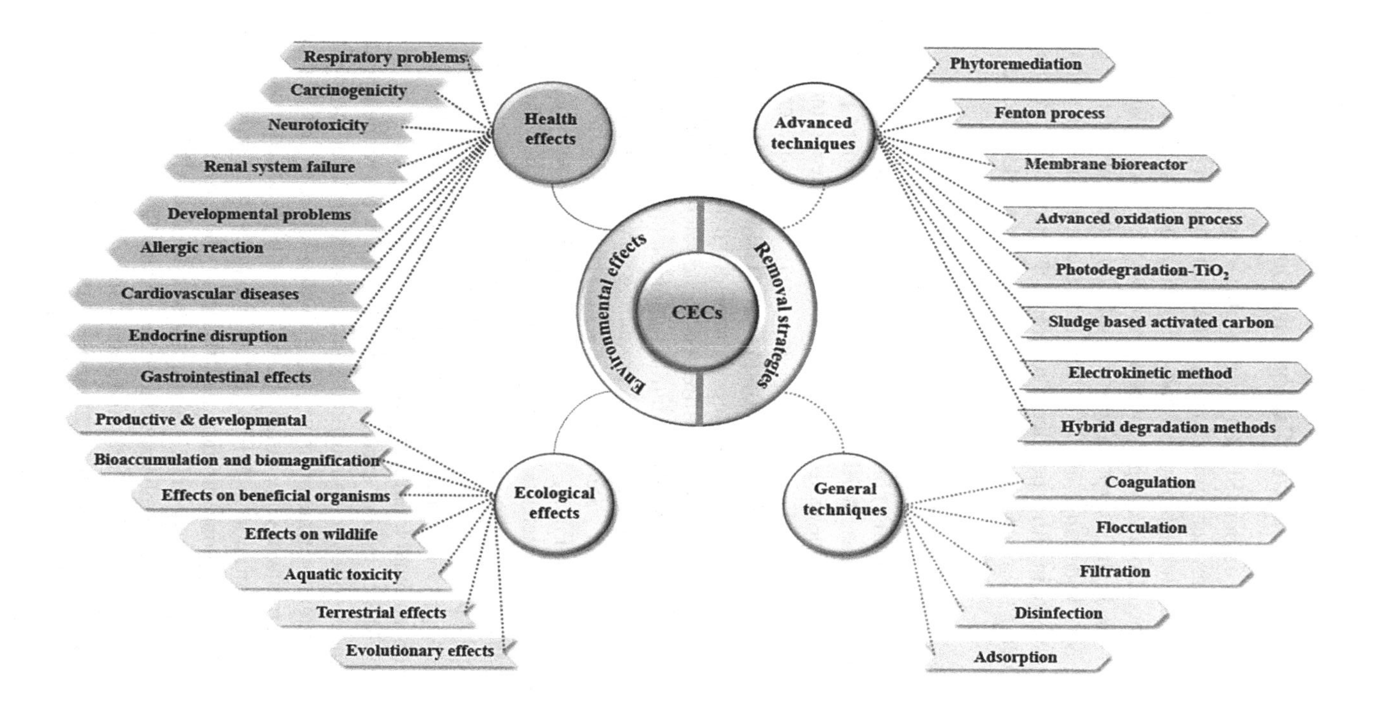

FIGURE 19.3 The environmental effects and removal strategies of CECs.

these treatment options, we can gain insights into the advances in CEC remediation and the potential for sustainable and efficient treatment solutions.

19.4.1 Sludge-based Activated Carbon Method

Sludge-based activated carbon (SBAC) materials have high adsorption performance and can effectively remove environmental pollutants, including typical organic matter and heavy metals, through physical and chemical processes. The SBAC method uses activated carbon from sewage sludge to remove CECs from wastewater. This method utilizes the high adsorption capacity of activated carbon to capture and effectively remove a wide range of CECs. The activated carbon is incorporated into the treatment process, in either powdered or granular form, to adsorb the contaminants, thus reducing their concentration in the treated water (Mohamed et al., 2023).

19.4.2 Membrane Bioreactors

Membrane bioreactors (MBRs) have emerged as a promising technology for removing CECs from wastewater. MBRs combine biological treatment processes with membrane filtration to achieve high-quality effluent and efficient removal of CECs. In an MBR system, a biological process, such as activated sludge or biofilm treatment, is integrated with a membrane filtration unit. The wastewater passes through the biological reactor, where microorganisms break down organic matter and metabolize pollutants, including CECs. The treated wastewater then flows through the membrane module, which consists of ultrafiltration or microfiltration membranes. These membranes act as a physical barrier, retaining suspended solids, bacteria, and CECs while allowing clean water to pass through. The use of membranes in MBRs provides several benefits. Firstly, the membrane barrier enhances the separation of solids and suspended particles, resulting in higher water quality and reduced turbidity. This improves the removal efficiency of CECs, as these contaminants are often present in the particulate form. Secondly, the small pore size of the membranes prevents the passage of bacteria and viruses, ensuring a higher level of microbial removal than conventional treatment processes.

Furthermore, MBRs exhibit excellent control over the hydraulic retention time, allowing for longer solids retention time and better biodegradation of CECs. The extended sludge retention time promotes the growth of diverse microbial communities capable of degrading a wide range of CECs. The high mixed-liquor suspended solids (MLSS) concentration in MBRs further enhances the removal efficiency by increasing the biomass concentration and reducing the hydraulic residence time required for treatment. However, it's important to note that MBRs also have some challenges. Fouling of the membranes due to the accumulation of solids and organic matter can occur, requiring regular maintenance and cleaning. This can impact the permeability and efficiency of the membranes. Additionally, the initial investment and operating costs of MBR systems can be higher than conventional treatment processes (Sengupta et al., 2022).

19.4.3 Advanced Oxidation Processes

AOPs have attracted significant attention as effective techniques for removing CECs from water and wastewater. AOPs involve the generation of highly reactive hydroxyl radicals (•OH) or other oxidative species to degrade and eliminate CECs through oxidation reactions. These processes offer several advantages regarding their versatility, effectiveness, and ability to target a wide range of CECs. AOPs typically employ different methods to generate reactive species, including chemical reactions, photolysis, and catalysts. Some common AOPs used for CECs removal include ozone-based processes, UV/H_2O_2, UV/TiO_2 photocatalysis, and Fenton or Fenton-like reactions. These processes can be applied individually or in combination to enhance the degradation efficiency.

AOPs offer several advantages for CEC removal. Firstly, they can effectively degrade various CECs, including pharmaceuticals, personal care products, pesticides, and other organic pollutants. This makes AOPs suitable for treating complex mixtures of CECs present in water and wastewater. Secondly, AOPs can target recalcitrant and persistent CECs that are not easily removed by conventional treatment methods. Furthermore, AOPs can be applied to different water and wastewater treatment scenarios, including drinking water treatment plants, industrial effluent treatment, and decentralized treatment systems. AOPs can be tailored and optimized to specific CECs in different water sources, ensuring more efficient removal and minimizing the formation of harmful by-products. However, it is important to note that AOPs also have certain limitations. The effectiveness of AOPs can be influenced by factors such as water quality, pH, temperature, and the presence of other compounds that can scavenge the •OH radicals. Additionally, the high energy requirements and operational costs associated with AOPs may pose challenges for large-scale implementation (Khan et al., 2020).

19.4.4 Photodegradation by Nano-scale TiO_2

Photodegradation using nano-scale titanium dioxide (TiO_2) involves using photocatalysts to degrade CECs under ultraviolet (UV) light. When exposed to UV radiation, the TiO_2 photocatalyst generates reactive oxygen species, which can break down organic contaminants. This method effectively degrades CECs with low biodegradability and is commonly used in advanced oxidation processes (Khan et al., 2020).

19.4.5 Fenton-Based Processes

Fenton-based processes (FBPs) utilize the Fenton reaction, which combines hydrogen peroxide and iron catalysts to generate hydroxyl radicals. These hydroxyl radicals have strong oxidizing properties and can effectively degrade a wide range of CECs. FBPs are often employed as an advanced oxidation technique, either alone or in combination with other processes, to enhance the removal of CECs from contaminated water (Munoz et al., 2012).

19.4.6 Phytoremediation

Phytoremediation is an eco-friendly and sustainable approach that utilizes plants to remove CECs from the environment. This natural remediation technique takes advantage of the unique abilities of plants to absorb, metabolize, and detoxify various contaminants, including CECs, present in soil, water, and air (Wu et al., 2021).

In phytoremediation, specific plant species, known as hyperaccumulators or tolerant plants, are selected based on their ability to accumulate high concentrations of CECs in their tissues without experiencing significant adverse effects. These plants can absorb CECs from the environment through their roots and transport them to their above-ground parts, such as leaves and stems. Once accumulated in the plant tissues, the contaminants can be removed through various mechanisms, including phytoextraction, rhizofiltration, phytodegradation, and phytostabilization (Shi et al., 2023).

Phytoextraction involves the uptake and accumulation of CECs in the plant's aerial parts. After a certain period of growth, the plants are harvested, and the contaminated biomass is removed, effectively removing the contaminants from the environment. Rhizofiltration, on the other hand, focuses on the filtration of CECs by the plant roots. The plant roots are surrounded by contaminated water, and the roots selectively absorb and retain the contaminants, purifying the water in the process.

Phytodegradation refers to the ability of certain plants and their associated microorganisms to degrade and detoxify CECs through metabolic processes. These plants produce enzymes that break down the contaminants into less toxic or non-toxic byproducts. Phytostabilization uses plants to immobilize and reduce the mobility of CECs in the environment, preventing their spread and further contamination. The plants create a physical barrier or bind the contaminants, reducing their bioavailability and potential for other organisms to be exposed.

Phytoremediation offers several advantages. It is a cost-effective and sustainable method that can be applied in a variety of contaminated environments, including industrial sites, agricultural lands, and wastewater treatment systems. Phytoremediation also has a minimal negative impact on the environment compared to traditional remediation methods that involve excavation, transport, and disposal of contaminated materials. Moreover, it can be esthetically pleasing and contribute to restoring and beautifying degraded areas.

However, phytoremediation does have some limitations. It can be a slow process, requiring long-term monitoring and management to ensure effective remediation. The efficiency of phytoremediation can also be influenced by factors such as plant species selection, environmental conditions, and the availability of CECs in the soil or water. Additionally, some CECs may not be easily absorbed or metabolized by plants, limiting the applicability of phytoremediation to certain contaminants (Tan et al., 2023).

19.4.7 Electrokinetic Remediation

It involves the application of an electric field to enhance the movement of CECs through the soil and toward collection electrodes, allowing for their removal from the environment. Electrodes are inserted into the contaminated soil in electrokinetic remediation, creating an electric circuit. Applying a direct current causes several electrochemical processes to occur, which drive the movement of CECs in the soil. The electric field mobilizes charged contaminants, such as heavy metals and ionic organic compounds, by attracting them toward the electrodes (Guedes et al., 2019).

Various electrokinetic phenomena, including electromigration, electro-osmosis, and electrophoresis, facilitate the migration of CECs. Electromigration involves the movement of charged CECs toward the oppositely charged electrode, while electro-osmosis refers to the movement of pore water carrying dissolved contaminants in response to the electric field. Electrophoresis occurs when charged particles are attracted to an electrode due to the electric field. As the contaminants migrate toward the electrodes, they can be collected and removed for further treatment or disposal. The collected CECs can undergo additional treatment processes to ensure proper disposal or recovery of valuable resources, if applicable (Guedes et al., 2021).

Electrokinetic remediation offers several advantages. It can effectively remove a wide range of CECs from soil and sediment, including heavy metals, inorganic ions, and organic compounds. It can be applied to cohesive and non-cohesive soils, making it suitable for diverse site conditions. Furthermore, it provides a uniform treatment throughout the soil volume, reaching contaminants that may be inaccessible to other remediation methods. However, electrokinetic remediation also has limitations. It can be time-consuming, requiring weeks to months to achieve significant contaminant removal, depending on the soil characteristics and initial contaminant concentrations. The energy consumption and associated electrical system running costs should also be considered (Guedes et al., 2021).

19.4.8 Hybrid Systems

Hybrid systems combine multiple treatment technologies to maximize the removal efficiency of CECs. These systems often integrate different methods, such as membrane filtration, activated carbon adsorption, and advanced oxidation processes, to achieve comprehensive and effective treatment. Hybrid systems offer the advantage of tailored treatment approaches, ensuring the removal of a wide range of CECs in complex wastewater streams (Ahmed et al., 2022).

19.5 CONCLUSION

The circular economy, which offers realistic long-term resource sustainability, is a potential approach for managing industrial waste. Circularity is a viable strategy for preserving resources, money, and energy due to the rising costs of various industries. In order to ensure a sustainable future for future generations, there is a significant focus on enhancing circular economy practices within the industrial sector and water management. Preserving the quality of waterbodies and mitigating the adverse impacts of contaminants on both human health and the environment rely heavily on strategies such as pollution prevention, source control, and efficient wastewater treatment. By safeguarding water quality and recognizing the inherent value of water and its diverse applications, we can work toward preserving its integrity and securing a better future. The uses of circular economy in various industrial activities and processes should therefore be the subject of additional efforts and research. By recognizing the significance of contaminants of emerging concern, encouraging the development of greener alternatives, and adopting sustainable practices, we can effectively address all the environmental challenges and work toward ensuring a cleaner and healthier environment for current and future generations.

REFERENCES

Ahmed, M., Mavukkandy, M. O., Giwa, A., Elektorowicz, M., Katsou, E., Khelifi, O., Naddeo, V., and Hasan, S. W. (2022) 'Recent developments in hazardous pollutants removal from wastewater and water reuse within a circular economy', *NPJ Clean Water*, 5(1), pp. 1–25. https://doi.org/10.1038/s41545-022-00154-5.

Englande, A. J., Krenkel, P., and Shamas, J. (2015) 'Wastewater treatment & Water reclamation', *Reference Module in Earth Systems and Environmental Sciences*. https://doi.org/10.1016/B978-0-12-409548-9.09508-7.

Gagliano, E., et al. (2020) 'Removal of poly- and perfluoroalkyl substances (PFAS) from water by adsorption: Role of PFAS chain length, effect of organic matter and challenges in adsorbent regeneration', *Water Research*, 171, p. 115381. https://doi.org/10.1016/j.watres.2019.115381.

Gaurav, G. K., et al. (2021) 'Review on polycyclic aromatic hydrocarbons (PAHs) migration from wastewater', *Journal of Contaminant Hydrology*, 236, p. 103715. https://doi.org/10.1016/j.jconhyd.2020.103715.

Ghimire, U., Sarpong, G., and Gude, V. G. (2021) 'Transitioning wastewater treatment plants toward circular economy and energy sustainability', *ACS Omega*, 6, pp. 11794–11803. https://doi.org/10.1021/acsomega.0c05827.

Guedes, P., et al. (2019) 'Electrokinetic remediation of contaminants of emergent concern in clay soil: Effect of operating parameters', *Environmental Pollution*, 253, pp. 625–635. https://doi.org/10.1016/j.envpol.2019.07.040.

Guedes, P., Dionísio, J., et al. (2021) 'Electro-bioremediation of a mixture of structurally different contaminants of emerging concern: Uncovering electrokinetic contribution', *Journal of Hazardous Materials*, 406, p. 124304. https://doi.org/10.1016/j.jhazmat.2020.124304.

Guillette Jr., L. J., et al. (2000) 'Alligators and endocrine disrupting contaminants: A current perspective 1', *American Zoologist*, 40(3), pp. 438–452. https://doi.org/10.1093/icb/40.3.438.

Gwenzi, W., et al. (2018) 'Sources, behaviour, and environmental and human health risks of high-technology rare earth elements as emerging contaminants', *Science of The Total Environment*, 636, pp. 299–313. https://doi.org/10.1016/j.scitotenv.2018.04.235.

Hoekstra, A. Y., and Mekonnen, M. M. (2012) 'The water footprint of humanity', *Proceedings of the National Academy of Sciences*, 109(9), pp. 3232–3237. https://doi.org/10.1073/pnas.1109936109.

Impellitteri, F., et al. (2023) 'Exploring the impact of contaminants of emerging concern on fish and invertebrates physiology in the mediterranean sea', *Biology*. https://doi.org/10.3390/biology12060767.

Kasonga, T. K., et al. (2021) 'Endocrine-disruptive chemicals as contaminants of emerging concern in wastewater and surface water: A review', *Journal of Environmental Management*, 277, p. 111485. https://doi.org/10.1016/j.jenvman.2020.111485.

Khan, J. A., et al. (2020) *Chapter 9 - Advanced Oxidation Processes for the Treatment of Contaminants of Emerging Concern* (pp. 299–365). Butterworth-Heinemann. https://doi.org/10.1016/B978-0-12-813561-7.00009-2.

Kim, M., et al. (2017) 'Review of contamination of sewage sludge and amended soils by polybrominated diphenyl ethers based on meta-analysis', *Environmental Pollution*, 220, pp. 753–765. https://doi.org/10.1016/j.envpol.2016.10.053.

Kumar, R., et al. (2022) 'A review on emerging water contaminants and the application of sustainable removal technologies', *Case Studies in Chemical and Environmental Engineering*, 6, p. 100219. https://doi.org/10.1016/j.cscee.2022.100219.

Kümmerer, K. (2009) 'The presence of pharmaceuticals in the environment due to human use – Present knowledge and future challenges', *Journal of Environmental Management*, 90(8), pp. 2354–2366. https://doi.org/10.1016/j.jenvman.2009.01.023.

Lipinski, C. A., et al. (1997) 'Experimental and computational approaches to estimate solubility and permeability in drug discovery and development settings', *Advanced Drug Delivery Reviews*, 23(1), pp. 3–25. https://doi.org/10.1016/S0169-409X(96)00423-1.

Maddela, N. R., et al. (2022) 'Major contaminants of emerging concern in soils: A perspective on potential health risks', *RSC Advances*, 12(20), pp. 12396–12415. https://doi.org/10.1039/D1RA09072K.

Matei, M., et al. (2023) 'Persistent organic pollutants (POPs): A review focused on occurrence and incidence in animal feed and cow milk', *Agriculture*. https://doi.org/10.3390/agriculture13040873.

Mohamed, B. A., et al. (2023) 'Circular economy in wastewater treatment plants: Treatment of contaminants of emerging concerns (CECs) in effluent using sludge-based activated carbon', *Journal of Cleaner Production*, 389, p. 136095. https://doi.org/10.1016/j.jclepro.2023.136095.

Mulchandani, A. and Westerhoff, P. (2016) 'Recovery opportunities for metals and energy from sewage sludges', *Bioresource Technology*, 215, pp. 215–226. https://doi.org/10.1016/j.biortech.2016.03.075.

Munoz, M., et al. (2012) 'Chlorinated byproducts from the fenton-like oxidation of polychlorinated phenols', *Industrial & Engineering Chemistry Research*, 51(40), pp. 13092–13099. https://doi.org/10.1021/ie3013105.

Pastorino, P. and Ginebreda, A. (2021) 'Contaminants of emerging concern (CECs): Occurrence and fate in aquatic ecosystems', *International Journal of Environmental Research and Public Health*. https://doi.org/10.3390/ijerph182413401.

Pesqueira, J. F. J. R., Pereira, M. F. R., and Silva, A. M. T. (2020) 'Environmental impact assessment of advanced urban wastewater treatment technologies for the removal of priority substances and contaminants of emerging concern: A review', *Journal of Cleaner Production*, 261, p. 121078. https://doi.org/10.1016/j.jclepro.2020.121078.

Rivera-Utrilla, J., et al. (2013) 'Chemosphere pharmaceuticals as emerging contaminants and their removal from water: A review', *Chemosphere*, 93(7), pp. 1268–1287. https://doi.org/10.1016/j.chemosphere.2013.07.059.

Rubio-Armendariz, C., et al. (2022) 'Microplastics as emerging food contaminants: A challenge for food safety', *International Journal of Environmental Research and Public Health*, 19, p. 1174.

Sauvé, S., et al. (2021) 'Circular economy of water: Tackling quantity, quality and footprint of water', *Environmental Development*, 39. https://doi.org/10.1016/j.envdev.2021.100651.

Sauvé, S., Bernard, S., and Sloan, P. (2016) 'Environmental sciences, sustainable development and circular economy: Alternative concepts for trans-disciplinary research', *Environmental Development*, 17, pp. 48–56. https://doi.org/10.1016/j.envdev.2015.09.002.

Sengupta, A., et al. (2022) 'Removal of emerging contaminants from wastewater streams using membrane bioreactors: A review', *Membranes*. https://doi.org/10.3390/membranes12010060.

Shah, S., et al. (2023) 'Microplastics and nanoplastics effects on plant–pollinator interaction and pollination biology', *Environmental Science & Technology*, 57(16), pp. 6415–6424. https://doi.org/10.1021/acs.est.2c07733.

Shi, Q., Kaur, P., and Gan, J. (2023) 'Harnessing the potential of phytoremediation for mitigating the risk of emerging contaminants', *Current Opinion in Environmental Science & Health*, 32, p. 100448. https://doi.org/10.1016/j.coesh.2023.100448.

Shi, Y., et al. (2015) 'Tissue distribution and whole body burden of the Chlorinated Polyfluoroalkyl Ether Sulfonic Acid F-53B in Crucian Carp (Carassius carassius): Evidence for a HIGHLY BIOACCUMULATIVE CONTAMINANT OF EMERGING CONCERN', *Environmental Science & Technology*, 49(24), pp. 14156–14165. https://doi.org/10.1021/acs.est.5b04299.

Siddiqi, M. A., and Clinic, M. (2003) 'New pollutants – old diseases', *Clinical Medicine*, 1(4), pp. 281–290.

Sirés, I., and Brillas, E. (2012) 'Remediation of water pollution caused by pharmaceutical residues based on electrochemical separation and degradation technologies: A review', *Environment International*, 40, pp. 212–229. https://doi.org/10.1016/j.envint.2011.07.012.

Stahel, W. R. (1997) 'The service economy: 'Wealth without resource consumption'?', *Philosophical Transactions of the Royal Society A: Mathematical, Physical and Engineering Sciences*, 355(1728), pp. 1309–1319. https://doi.org/10.1098/rsta.1997.0058.

Tan, H. W., et al. (2023) 'A state-of-the-art of phytoremediation approach for sustainable management of heavy metals recovery', *Environmental Technology & Innovation*, 30, p. 103043. https://doi.org/10.1016/j.eti.2023.103043.

Vassallo, A., et al. (2021) 'Antibiotic-resistant genes and bacteria as evolving contaminants of emerging concerns (e-CEC): Is it time to include evolution in risk assessment?', *Antibiotics*. https://doi.org/10.3390/antibiotics10091066.

Stahel, A. (2010) *The Performance Economy*. 2nd edn. Palgrave McMillan.

Wu, Y., et al. (2021) 'Phytoremediation of contaminants of emerging concern from soil with industrial hemp (Cannabis sativa L.): A review', *Environment, Development and Sustainability*, 23(10), pp. 14405–14435. https://doi.org/10.1007/s10668-021-01289-0.

Yadav, D., et al. (2021) 'Environmental and health impacts of contaminants of emerging concerns: Recent treatment challenges and approaches', *Chemosphere*, 272, p. 129492. https://doi.org/10.1016/j.chemosphere.2020.129492.

Index

F

G

H

L

M

N

Printed in the United States
by Baker & Taylor Publisher Services